Nanoindentation of Brittle Solids

Nanoindentation of Brittle Solids

Arjun Dey

Anoop Kumar Mukhopadhyay

CRC Press
Taylor & Francis Group
Boca Raton London New York

CRC Press is an imprint of the
Taylor & Francis Group, an **informa** business

The figures on the cover are from experiments conducted at the 'Nanoindentation Laboratory' of the Mechanical Property Evaluation Section under the Materials Characterization Division of CSIR-Central Glass and Ceramic Research Institute, Kolkata.

Top row, left to right: SPM images of 4 × 4 square array of Berkovich nanoindentation on fused quartz, SPM images of array of scratch trails on DLC based composite thin film and SPM image of a Berkovich nanoindent on soda-line-silica glass.

Bottom row, left to right: FESEM image of a Berkovich nanoindent on soda-line-silica glass, optical micrograph of 4 × 4 square array of Berkovich nanoindentation on human cortical bone at cross section and FESEM image of a Berkovich nanoindent on microplasma sprayed bioactive hydroxyapatite coating.

CRC Press
Taylor & Francis Group
6000 Broken Sound Parkway NW, Suite 300
Boca Raton, FL 33487-2742

First issued in paperback 2017

© 2014 by Taylor & Francis Group, LLC
CRC Press is an imprint of Taylor & Francis Group, an Informa business

No claim to original U.S. Government works

Version Date: 20140404

ISBN 13: 978-1-138-07653-2 (pbk)
ISBN 13: 978-1-4665-9690-0 (hbk)

Visit the Taylor & Francis Web site at
http://www.taylorandfrancis.com

and the CRC Press Web site at
http://www.crcpress.com

Dedicated to

Manju, Brishti, Tutum

and

Omprita

Contents

Section 1   Contact Mechanics

Section 2   Journey towards Nanoindentation

Section 3   Static Contact Behavior of Glass

Section 4   Dynamic Contact Behavior of Glass

Section 5   Static Contact Behavior of Ceramics

Section 6   Static Behavior of Shock-Deformed Ceramics

Section 8 Nanoindentation Behavior of Functional Ceramics

Section 9   Static Contact Behavior of Ceramic Coatings

Section 11   Nanoindentation Behavior on Ceramic-Based Natural Hybrid Nanocomposites

Section 12 Some Unresolved Issues in Nanoindentation

Prologue

How Did it All Happen?

Research in brittle solids like glass and ceramics and their nanoscale characterization are being driven today both by market demand for better engineered materials and the academic need for understanding how the damages initiate, incubate and propagate in brittle solids. The global advanced ceramics market is projected to reach *US$56.4 Billion* by 2015 and *US$68 Billion* by the year 2018. The Asian markets of China, India, South Korea, Taiwan and Thailand promise substantial growth potential at a robust pace of about 8% through 2018. Even for an apparently simple amorphous brittle solid like advanced flat glass, the market was worth *US$12.9 Billion in 2012* and projected to grow at a rate of 9.1% annually through 2014. The global market for LCD glass substrates was about *US$14 Billion* last year and by 2017 the requirement of LCD glass substrate is projected to be about 500 million square meters. The Indian glass market size is about *US$33 Million* today with a *CAGR of about 14.7%* projected for the next decade. It is expected to grow in glass consumption by about 9% in construction, 19% in automotive, 10–12% in consumer goods and 12–15% in pharmaceuticals sectors. The proposed investment by major players in *next decade toward capacity expansion and new technologies will be around US$70 Million* in India alone. Similar is the situation with synthetic biomaterials. It has been noted that the U.S. market is the largest geographical segment for biomaterials, and is expected to be worth *US$22.8 Billion by 2014 with a CAGR of 13.6% from 2009 to 2014.* Europe is the second largest segment and is expected to reach *US$17.7 Billion by 2014, with a CAGR of 14.6%.* Similarly, the *Asian market size is estimated to increase at the highest CAGR of 18.2% in the same period.* Further, the forecast is that the global dental implants market will reach *US$4.07 Billion in 2014.* The global dental implants market has also been witnessing technological advancements in the quality and hence the lifetime of such implants. In fact, forecast for the *dental implants market in India is to reach US$101.2 Million by 2014.*

Why is so much money going into this field of brittle solids? It is obvious, then, that there exists a huge market demand and it is projected to grow further, at a very fast pace in days to come. That is why the improvement in fabrication technology and new product development at competitive prices will be the key to future market growth and capturing the same.

However, all the common and advanced applications of glass, advanced structural and functional ceramics, soft coatings, hard thin films as well as synthetic bio-materials will demand the development of engineered materials with better *"contact damage resistance"* properties to enhance their marketability. But the problem is that, being fundamentally brittle materials, both glass and ceramics are prone to suffer from contact damage–induced brittle fracture. Akin to the cases of glass and ceramics, in the cases of both natural and artificial bio-materials that are being used today or are contemplated to replace the natural bio-materials in days to come, the physics of deformation at the microstructural length scale are far from well understood.

Hence, it is urgently needed to develop better microstructurally engineered glass, ceramics, bio-materials etc., for various applications. This need therefore demands a thorough understanding of basic contact-induced deformation mechanisms and damage initiation as well as growth mechanisms at macro-, micro- and/or nanostructural length scales of the glass and ceramic microstructures. However, it is not always very easy. The reason is that the processing of especially ceramics and glasses with zero defects is very difficult. The processing-induced volume or surface flaws are almost omnipresent. In addition to their presence, in some more complicated situations, e.g., in the case of ceramic coatings, the characteristically heterogeneous microstructure may offer more hindrance than support to evaluation of the nanomechanical properties. And that is from where the essential challenge of the subject matter, e.g., the "nanoindentation of brittle solids" of the present book sprouts out.

It is to be appreciated further that evaluation of mechanical properties and understanding of deformation behaviour of brittle materials, more specifically ceramics in bulk, coatings and thin film forms as well glass, is really a challenging task for any researcher. Once the student simply raised the question: why is it so challenging? Every material should have a definite failure point. Why do we have to worry so much about the mechanical behaviour of brittle materials? Then, the teacher explained in a simple manner that classical brittle materials do not show delayed failure like ductile metals and because of this, a catastrophic failure is predominant in them. When we are using those brittle materials in practical applications, they may fail catastrophically in a very premature manner and hence can cause huge loss in terms of both money and human life. Now, if the tip of your pencil breaks (interestingly, that is also a case of sudden failure!), you can sharpen it, at least once or twice. But, in the case of a floor tile or the armour of an aircraft or ship, the bullet-proof jacket of a soldier, biomedical implants, ceramic coatings on cutting tools, silicon nitride ball bearings or non-oxide ceramic-based radomes or superhard ceramic thin films in special tribological applications, if catastrophic failure happens it is impossible to stop it at that point of time. We all may remember, for example, what happened due to the failure of one small insulation tile in the case of the Space Shuttle Columbia. It disintegrated during re-entry over Texas on February 1, 2003.

However, we can predict the in-service behaviour if we know its mechanical properties and understand its deformation behaviour in-depth in responses to different types of loading, e.g., mode I, mode II, mode III and mixed mode loadings as well as strain rates. When we see the failure at the macroscopic length scale, it must have originated a few moments back, at its microstructural length scale. Therefore, it's not surprising that the mechanical properties measurement at local micro-/nanostructural length scale is gaining huge popularity.

Today, it is the era of engineered multifunctional materials. The buzz word today is research in nanomaterials of 1D/2D structures. Therefore, it is required to evaluate the micro-/nanomechanical properties at the length scale; that is, of the order of the microstructural unit blocks. That is why the depth sensing indentation or nanoindentation technique has evolved in early 1990s and is continuing to grow today with huge popularity at a very fast pace. It gives us the chance to probe at such small length scales as mentioned above, as if we are sitting there virtually, nanomechanically disturbing the structure in a controlled manner and then recording the system's response.

Well then the question is what are these brittle solids that we have looked at? The list includes the SLS glass, alumina and ZTA ceramics, different bulk and fiber reinforced ceramic matrix composites like C/C, C/SiC, alumina/ monazite, hydroxyapatite/mullite, soft coatings like hydroxyapatite coating, hard anodic coating like micro arc oxidized coating, soft and hard thin films like $Mg(OH)_2$, TiN, DLC as well as the natural nanobiological hybrid composites like tooth, bone and fish scale, etc. Both static and dynamic contact situations were considered. Many technical issues like the effects of load, depth, loading rate, etc. have been studied. In this book, we have shown that nanoindentation is a potential tool for characterizing a large domain of materials, especially for those which are brittle in nature. In every chapter, we have tried to correlate microstructural issues with the micro-/nanomechanical responses. Further, I hope that this book will explain the importance and flexibility of the nanoindentation technique for multifaceted nanomechanical characterization of brittle materials.

Arjun Dey

The need and the importance to understand the nanoscale contact and deformation behaviour of bulk ceramics was noticed by me during my postdoctoral research work at the Department of Mechanical and Mechatronic Engineering, Sydney University, Australia where I had an opportunity to work with world famous professors Yiu Wing Mai and Michael V Swain. After a decade, when I worked with Dr. R.W. Steinbrech at Forshungszentrum, Juelich, Germany, I got further exposure to contact mechanics of thermal barrier coatings where heterogeneous microstructure can play a crucial role during evaluation of its micro-/nanomechanical properties by the depth-sensitive indentation or the nanoindentation technique. Afterwards, the Nanoindentation Laboratory was established around 2003–2004 under my leadership at CSIR-CGCRI, Kolkata. The research presented in this book is the fruit of our work done at this laboratory over a period of about a decade or so by myself and my students, especially Dr. Dey. We have collaborated quite extensively with fellow researchers in and outside of CSIR-CGCRI. In the process, we have tried to understand the physics of deformation at micro-/nanoscale contacts of several brittle solids as mentioned above. Presently, the area of research in the nanoindentation domain of brittle solids is a prime area of research importance due to the wide variety of applications of such brittle materials. I sincerely hope that this book will give the reader an idea about how the nanoindentation technique can be used to correlate the nano-mechanical properties and microstructural parameters. To the best of the author's knowledge, this is the first attempt to publish a book on the application of the nanoindentation technique especially dedicated to brittle materials like glass and ceramics.

Anoop Kumar Mukhopadhyay

Preface

What Is in it for Us?

This book is divided into 12 Sections. Section 1 is on contact mechanics. It comprises two chapters. Chapter 1 deals with contact issues in brittle solids, while Chapter 2 concentrates more upon the mechanics of elastic and elastoplastic contact. The importance of elastoplastic contact is well recognized in the case of brittle solids in general, and glass and ceramics in particular.

Section 2 begins with a journey toward the main topic of this book, that is the science and technology of nanoindentation, especially in the backdrop or purview of brittle solids. This section consists of 7 chapters. Chapter 3 gives a brief history of indentation. In Chapter 4, we have discussed the concepts of hardness and elastic modulus of a material. Next in Chapter 5, the basic ideas of nanoindentation have been put forward, with a special emphasis on why it is necessary and where its applications lie altogether. But it is also important to know about the nanoindentation data analysis methods which have been discussed in Chapter 6. The various nanoindentation techniques are elaborated in Chapter 7. But one can recognize that it is not only important to know theoretically how the whole technique works; it is also important to understand how it actually translates to real life practice. For instance, what are the basic components of a nanoindentation machine and how do the different components work? It is also important to know about the ranges of different commercially available machines and their resolutions. These issues are briefly discussed in Chapter 8. Now, in this book we have discussed results obtained from the nanoindentation experiments conducted on a truly wide variety of brittle solids, e.g., glass, ceramics, shock-loaded ceramics, different types of ceramic matrix composites, structural and functional ceramics, bioactive thick ceramic coatings as well as hard thin ceramic films. We have also included nanoindentation results from our recent research on natural biomaterials like tooth, bone and fish scale materials. These aspects are briefly dealt with in Chapter 9.

Section 3 discusses the static contact behaviour of glass. It actually comprises three chapters; namely 10, 11, and 12. In Chapter 10, we have discussed the nanoindentation response if the contact is made too quickly in glass. Chapter 11 actually poses the question of whether if it is possible to enhance the nanohardness of glass. To the best of our knowledge this is the very first such attempt. Chapter 12 discusses the energy issues related to the nanoindentation of glass.

Section 4 deals with the dynamic contact behaviour of glass spanning Chapters 13 to 15. What happens if a specimen like glass is damaged in microscale dynamic contact events? This issue is discussed in Chapter 13. The next question that naturally appears is how it matters whether such a microscale contact is slow or fast. As elaborated in Chapter 14, it really matters very much for dynamic damage evolution in glass. When the speed of the dynamic contact is varied, the consequences are portrayed in Chapter 14. We also wanted to ask how much is the damage inside a scratch groove in glass? How can we quantify the damage in terms of the nanomechanical properties evaluated at the local microstructural length scale? To the best of our knowledge, this is the very first such attempt. The results of our related experimental observations are summarized in Chapter 15.

Section 5 that spans Chapters 16 to 18 concentrates mainly on the nanoscale static contact behaviour of a typical brittle ceramic like alumina. Chapter 16 not only describes the nanoindentation response of a coarse grain alumina, it also questions how relevant the grain size is as far as the intrinsic capability against contact-induced damage of a ceramic is concerned. The results depicted in Chapter 17 raise two very important questions, e.g., if the energy dissipation rate from the loading train into the microstructure of a given ceramic really matters in its response against contact-induced deformation and/or micro damage evolution and if it does, what will be the mechanisms of real relevance? To the best of our knowledge, this is the very first such attempt. In a kind of self-motivated manner, we have tried in Chapter 18 to devise a rational picture that addresses the issues raised in Chapter 17, but admittedly it opens up possibly more areas of concern for future research than were possible to be addressed in our humble effort made in this book.

Coming up to Section 5 we had somewhat learnt about the very interesting aspects of an interaction triangle that possibly exists between the microstructural units, e.g., grain size, the probe length scale and the rate of probing the load and the loading rate and the most important one, the spatial extent of interaction between these two factors on the one hand and the length scales of the naturally present pre-existing defects on the other hand. Now we wanted to pose another very important constraint on this scenario, i.e., what happens to its nanoindentation response, if the ceramic is already highly damaged; say, due to very high strain rate or high pressure impact from a projectile!

Thus, Section 6 that spans Chapters 19 to 22 tries to put forward a critical look in this complicated scenario. To the best of our knowledge this is the very first such attempt. Through these chapters we have shown how the nanoscale contact deformation resistance of a typical coarse grain ceramic, e.g., 10 μm grain size alumina would be affected under such a scenario, whether the process will be load- and loading rate-dependent and how the interaction picture changes, as a function of the nanoindentation zone of influence and the extent of pre-shock history that the sample has gone through.

Section 7 spans Chapters 23 to 25 and asks what happens to the nanoindentation response in different kinds of ceramic matrix composites (CMCs), e.g., C/C and C/C-SiC composites, HAp-based biological-composites, tape cast multilayered composites as well as particulate reinforced CMCs. Similarly, the nanoindentation behaviour of a wide variety of functional ceramics is covered in Chapters 26 to 31 of Section 8. Chapter 26 concentrates on the nanoindentation study on silicon that phase transforms under nanoindentation-induced pressure, while Chapter 27 elaborates on the nanomechanical behaviour of ZTA where the tetragonal zirconia can phase transform under appropriate stress to the monoclinic phase. In Chapter 28 we have tried to address the nanoindentation responses of two actuator ceramics, e.g., $(Pb_{0.88}Ba_{0.12})[(Zn_{1/3}Nb_{2/3})_{0.88}Ti_{0.12})]O_3$ (PZN-BT) and $(Pb_{0.8}Ba_{0.2})[(Zn_{1/3}Nb_{2/3})_{0.8}Ti_{0.2})]O_3$ (PZN-BT-PT) ceramics and show how the results correlate with the hysteresis loop measurements in the polar region of these two actuator materials. Probably the very first results ever obtained on nanoindentation response of sol-gel derived, green compacted nano bismuth ferrite multiferroic ceramics are depicted in Chapter 29. Encouraged by the interesting results obtained for many functional materials as depicted above, it was decided to extend the efforts to the realm of materials for renewable energy or greener energy sources. It is in this perspective that the nanoindentation experiments were conducted on all three components—anode, electrolyte and cathode of a ceramic solid oxide fuel cell (SOFC) and also on the glass-ceramic sealants used to connect the different components of a SOFC stack in a condition that prevents leakage of any kind. To the best of our knowledge, these are the very first such efforts made in this field and the results are depicted in Chapters 30 and 31.

Section 8, spread across Chapters 32 to 37, actually puts forward the question of whether it is possible to utilize the nanoindentation technique to investigate the nanoscale contact deformation behaviour of ceramic coatings which are thick, porous, and highly micro-cracked and hence possess truly heterogeneous microstructures. It has been shown that the same really can be achieved. Two coatings were investigated. One is a bioactive ceramic microplasma sprayed hydroxyapatite (MIPS-HAp). The other is a protective ceramic oxide coating formed on magnesium alloy by micro arc oxidation (MAO) coatings. Thus, Chapter 32 studies the nanoindentation on MIPS-HAp coatings, while the Weibull modulus of nanohardness and elastic modulus of the same has been discussed in Chapter 33. Keeping in mind the anisotropic microstructure of the MIPS-HAp coatings, the anisotropy in nanohardness has been examined in Chapter 34. To the best of our knowledge, in Chapter 36 the very first attempt to evaluate the microstructural scale relevant fracture toughness of the same MIPS-HAp coatings has been explored. Next in Chapter 36, the nanomechanical efficacy of the same was critically examined after immersion in simulated body fluid. Further, the nanoscale contact response of the important protective MAO coatings was examined in Chapter 37.

The cases of the soft $Mg(OH)_2$ and hard TiN, Al_2O_3 and metal doped as well as undoped DLC thin ceramic thin films in terms of nanoindentation behavior as well as their nanotribological behavior have been briefly exposed in Chapters 39 to 41 which comprise Section 10.

So far we have discussed nanoindentation responses of glass, ceramics, different types of CMCs, functional ceramics, thick coatings as well as soft and hard ceramic thin films of tremendous biological, functional, structural, industrial, tribological as well as commercial importance. But one point has not been explored. That is the aspect of structure and nanomechanical property relationships of the natural biomaterials, e.g., human teeth, bone, fish scales etc. developed by mother nature over the ages in such a way that they possess nanoscale to macroscale functionally graded microstructure. Thus, Section 11 spanning Chapters 42 to 45 focuses on the nanoindentation behaviour of ceramic based natural hybrid nanocomposites.

In Chapters 46 to 51 in Section 12 the global, yet unresolved issues like the forward and reverse indentation size effect (ISE, RISE), pop-in, loading rate, substrate and residual stress effects, as well as the data reliability in brittle solids have been touched upon. A brief note on scope and direction for future research in this exciting, ever growing field of the science and technology of nanoindentation in general, and brittle materials in particular has been sketched in Chapter 52. Finally, the book ends with a summary of the major findings that emerge out of the results presented in this compendium of our research results.

Acknowledgments

At the onset, we express our sincere thanks to the directors of CSIR-Central Glass and Ceramic Research Institute (CSIR-CGCRI), Kolkata and ISRO Satellite Centre, Bangalore for their constant encouragements and use of their infrastructural facilities. We are grateful to Mr. Srikanta Dalui, Mr. Manas Raychoudury, Mr. Haradhan Das, Mr. Ram Narayan Kumar, Mr. Kartik Banerjee, and Mr. Biplab Naskar of the Mechanical Property Evaluation Section, CSIR-CGCRI for help received during this work. We are especially indebted to the late Dr. D. Basu, Dr. M. K. Sinha, Mr. B. Kundu, and Mr. Anath Karmakar from the Bio-Ceramics and Coatings Division of CSIR-CGCRI. We also express our sincere thanks for the incredible help and support received from all the former students and research scholars of the Nanoindentation Laboratory at CSIR-CGCRI.

What Arjun Says....

It is now my turn to thank the people who were involved behind the curtain for the completion of the present book. First of all, I am really grateful to my beloved parents, Mr. Dilip Kumar Dey and Ms. Minati Dey and to my dear wife Omprita Chatterjee who took on all the responsibilities for family matters and gave me the encouragement to publish this book. I am really grateful to my dearly loved father-in-law Er. Santi Priya Chatterjee, mother-in-law Mrs. Shubhra Chatterjee, my spouse's sister Mrs. Epsita Chatterjee, her husband Dr. Sirshak Dutta and their young son Master Atmabhash. I am also most thankful to my dear boudi (sister-in-law) Ms. Manjulekha Mukherjee (spouse of Dr. Mukhopadhyay) for providing continuous mental support and caring for almost every need during the publication of this book.

Very special thanks are due to Dr. A. K. Sharma, Group Director, Thermal System Group, ISAC-ISRO; Prof. Bikramjit Basu, IISc Bangalore; Prof. Tapas Laha, IIT Kharagpur; Prof. Srinivasa Rao Bakshi, IIT Madras; Prof. Vikram Jayram, IISc; Prof. Rabibrata Mukherjee, IIT Kharagpur; Dr. Harish Barshilia, CSIR-NAL; Prof. Niloy K. Mukhopadhyay, IIT-BHU; Prof. Kuruvilla Joseph and Prof. K. Prabhakaran, IIST, Trivundrum; Prof. Subroto Mukherjee, IPR, Gandhinagar; Dr. Arvind Sinha, CSIR-NML; Prof. Ajoy Kumar Ray, Vice Chancellor, BESU, Shibpur; and Prof. Gerhard Wilde, Director, Institute of Materials Physics, University of Muenster. Moreover, I have always received enormous supports from Prof. Nil Ratan Banyopadhyay and

Prof. Subhabrata Datta from BESU, Shibpur, without whose support my academic and research career would not, perhaps, have grown to where it is today. Before I finish, I am really fortunate to meet Dr. Anoop who is not only my Master's and PhD guide and co-author of the present book but also he is the path finder of my research life.

Arjun Dey

What Anoop Says....

I thank my late, beloved parents Mr. Girija Bhusan Mukherjee and Mrs. Kamala Mukherjee and my late in-laws Mr. Beni Madhab Bhattacharya and Mrs. Bijaya Bhattacharya for making me what I am today. I wish all of them were alive today! The huge support received from Mr. and Mrs. S. K. Mukherjee of Hindustan Aeronautics Limited, Bangalore, Prof. S. Bhattacharya of Institute of Management Technology, Ghaziabad, and Dr. Mrs. A. Bhattacharya of RKDIT, Ghaziabad, Mr. and Mrs. A. Bhattacharya of Silvasa, Mumbai is also gratefully acknowledged. It would be an oversight not to recognize the tremendous contributions of all four of my sister-in-laws, four late brothers and one late sister, two late uncles and aunts to my upbringing and their huge encouragement of my academic efforts.

I am also more than grateful for my entire life to my dear, beloved wife Mrs. Manjulekha Mukherjee and to my two sweet daughters Miss Roopkatha Mukhopadhyay, aged 16, and Miss Sanjhbati Mukhopadhyay, aged 11; without their tremendous patience, love, care and mental support, this book would not have seen the light of the day. It is a joy for me to recognize the kind encouragement received from my nephew and his wife Mr. N. Mukherjee, Mrs. M. Mukherjee and my niece, Miss L. Mukherjee.

I also appreciate very much the kind contributions made by Profs. Y-W. Mai, M.V. Swain and Mark Hoffman of Australia; Dr. Brian Lawn of NIST, USA; Prof. S. Priyadarshy of USA; Dr. R. W. Steinbrech of Germany; Mr. I. Dean of South Africa; Prof M. K. Sanyal, Director, SINP; Dr. D. Chakraborty, my own PhD guide; Dr. S. Kumar, Dr. B. K. Sarkar, Dr. C. Ganguly, Dr. H. S. Maity, Dr. K. K. Phani, Dr. D. K. Bhattacharya, late Drs. A. P. Chatterjee, D. Basu and Mr. D. Chakraborty, all of CSIR-CGCRI; Dr. S. Tarafder of CSIR-National Metallurgical Laboratory, Jamshedpur; Prof. B. Basu of IISc, Bangalore; Prof. N. R. Bandyopadhyay of BESUS; Prof. A. N. Basu of Jadavpur University; and, of course, Prof. I. Manna of IIT Kanpur to my development as a researcher. I especially thank Prof. R. Mukherjee of IIT Kharagpur and the late Mr. P. Basu Thakur of Icon Analytical Pvt. Ltd., India, without whose very active support the Nanoindentation Laboratory would not have been created at CSIR-CGCRI. I also thank my colleagues Mr. D. Sarkar,

Mr. I. Biswas, Drs. D. Bandyopadhyay and G. Banerjee, Miss S. Datta, Mr. S. Dey, Mrs. R. Chakraborty and Mr. S. Biswas of the Project Management Division, and all my other colleagues from the Publication Section and the Non-Oxide Ceramic and Composite Division of CSIR-CGCRI for their help and cooperation during the course of writing this book.

I am also very grateful to my dear brothers Mr. S. Acharya and Mr. D. Moitra of CSIR-CGCRI and to Mr. R. Das as well as Dr. Anup Khetan of RTIICS, without whom I would not have been able to write even the very first page of this book. I also thank Dr. S. K. Bhadra, Chief Scientist and Mr. K. Dasgupta, Director of CSIR-CGCRI for their kind encouragement, support, and advice at all stages. Very special thanks are due to my colleagues Dr. R. N. Basu, Dr. D. Kundu, Dr. G. De, Dr. H. S. Tripathy, Dr. S. Dasgupta and Mr. A. K. Chakraborty of CSIR-CGCRI. I also thank the Almighty to have kindly provided me a student like Dr. Arjun Dey, the co-author of this book. He and his wife, Omprita, are part of my life and definitely much more than mere students or junior colleagues to me. I feel very proud to have the opportunity to be his thesis supervisor, just by chance, as they say, by the decree of God. I acknowledge all the near and distant relatives, teachers, friends, colleagues, former and present students and well wishers for their support and kind encouragement received at every bend of my life and particularly during the execution of this book writing project. Last but not the least, I express my gratitude to my dear friend, the poet, Joy Goswami, who taught me how to live through the hours of grief and pain, through the hours of tortures and torments, just to see through the apparently endless night that ends in a new whisper of love, a new breath of life, and the sunrise of a new day.

Anoop Kumar Mukhopadhyay

About the Authors

Dr. Arjun Dey is presently working as a scientist at the Thermal Systems Group of ISRO Satellite Centre, Indian Space Research Organisation, Department of Space, Bangalore. Dr. Dey earned a bachelor's degree in mechanical engineering in 2003; followed by a master's degree in materials engineering from Bengal Engineering and Science University, Shibpur, Howrah in 2007. While working at CSIR-Central Glass and Ceramic Research Institute (CSIR-CGCRI), Kolkata, he earned his doctoral degree in materials science and engineering in 2011 from the Bengal Engineering and Science University, Shibpur, Howrah.

He has obtained many prestigious awards such as the "Dr. R. L. Thakur Memorial Award" from The Indian Ceramic Society in 2012, "DST-Fast Track Young Scientist Scheme Project Grant Award" from Department of Science and Technology, India in 2011, "Young Engineers Award" and "Metallurgical and Materials Engineering Division Prize" from The Institution of Engineers (India) in 2010–2011 and "Young Scientist Award" from Materials Research Society of India, Kolkata chapter in 2009 for significant contributions in the field of materials science and engineering. Arjun also obtained the prestigious CSIR-Senior Research Fellowship Award, from the Council of Scientific and Industrial Research (CSIR) to accomplish his PhD.

The research work of Dr. Dey culminated in more than 125 publications to his credit. He is a member of prestigious professional and statutory bodies including the Institution of Engineers (India), Materials Research Society of India (MRSI), Astronautical Society of India, Indian Science Congress Association, The Indian Physical Society and the Society of Biomaterials and Artificial Organs. His current research interests cover a truly diverse span, e.g., oxide and metallic thin film and micro arc oxidation coating for spacecraft application, physics of nanoscale deformation of brittle solids, very high strain rate shock physics of ceramics, tribology of ceramics, nanotribology of ceramic coatings and thin films, microstructure mechanical and/or functional property correlation of (a) structural and bio-ceramics, bio-ceramic coatings, biomaterials, (b) multilayer composites, and (c) thick/thin hard ceramic coatings.

He serves as reviewer for many National and International Journals. Recently, Dr. Dey has chaired in technical session on "Surface Engineering, Thin Films and Coatings" in 3rd Asian Symposium on Materials & Processing (ASMP – 2012) held at IIT Madras during August 30-31, 2012, Chennai.

Dr. Anoop Kumar Mukhopadhyay is a chief scientist and head of the mechanical property evaluation section in the Materials Characterization Division of CSIR-CGCRI, Kolkata, India. He also heads the Program Management Division and Business Development Group of CSIR-CGCRI. He obtained his bachelor's degree with honours in physics from Kalyani University, Kalyani in 1978 followed by a master's degree in physics from Jadavpur University, Kolkata in 1982. In 1978, he initiated in India the research work on evaluation, analysis and microstructure mechanical properties correlation of non oxide ceramics, for high temperature applications prior to joining CSIR-CGCRI, Kolkata, India in 1986, as a staff scientist. Working on the critical parameters that control the high temperature fracture toughness of silicon nitride and its composites, he earned his Ph D. degree in science, in 1988 from the Jadavpur University, Kolkata.

During 1990–1992 he was awarded the prestigious Australian Commonwealth Post Graduate Research Fellowship and made pioneering contributions about the role of grain size in wear of alumina ceramics during his post doctoral work on development of wear and fatigue resistant oxide ceramics with world renowned Prof. Yiu-Wing Mai and Prof. Michael V. Swain at the University of Sydney, Australia.

At CSIR-CGCRI, Kolkata, Dr. Mukhopadhyay established an enthusiastic research group on evaluation and analysis of mechanical and nanomechanical properties of glass, ceramics, bioceramic coatings and biomaterials, thin films and natural biomaterials. Dr. Mukhopadhyay wrote more than 200 publications. He wrote 7 patents with 3 of them already granted, 2 book chapters already published and two books (in progress) to his credit. He has supervised seven doctoral students including one candidate who has already earned, a PhD at Bengal Engineering and Science University, Shibpur, Howrah in 2011. He contributed three chapters in "Handbook of Ceramics" edited by Dr. S. Kumar, internationally famous glass technologist and former director of CSIR-CGCRI, Kolkata, India and published by Kumar and Associates, Kolkata. He serves on the editorial board of *Soft Nanoscience Letters*.

In 2008, he won the Best Poster Paper Award at the 53rd DAE Solid State Physics Symposium. He also won in 2000 the Sir C V Raman Award of the Acoustical Society of India. In the same year, he also won the Best Poster Paper Award of the Materials Research Society of India. He was also awarded in 2000 the Visiting Scientist Fellowship to work on the fracture and nanoindentation behaviour of ceramic thermal barrier coatings with the world renowned scientist, Dr. R. W. Steinbrech at the Forschungszentrum, Juelich, Germany. He was awarded in 1997 the Outstanding Young Person Award for Science and Innovation by the Outstanding Young Achievers

Association, Kolkata and won Lions Club of India award in 1996. His work was recognized in 1995 through the best Best Poster Paper Award of the Materials Research Society of India.

Recently in 2010, his paper won the Best Research Paper Award at the Diamond Jubilee Celebration Ceremony of CSIR-CGCRI, Kolkata. His current research interests cover a truly diverse span, e.g., physics of nanoscale deformation for brittle solids, very high strain rate shock physics of ceramics, tribology of ceramics, nanotribology of ceramic coatings and thin films, microstructure mechanical and/or functional property correlation as well as ultrasonic characterisation and fatigue of (a) structural and bio-ceramics, bio-ceramic coatings, bio-materials (b) multilayer composites (c) thick/thin hard ceramic coatings. He also has a very active interest in microwave processing of ceramics, ceramic composites and ceramic metal or ceramic/ceramic joining.

Contributors

Saikat Acharya
Mechanical Property Evaluation
  Section
Central Glass and Ceramic Research
  Institute
Kolkata, India

Nil Ratan Bandyopadhyay
School of Materials Science and
  Engineering
Bengal Engineering and Science
  University
Howrah, India

Payel Bandyopadhyay
Central Glass and Ceramic Research
  Institute
Kolkata, India

Sujit Kumar Bandyopadhyay
Variable Energy Cyclotron Centre
Kolkata, India

Bikramjit Basu
Materials Research Center
Indian Institute of Science
Bangalore, India

Rajendra Nath Basu
Fuel Cell and Battery Division
Central Glass and Ceramic Research
  Institute
Kolkata, India

Manjima Bhattacharya
Central Glass and Ceramic Research
  Institute
Kolkata, India

Nilormi Biswas
Central Glass and Ceramic Research
  Institute
Kolkata, India

Manaswita Bose
Department of Energy Science and
  Engineering
Indian Institute of Technology
Mumbai, India

Himel Chakraborty
School of Materials Science and
  Engineering
Bengal Engineering and Science
  University
Howrah, India

Riya Chakraborty
Central Glass and Ceramic Research
  Institute
Kolkata, India

Pradip Sekhar Das
Mechanical Property Evaluation
  Section
Central Glass and Ceramic Research
  Institute
Kolkata, India

Probal K. Das
Non-Oxide Ceramic and Composite
  Division
Central Glass and Ceramic Research
  Institute
Kolkata, India

Tapobrata Dey
Fuel Cell and Battery Division
Central Glass and Ceramic Research
  Institute
Kolkata, India

Prakash C. Ghosh
Department of Energy Science and
  Engineering
Indian Institute of Technology
Mumbai, India

Saswati Ghosh
Fuel Cell and Battery Division
Central Glass and Ceramic Research
  Institute
Kolkata, India

Debkalpa Goswami
Department of Production
  Engineering
Jadavpur University
Kolkata, India

A. K. Himanshu
Variable Energy Cyclotron Centre
Kolkata, India

Devashish Kaushik
Department of Ceramic Engineering
Indian Institute of Technology
Varanasi, India

Anil Kumar
Ceramic Matrix Composite Division
Advanced Systems Laboratory
Defence Research and Development
  Organisation
Hyderabad, India

P. Kundu
Fuel Cell and Battery Division
Central Glass and Ceramic Research
  Institute
Kolkata, India

Prafulla K. Mallik
Department of Metallurgical and
  Materials Engineering
Indira Gandhi Institute of
  Technology
Orissa, India

Sekhar Nath
Applied Interfacial and Materials
  Science
CavinKare Research Centre
Chennai, India

A. K. Pal
Department of Instrumentation
  Science
Jadavpur University
Kolkata, India

Rajib Paul
Department of Instrumentation
  Science
Jadavpur University
Kolkata, India

Anju M. Pillai
Thermal Systems Group
ISRO Satellite Centre
Bangalore, India

A. Rajendra
Thermal Systems Group
ISRO Satellite Centre
Indian Space Research Organisation
Bangalore, India

R. Uma Rani
Thermal Systems Group
ISRO Satellite Centre
Bangalore, India

V. Sasidhara Rao
Thermal System Group
ISRO Satellite Centre
Bangalore, India

I. Neelakanta Reddy
Thermal System Group
ISRO Satellite Centre
Bangalore, India

V. R. Reddy
Department of Physics
Sri Venkateswara University
Tirupati, India

Sadanand Sarapure
Department of Mechanical
Engineering
Acharya Institute of Technology
Bangalore, India

Soumya Sarkar
Non-Oxide Ceramic and Composite
Division
Central Glass and Ceramic Research
Institute
Kolkata, India

Pintu Sen
Variable Energy Cyclotron Centre
Kolkata, India

A. Das Sharma
Fuel Cell and Battery Division
Central Glass and Ceramic Research
Institute
Kolkata, India

Anand Kumar Sharma
Thermal Systems Group
ISRO Satellite Centre
Bangalore, India

Jyoti Kumar Sharma
Department of Ceramic
Engineering
Indian Institute of Technology
Varanasi, India

Arnab Sinha
School of Materials Science and
Engineering
Bengal Engineering and Science
University
Howrah, India

Tripurari Prasad Sinha
Department of Physics
Bose Institute
Kolkata, India

N. Sridhara
Thermal Systems Group
ISRO Satellite Centre
Bangalore, India

Hari Krishna Thota
Thermal Systems Group
ISRO Satellite Centre
Bangalore, India

Section 1

Contact Mechanics

1

Contact Issues in Brittle Solids

Payel Bandyopadhyay, Debkalpa Goswami, Nilormi Biswas,
Arjun Dey, and Anoop Kumar Mukhopadhyay

1.1 Introduction

The purpose of this chapter is to provide an easy conceptual picture of
the contact-induced deformation of a brittle solid. Contact issues are very
important in our daily life. If we walk on the road or a car runs on the
path, the activity involves contact issues. For many advanced applications,
the grinding and polishing of the components made of brittle solids is a
major issue of technical, scientific, and academic interest. But, basically,
what are these processes? These are nothing but contact processes between
two solids. Here, we take a very humble approach to simulate the everyday
contact issues by using two solid bodies to understand the damage evolution. The solid body that actually applies the load is called the *indenter*, and
the other solid body that actually undergoes deformation under the loaded
indenter is called the *sample*. The deformation of this sample is the area of
interest of this book.

1.2 Elasticity and Plasticity

We are about to study the deformation of solids under loads in this book.
Do you ever think what actually happens inside a solid body when you
apply force on it? Take a very simple example. When you feel some pressure about your body, how do you react? At first, you try to tolerate the
pressure. If the pressure further increases, then you feel very stressed and
you are in a strained condition. Similar things actually happen for a solid
body. When you apply force to a body, the force actually acts on a uniquely
defined area of a surface. The solid body undergoes deformation, and a
reaction force generates inside the material to resist the deformation. Stress
is the reaction force per unit area. It is usually measured by the applied

force, as the applied force is proportional to the reaction force. The term *strain* actually relates to the deformation. Thus, it is the deformation per unit dimension. The typical stress-versus-strain curve is illustrated in Figure 1.1a. At first, the stress is proportional to the strain up to a certain limit. This is the proportionality limit, i.e., the point A in Figure 1.1a. Within this limit, stress is proportional to the strain, which is what the Hooke's law states. Further, the constant of proportionality is known, simplistically, as the elastic modulus. The point B in Figure 1.1a is called the elastic limit. The elastic limit of a material is the lowest stress at which permanent deformation could be measured. For elastomers, the elastic limit is much higher

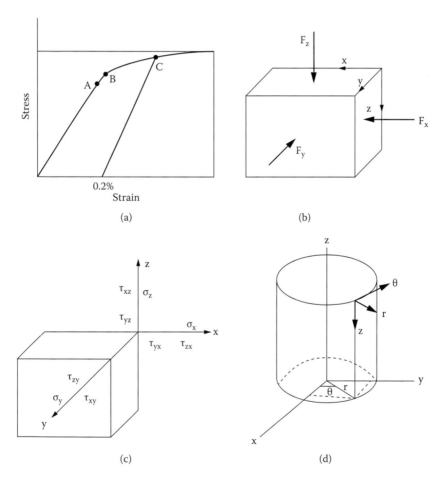

FIGURE 1.1
(a) Typical stress–strain curve; (b) applied forces acting on a solid body; (c) directions of the components of the generated stress on different planes in a solid body; and (d) cylindrical coordinate system. (Reprinted with permission of Bandyopadhyay and Mukhopadhyay [6] from Elsevier.)

than the proportionality limit. The point C in Figure 1.1a is the offset yield point. The yield point is a point in the stress-versus-strain curve where plastic deformation starts. When it is very difficult to define the yield point uniquely, then we define a point at about 0.2% of the strain. This is called the offset yield point. Moreover, within the elastic limit, the solid body deforms elastically, which implies that it can recover fully after withdrawal of the externally applied load. Beyond this point, the solid body starts to deform permanently. When applied load reaches a critical value, such that the stress experienced by the solid body is equal to or greater than the failure strength of the solid, it cannot tolerate further application of load and fails completely.

1.3 Stresses

The term *stress* has huge importance in this field of research. There are three types of stress. The first is the tensile stress. This type of stress tries to pull the surface apart into at least two parts in two opposite directions. Thus it always acts normal to a given plane. The second is the compressive stress. This type of stress tries to compress the surface from one given or from two given opposite directions. The compressive stress and/or stresses also act(s) normal to a given plane. Such stresses are usually expressed as σ_{xx}, σ_{yy}, and σ_{zz}.

Thus, the stresses are taken as tensile if the sign is positive and compressive if the sign is negative. Further, they act along x-, y-, and z-directions. In other words, they act on the yz, zx, and xy planes. It is well known that force is a vector quantity that has three components. Let the components be F_x, F_y, and F_z in the x-, y-, and z-directions, as shown in Figure 1.1b. F_x is perpendicular to the yz plane. So, $\sigma_{xx} = F_x$/(area of action of force on the yz plane). Similarly, the generated normal stresses on the xy and zx planes are respectively given by $\sigma_{zz} = F_z$/(area of action of force on the xy plane) and $\sigma_{yy} = F_y$/(area of action of force on the zx plane).

The third type of stress is the shear stress, which is totally different from the two other stresses because shear force is different from the force that acts along the direction normal to a surface. The shear force is a force that tries to slide past one part over another part of a given surface and/or a given surface over another given surface. Thus, the shear stress is basically the reaction force per unit area with which a given material tries to resist finally sliding out due to shearing action of the externally applied shear force. As the component F_x is parallel to the xy plane, it generates a shear stress component on the xy plane. The generated shear stress component on the xy plane is, $\tau_{zx} = F_x$/(area of action of force on xy plane). Similarly, F_y is also parallel to the xy plane, so it also generates shear stress in that plane.

Therefore, altogether nine components of stresses can act on a solid and may be expressed by a corresponding matrix notation as follows:

$$
\begin{bmatrix}
\sigma_{xx} & \tau_{xy} & \tau_{xz} \\
\tau_{yx} & \sigma_{yy} & \tau_{yz} \\
\tau_{zx} & \tau_{zy} & \sigma_{zz}
\end{bmatrix}
$$

The directions of the nine components of the generated stress are shown in Figure 1.1c.

The maximum tensile stress due to static contact $\left(\sigma_m^s\right)$ under a normal load (P) in the Hertzian contact situations (as will be described many times in this book) is calculated using the following equations [1, 2]:

$$
\sigma_m^s = \frac{(1-2v_s)p_0}{2} \tag{1.1}
$$

where $p_0 = P/\pi a_s^2$ is the unit of stress and the corresponding contact radius (a_s) is given by [1]

$$
a_s = \left(\frac{4}{3}kPrE_s\right)^{\frac{1}{3}} \tag{1.2}
$$

In equation (1.2), r is the radius of the indenter. It typically varies from 150 nm for a sharp Berkovich indenter to about 200 μm for a blunt spherical indenter. Here, E_s is the Young's modulus of the sample, and k is a factor given by [1]

$$
k = \frac{9}{16}\left[(1-v_s^2)+(1-v_i^2)\frac{E_s}{E_i}\right] \tag{1.3}
$$

In equation (1.3), v_i and E_i are the Poisson's ratio and the Young's modulus of the indenter, respectively, and v_s and E_s are the Poisson's ratio and Young's modulus of the sample. Further, the maximum tensile stresses due to dynamic contact $\left(\sigma_m^d\right)$ between a brittle solid (e.g., a glass surface) and the sliding indenter can be obtained from the following equation [1–3]:

$$
\sigma_m^d = (1+15.5\mu)\sigma_m^s \tag{1.4}
$$

where μ is the coefficient of friction between the glass sample and the sliding indenter, and σ_m^s is calculated using equations (1.1), (1.2), and (1.3). The position of the maximum tensile stress occurs for all coefficients of friction (μ) at the trailing edge of the indenter. For all applied normal load $P > P_c$, the critical load for crack initiation, a cone-shaped fracture is initiated [1, 2] because now $\left(\sigma_m^s\right)$ will be greater than the fracture strength σ_f of the brittle solid (e.g., a glass). The frustum of this fracture cone intersects the glass surface close to the circle of contact of radius (a_s), which now becomes the critical contact radius a_c. These cracks are termed as partial cone cracks [1] because each individual crack leaves behind an incomplete arcuate trace on the glass surface. The incomplete surface traces of these partial cone cracks on the glass surface are termed as *ring cracks* [1], which are nearly equispaced. It follows from this that introduction of the sliding contact enhances the possibility of producing manifold ring cracks.

The values of normalized shear stress (τ') can be estimated at various points in the (r,z) plane in a cylindrical coordinate system (Figure 1.1d) using the following equations [4–6]:

$$\tau' = \frac{\sigma_1' - \sigma_2'}{2} \tag{1.5}$$

Here the normalized principal stresses σ_1' and σ_3' are given by [4–6]:

$$\sigma_1' = \frac{\sigma_r' + \sigma_z'}{2} + \sqrt{\left(\frac{\sigma_r' + \sigma_z'}{2}\right)^2 + \sigma_{rz}'^2} \tag{1.6}$$

$$\sigma_3' = \frac{\sigma_r' + \sigma_z'}{2} + \sqrt{\left(\frac{\sigma_r' + \sigma_z'}{2}\right)^2 + \sigma_{rz}'^2} \tag{1.7}$$

The normalized radial stress σ_r' is given by [4–6]:

$$\sigma_r' = \frac{\sigma_r}{P_m} = \frac{3}{2}\left\{\frac{1-2v}{3r'^2}\left[1-\left(\frac{z'}{\sqrt{u'}}\right)^3\right]\right.$$

$$+\left(\frac{z'}{\sqrt{u'}}\right)^3 \frac{u'}{u'^2 + z'^2} + \frac{z'}{\sqrt{u'}}\left[u'\frac{(1-v)}{1+u'} + (1+v)\sqrt{u'}\tan^{-1}\left(\frac{1}{\sqrt{u'}}\right) - 2\right]\right\} \tag{1.8}$$

where $2v$ is the Poisson's ratio of the sample. The normal stress σ_z' is given by [4–6]:

$$\sigma_z' = \frac{\sigma_z}{P_m} = -\frac{3z'^3}{2\left(u'^2 + z'^2\right)\sqrt{u'}} \tag{1.9}$$

However, the stress component σ_{rz}' acting at the r,z plane is given by [4–6]:

$$\sigma_{rz}' = \frac{\sigma_r}{P_m} = -\frac{3r'z'^2\sqrt{u'}}{2(u'^2 + z'^2)(1+u')} \tag{1.10}$$

Here, the maximum contact pressure, P_m, is given by [6, 7]:

$$P_m = \left(\frac{6E_r^2 P_{eff}}{\pi^3 R^2}\right)^{1/3} \tag{1.11}$$

In equation (1.11), R is the indenter radius and P_{eff} is the effective normal load, given by

$$P_{eff} = P(1+\mu^2)^{0.5} \tag{1.12}$$

where $2P$ is the applied normal load and μ is the friction coefficient.
E_r is the reduced Young's modulus, given by

$$\frac{1}{E_r} = \frac{1-v_s^2}{E_s} + \frac{1-v_i^2}{E_i} \tag{1.13}$$

where $2v_s$ and E_s are, respectively, the Poisson's ratio and Young's modulus of the sample, and v_i and E_i are the Poisson's ratio and Young's modulus of the indenter.

In equations (1.8), (1.9), and (1.10), the normalized displacement u' is given by [4, 5]:

$$u' = \frac{1}{2}\left[r'^2 + z'^2 - 1 + \left\{\left(r'^2 + z'^2 - 1\right)^2 + 4z'^2\right\}^{1/2}\right] \tag{1.14}$$

In equations (1.8), (1.9), (1.10), and (1.14), the normalized radial distance r' and the normalized depth z' are given by [4–6]:

$$z' = \frac{z}{a_s} \tag{1.15}$$

$$r' = \frac{r}{a_s} \tag{1.16}$$

Here, the dynamic contact radius a_d is given by [4–6]:

$$a_d = \left(\frac{3P_{eff}R}{4E_r}\right)^{1/3} \tag{1.17}$$

The normalized stresses are functions of Poisson's ratio only. Figures 1.2a–g show the typical stress contours for a material of Poisson's ratio $v \approx 0.34$. The normal load is applied at (0, 0) position, and the various normalized values of the stresses are indicated with different colors. All the stress contours are distorted semicircles. The magnitudes of σ_1, σ_2, σ_3, σ_r, and σ_{rz} are maximum

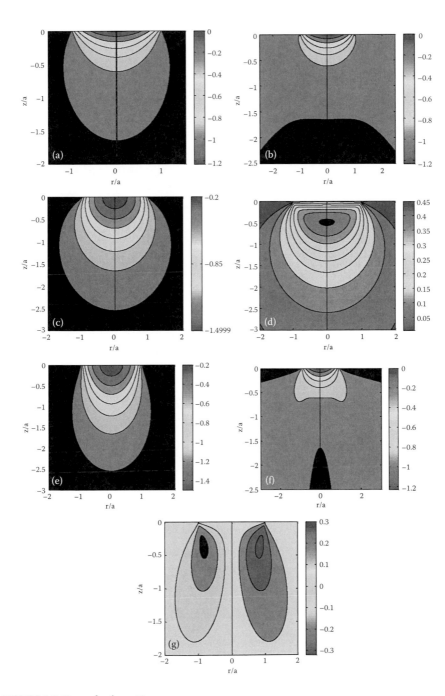

FIGURE 1.2 (See color insert.)
Normalized stress contours under a Hertzian contact for Poisson's ratio $v = 0.34$: (a) principal stress σ_1', (b) principal stress σ_2', (c) principal stress σ_3', (d) shear stress τ, (e) normal stress σ_z', (f) radial stress σ_r', and (g) principal stress acting on the *rz* plane, σ_{rz}.

at the point of contact, i.e., at (0, 0) point (Figures 1.2a–e). But the shear stress is maximum at a point slightly inside the material (Figure 1.2f). The contour of the component of the principal stress acting on the rz plane (σ_{rz}) is completely different in shape. It has two regions in the stress distribution plot that are mirror images of one another (Figure 1.2g). The value of the normalized stress is positive on one side and negative on another side, which implies that the stress is tensile on one side and compressive on another side. The damage evolution on the subsurface depends on both the principal and shear stresses. The magnitudes of the generated stresses actually control the damages in the subsurface. When the magnitudes exceed a critical value, they generate microcracks that further increase the removal of material. The maximum shear stress underneath the indenter is theoretically predicted using the following equation [5–8]:

$$\tau_{\max} = 0.445 \left(\frac{16 P_{\text{eff}} E_r^2}{9\pi^2 R^2} \right)^{1/3}. \tag{1.18}$$

1.4 Conclusions

This chapter presented the relevant aspects of contact deformation of brittle solids and the related mathematical formalisms. This is just a starting point for us to look into further details about the contact mechanics when the contact is purely elastic and when the contact is elastoplastic in nature. This is what we are going to do in Chapter 2. The reason for doing so is that in both glass and ceramics, which are brittle solids, both of these types of deformations can and do happen during the indentation process in general, and nanoindentation in particular.

References

1. Lawn, B. R. 1967. Partial cone crack formation in a brittle material loaded with a sliding spherical indenter. *Proceedings of the Royal Society of London A* 299:307–16.
2. Lawn, B. R., and F. C. Frank. 1967. On the theory of Hertzian fracture. *Proceedings of the Royal Society of London A* 299:291–306.
3. Hamilton, G. M., and L. E. Goodman. 1966. The stress field created by a circular sliding contact. *Journal of Applied Mechanics* 33:371–76.
4. Fischer Cripps, A. C. 2000. *Introduction to contact mechanics*. New York: Springer.
5. Packard, C. E., and C. A. Schuh. 2007. Initiation of shear bands near a stress concentration in metallic glass. *Acta Materialia* 55:5348–58.

6. Bandyopadhyay, P., and A. K. Mukhopadhyay. 2013. Role of shear stress in scratch deformation of soda-lime-silica glass. *Journal of Non-Crystalline Solids* 362:101–13.
7. Mao, W. G., Y. G. Shen, and C. Lu. 2011. Deformation behavior and mechanical properties of polycrystalline and single crystal alumina during nanoindentation. *Scripta Materialia* 65:127.
8. Shang, H., T. Rouxel, M. Buckley, and C. Bernard. 2006. Viscoelastic behavior of a soda-lime-silica glass in the 293–833 K range by micro-indentation. *Journal of Materials Research* 21:632–38.

2

Mechanics of Elastic
and Elastoplastic Contacts

Manjima Bhattacharya, Arjun Dey, and Anoop Kumar Mukhopadhyay

2.1 Introduction

In the previous chapter we looked into the contact issues in brittle solids. In this chapter, we shall therefore try to understand what the different types of contacts are and how the mechanics change with the type of contact. There are excellent reviews [1–4] available on this subject, particularly in connection with the science of deformation and fracture at the nanoscale, for a truly wide variety of materials. While that knowledge base is of extraordinary importance to the advanced researcher, here, in particular, our approach will be to initiate the understanding of contact mechanics for a beginner. Such an understanding will help us to make a brief scan of the different existing models used to describe the contacts in terms of experimentally measurable physical quantities. As such, there can be many possible types of contact that can happen theoretically in a solid. To start with, it must be borne in mind here that two materials making contact, are involved. One of these is the indenter, and the other is the sample, which is exposed to the indentation by the indenter. There can be situations when the sample is perfectly elastic while the indenter is incompressible and rigid. It may also happen that the sample may be considered as perfectly plastic up to a fixed stress, and the indenter may be still considered as perfectly rigid with a sharp or a worn tip. It may also happen that the indentation process can be likened to that of a physical expansion of an existing cavity in a given solid under the applied contact pressure. There may be many other possible combinations of identities of the idealized indenter and the idealized sample, but for the sake of briefness, we shall deal here with only those three models that are most extensively utilized in the literature [5–7]. These are

1. Elastic indentation model
2. Rigid perfectly plastic model

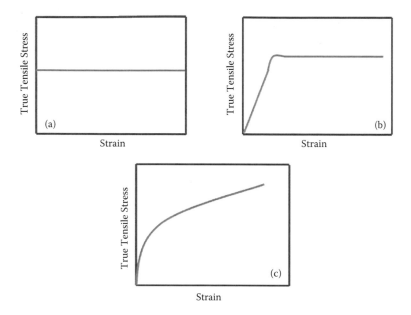

FIGURE 2.1
Stress-versus-strain behaviors corresponding to (a) the elastic indentation model, (b) the rigid perfectly plastic model, and (c) the spherical-cavity expansion model.

3. Spherical-cavity expansion model
4. Elastic and perfectly plastic model

The stress-versus-strain behaviors of solids pertinent to these three models are shown in Figures 2.1a–c.

2.2 The Different Models

2.2.1 The Elastic Indentation Model

Hertz in the year 1882 [5] developed the elastic indentation model. Later on, other researchers [6, 7] had worked independently on the same problem. The basic premise of these models is simple. Here, the indenter is assumed to be a rigid cone. This means that any deformation of the conical indenter will be insignificant in comparison to that of the sample. The sample is assumed to be linearly elastic (Figure 2.2). The corresponding pressure distribution, $p(r)$, for the contact circle of the cone can be expressed as:

$$p(r) = \frac{E \cot \theta}{2(1-v^2)} \cosh^{-1}\left(\frac{a}{r}\right) \tag{2.1}$$

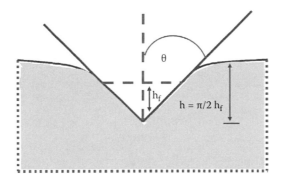

FIGURE 2.2
Schematic representation of the indentation in an elastic solid. (Adapted from Mukhopadhyay and Paufler [4].)

where r is the radial distance measured from the center of the contact circle of diameter a created by the indentation process on the sample, E is the Young's modulus, v is the Poisson's ratio, and θ is the semi-apical angle of the cone. From equation (2.1), using the boundary condition $a = r$, the mean contact pressure can be written as:

$$p_m = H = \frac{E \cot \theta}{2\left(1 - v^2\right)} \tag{2.2}$$

This quantity, p_m, then can be likened to be the measure of hardness (H) of a linearly elastic solid due to indentation by a perfectly rigid indenter. However, the indenter shape can be of many types, e.g., wedge, spherical, and cylindrical with a flat contact end. The solutions for these and many other geometries are readily available today in the literature [7, 8] for the interested student and researcher. In particular, for a spherical indenter of radius R, H can be expressed as a function of the load (P) by equation (2.3):

$$H = \frac{P^{\frac{1}{3}}}{\pi}\left(\frac{4E}{3R\left(1 - v^2\right)}\right)^{\frac{2}{3}} \tag{2.3}$$

In deriving equation (2.3), we need to assume that the friction and the elastic deformation of the indenter are negligible. Now, the impact of the indentation disappears completely after unloading for a perfectly elastic material. Therefore, the depth of penetration can be measured from the instrumented hardness testing equipment so that the area can be found from the depth measurement.

2.2.2 The Rigid Perfectly Plastic Model

Prandtl [9] advocated this model, which was based on the slip-line field theory. This approach was later put on a more generalized mathematical framework [10]. Here also, the indenter was assumed to be perfectly rigid. But, in contrast to the elastic model, here the sample is considered to be rigid and perfectly plastic. It means that no plastic deformation will occur until a limiting or cutoff stress Y is attained. The uniform pressure p across the punch is given by:

$$p = cY = 2k\left(1 + \frac{\pi}{2}\right) \tag{2.4}$$

This uniform pressure p can be likened to the hardness, H. Therefore, one can safely assume that, in general, $p = H = cY$, where c is the appropriate constraint factor, depending on the geometry of the indenter and the interfacial friction. For instance, when the stress condition fulfills the Tresca criterion, c will assume the value 2.6. Similarly, when the stress condition fulfills the von Mises criterion, c may be as high as 3 [11–13]. The most interesting aspect of this model is that even for a sharp contact situation, it can predict hardness values for materials having a high magnitude of the ratio E/Y. However, when the contact situation to be tackled is that of a blunt indenter, the model is not successful in predicting the hardness values of those materials having a relatively lower magnitude of E/Y. In such cases, where there is no friction at the interface between the sample and the rigid wedge-shaped indenter of semi-apical angle θ, the pressure across the face of the indenter is given by:

$$p = 2k(1 + \theta) \tag{2.5}$$

In other words, the material's hardness, H, can be equivalently expressed as

$$H = 2k(1 + \theta) \tag{2.6}$$

2.2.3 The Spherical-Cavity Expansion Model

In this model, the indentation process has been imagined to be like that of a physical expansion of an existing cavity in a given solid under the applied contact pressure. Johnson's analysis [11, 12] has been the most successful one for the spherical-cavity model (Figure 2.3). The situation closely resembles the indentation of elastic–plastic solids. The results of this model could easily fit for the materials having higher as well as lower E/Y values. As indicated in Figure 2.3, the radial expansion model is the bottom half of the spherical cavity. It follows that the picture assumes that there is no

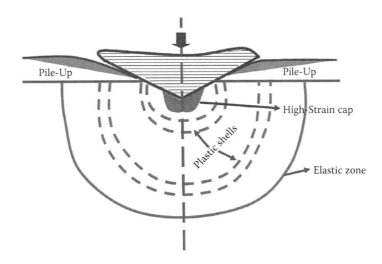

FIGURE 2.3
Schematic representation of the indentation process according to the spherical-cavity expansion model. (Adapted from Mukhopadhyay and Paufler [4].)

piling up of any displaced material outside the indentation. It is therefore obvious that indentation volume is taken care of in the region well behind the plastic shell where the material is considered to be totally elastic. Thus, to take care of the indentation volume, the shape of the indenter comes into the picture that represents the ratio of hardness, H, to the yield stress, Y. For instance, when the indenters are of conical and pyramidal shapes, the relation is given by:

$$\frac{H}{Y} = \frac{2}{3}\left[1 + \ln\left(\frac{E\cot\theta}{3Y}\right)\right] \qquad (2.7)$$

In the case of spherical indenters,

$$\frac{H}{Y} = \frac{2}{3}\left[1 + \ln\left(\frac{E(a/R)}{3Y}\right)\right] \qquad (2.8)$$

This treatment assumes that the material has a constant yield stress, Y, and that there is no work hardening produced by the indentation process itself. According to Hill, Lee, and Tupper [10], when the elastic–plastic boundary is at a distance c from the center of the contact, the cavity pressure is given by:

$$P = Y\left(\frac{2}{3} + 2\ln\frac{c}{a}\right) \qquad (2.9)$$

It follows from equation (2.9) that when the elastic–plastic boundary coincides with the boundary of the cavity (i.e., at $c = a$), the contact pressure is given by:

$$P_C = \frac{2}{3}Y \qquad (2.10)$$

Thus, this value of P defines a critical applied pressure P_c required for plastic deformation to occur at all. If the applied pressure P is less than P_c, no plastic deformation occurs. However, it automatically follows that if the indentation pressure is less than Y, no plastic indentation usually takes place.

2.2.4 The Elastic and Perfectly Plastic Model

Recently, the analytical relationship for calculating hardness from the uniaxial material properties and indenter geometry for a wide variety of elastic and plastic materials has been developed [14] using an elastic, perfectly plastic model. From this model, the hardness of a material due to the conical indentation is given by:

$$H = \frac{E\tan\theta}{2(1-v^2)}\tanh\left(\frac{2(1-v^2)YC_\theta}{E\tan\theta}\right) \qquad (2.11)$$

where $C_\theta = \frac{2}{\sqrt{3}}(2.845 - 0.002\,\theta)$ and the semi-apical angle of the indenter is in the range of $0° < \theta \le 37.5°$.

2.3 Conclusions

We have tried here to understand how the elastic indentation model, the rigid perfectly plastic model, and the spherical-cavity expansion model can be utilized to determine hardness of a given material. In addition, the use of an elastic, perfectly plastic model for evaluation of a material's hardness was also briefly discussed. It is not only important to understand the mechanics part of the whole story, it is also imperative that we learn a little bit about the history of indentation, per se. Therefore, in Chapter 3 we present a brief history of the science and technology of the indentation technique itself.

References

1. Fischer-Cripps, A. C. 2006. Critical review of analysis and interpretation of nanoindentation test data. *Surface and Coatings Technology* 200:4153–65.
2. Palacio Manuel, L. B., and B. Bhushan. 2013. Depth-sensing indentation of nanomaterials and nanostructures. *Materials Characterization* 78:1–20.
3. Hussein, N., K. Kalantar-zadeh, M. Bhaskaran, and S. Sriram. 2013. In situ nanoindentation: Probing nanoscale multifunctionality. *Progress in Materials Science* 58:1–29.
4. Mukhopadhyay, N. K., and P. Paufler. 2006. Micro- and nanoindentation techniques for mechanical characterisation of materials. *International Materials Reviews* 51:209–45.
5. Hertz, H. 1882. Ueber die Berührung fester elastischer Körper. *Journal für die reine und angewandte Mathematik* 92:156–71.
6. Boussinesq, J. 1885. *Applications des potentials a l'etude de equilibre et du mouvement des solides elastiques*. Paris: Gauthier Villars.
7. Sneddon, I. N. 1965. The relation between load and penetration in the axisymmetric Boussinesq problem for a punch of arbitrary profile. *International Journal of Engineering Science* 3:47–57.
8. Timoshenko, S., and J. N. Goodier. 1951. *Theory of elasticity*. New York: McGraw-Hill.
9. Prandtl, L. 1920. Nachrichten von der Koniglichen Gesellschaft der Wissenschaften zu Gottingen, Mathematisch-Physikalisch. *Klasse* 37:74–85.
10. Hill, R., E. H. Lee, and S. J. Tupper. 1947. The theory of wedge indentation of ductile materials. *Proceedings of the Royal Society A* 188: 273–89.
11. Johnson, K. L. 1970. The correlation of indentation experiments. *Journal of the Mechanics and Physics of Solids* 18:115–26.
12. Johnson, K. L. 1985. *Contact mechanics*. Cambridge, U.K.: Cambridge University Press.
13. Yu, W., and J. P. Blanchard. 1996. An elastic–plastic indentation model and its solution. *Journal of Materials Research* 11:2358–67.
14. Fischer-Cripps, A. C. 2007. *Introduction to contact mechanics*. New York: Springer.

Section 2

Journey towards Nanoindentation

3

Brief History of Indentation

Nilormi Biswas, Arjun Dey, and Anoop Kumar Mukhopadhyay

3.1 Introduction

Nanoindentation today is a well-established experimental means to investigate the deformation of bulk materials, thick coatings, biomaterials, and thin films [1–5]. It is almost obvious, therefore, that to understand the nanoindentation behavior and mechanisms of different materials, it is equally important also to know a little about the path that has been charted by numerous researchers [6–35] whose phenomenal contributions have carried us to where we are today. Thus, our focus in this chapter will be the history of indentation.

3.2 How Did It All Happen?

Well, prehistoric humans also used indentation, albeit unknowingly, by using a sharp pointed stone or an arrow to intimidate, attack, or kill animals and/or enemies. Also, there is the popular comic-strip story about the "walking ghost" who used to leave "a danger signal" impression on the "evil fellow's" cheek or forehead—a permanent dent or impression on the body. Thus indentation is supposed to leave a permanent mark. It is permanent because it is causing deformation beyond that acceptable by the purely classical elastic limit as provided by Hooke's law.

3.3 And Then There Was a...

This is how a chapter in your school history book would often start, isn't it? The earliest report regarding the study of hardness of materials was given by Huygens. He wrote about this in his *Traité de la Lumière* [6]. What he had found out is the following: while trying to scratch an Iceland spar by a knife held at two different angles to the sliding direction, he observed that the performance of the same knife had differed [6, 7]. So, one direction was an easier direction or softer direction of scratching, and the other direction was a relatively harder direction of scratching. Similarly, none other than Sir Isaac Newton had noticed that a ductile metal was more difficult to polish than a brittle glass [8, 9].

In 1822, Mohs introduced a 10-stage hardness scale and instructions for its realization [10]. The first machines for measuring indentation hardness were reported around the mid-1850s [11, 12]. Around 1900, machines for measuring indentation hardness under both static and dynamic conditions became commercially available [13–15]. This technological growth was driven by the market demand. It was a prime time in the history of industrial developments in human civilization. Hence, there was a huge market demand for high-strength, high-hardness iron and steel as well as for wear-resistant tools that would be required to machine them [16, 17].

3.4 Modern Developments: Nineteenth-Century Scenario

In the early 1900s, the Brinell ball test [18] was the most widely used hardness-measuring technique. It originally consisted of pressing a hard steel ball under a known load into the material of interest. The Brinell hardness (H_c) was then calculated by dividing the load P by the surface area of the indentation [19]. The same method was also suggested by Martens in 1898 [13, 20], i.e.,

$$H = P/\pi Dd \tag{3.1}$$

where D is the diameter of the ball indenter and d is the depth of the indentation. Later on, Meyer [21] found that the indentation diameter, d_i, is a function of the load, i.e.,

$$P = ad_i^n \tag{3.2}$$

where a and n are numbers that both depend on the material being tested. The quantity a also depends on the size of the ball.

One problem with the Brinell test is that the hardness number tended to increase with increasing load due to a combination of work hardening and an increase in the diameter of the indentation. To counter this problem, the "scleroscope" was developed in 1907. It measured the rebound of a steel ball or pointed cylinder dropped down a graduated glass tube onto the material of interest [22]. It helped to make the pioneering observation that the "Shore hardness" (the name derived from the inventor of the scleroscope machine) was proportional to elasticity. Another dynamic method of measuring hardness was patented in the same year [23]. Repetitive Shore hardness tests on the same spot led to hardness enhancement due to work hardening with the number of tests [24].

3.5 Comparison of Techniques

To address the issue of the differences between the static and dynamic mechanical properties, Shore and Hadfield performed a large number of experiments [25]. They identified that the Shore hardness gave results closer to the elastic [24] limit of the material, whereas Brinell hardness measured properties closer to the ultimate strength. Thus, it was identified from the results of these experiments that the results from these two tests would be comparable if, and only if, low indentation loads were applied in the Brinell hardness measurement experiments [26, 27].

3.6 Major Developments beyond 1910

In 1919, a method of hardness testing was patented by Rockwell and Rockwell. It had the ability to subtract the elastic response of the measured data so that only the plastic hardness could be measured. Before the invention of this machine, there was no means to measure the hardness of materials harder than those that the Brinell test could accurately measure. This problem was solved by the invention of the Rockwell C hardness tester, which used a diamond cone [28].

The next stage of development followed during the mid-1920s to the mid-1930s [29–31]. This period saw the development of the Vickers hardness testers. The Vickers hardness testers used a square-section diamond pyramid known as the Vickers pyramid, such that the geometrical similarity is preserved during an indentation test [29–31]. It was already known that the conical indenters also exhibited geometrical similarity [32].

The subsequent level of development was prompted by the urge to determine both the recovered and the unrecovered dimensions of the indentation impression. This need led to the development of the Knoop hardness testers by the end of 1940s. The Knoop hardness testers [33] used a diamond indenter with a specific geometry. Here, the diamond indenter had diagonals that differ from each other by a factor of 7. This inherent anisotropy in the shape of the indenter causes the maximum elastic recovery along the shortest diagonal and the minimum along the longest diagonal.

3.7 Beyond the Vickers and Knoop Indenters

A major technical as well as technological problem with both Vickers and Knoop diamond indenters was that it was always difficult to exactly ensure that all four sides of the corresponding four-sided pyramid would meet at a point. This problem was soon identified by Berkovich [34] within only about a decade after the Knoop diamond indenter was identified. How did Berkovich address this issue? Well, to overcome this problem, Berkovich introduced the triangular cross-section (or three faceted) pyramidal diamond indenter [34]. The introduction of the Berkovich indenters has certainly played a very important role in the subsequent development of the commercial nanoindentation machines that provide the facility for indentation testing at ultralow micronewton load ranges today [35]. Table 3.1 presents a

TABLE 3.1

Comparison of Hardness Evaluation for Four Different Indentation Tests

Method of Testing	Impression Shape	Contact Area	Hardness Value
Berkovich		$A = a^3\sqrt{(3/2)}$	$H_B = F/A$ $= F/a^3\sqrt{(3/2)}$
Knoop		$A = d^2/14.4$	$H_K = F/A$ $= F \times 14.4/d^2$
Vickers		$A = d^2/(2 \times \sin 68°)$ (theoretically, $d_1 = d_2 = d$)	$H_V = F/A$ $= 1.8544 \times F/d^2$
Brinell		$A = \{\pi D^2$ $[1 - (1 - (d/D)^2)]^{1/2}\}/2$	$H_C = F/A$ $= 2F/\{\pi D^2$ $[1 - (1 - (d/D)^2)]^{1/2}\}$

Note: F = applied force; A = projected contact area.

compendium of the quantitative yet comparative picture of how hardness values are measured by the different indentation testing techniques.

3.8 Conclusions

In this chapter we have tried to understand the genesis of indentation testing from near the end of the sixteenth century up to the 1980s. The paths from Brinell hardness testers to today's Berkovich nanoindenters were charted, along with a retrospective of the logical sequence of developments in the history of indentation. Now that we understand how the field of indentation testing has evolved, Chapter 4 will discuss the situations in which this technique can be used to derive extremely important surface mechanical properties such as nanohardness and the Young's modulus.

References

1. Sebastiania, M., C. Eberl, E. Bemporad, and G. M. Pharr. 2011. Depth-resolved residual stress analysis of thin coatings by a new FIB–DIC method. *Materials Science and Engineering A* 528:7901–8.
2. Li, X., and B. Bhushan. 2002. A review of nanoindentation continuous stiffness measurement technique and its applications. *Materials Characterization* 48:11–36.
3. Nili, H., K. Kalantar-zadeh, M. Bhaskaran, and S. Sriram. 2013. In situ nanoindentation: Probing nanoscale multifunctionality. *Progress in Materials Science* 58:1–29.
4. Roa, J. J., G. Oncins, J. Díaz, X. G. Capdevila, F. Sanz, and M. Segarra. 2011. Study of the friction, adhesion and mechanical properties of single crystals, ceramics and ceramic coatings by AFM. *Journal of the European Ceramic Society* 31:429–49.
5. Kurland, N. E., Z. Drira, and V. K. Yadavalli. 2012. Measurement of nanomechanical properties of biomolecules using atomic force microscopy. *Micron* 43:116–28.
6. Huygens, C. 1690. Ou sont expliquées les causes de ce qui luy arrive dans la réflexion et dans la refraction, et particulièrement dans l'étrange refraction du cristal d'Islande. In *Traité de la Lumière*, 95–96. Leiden, Netherlands: Pierre van der Aa.
7. Huygens, C. 1912. *Treatise on light*. London: Macmillan.
8. Newton, I. 1730. *Opticks: or a treatise of the reflections, refractions, inflections and colours of light*, Proposition 7, Theorem 6. London: William Innys.
9. Newton, I. 1931. *Opticks: or a treatise of the reflections, refractions, inflections and colours of light*, 104–7. London: G. Bell and Sons.
10. Mohs, F., and W. Haidinger. 1825. *Treatise on mineralogy or the natural history of the mineral kingdom*. Edinburgh: printed for A. Constable; London: Hurst, Robinson.

11. Wade, W. 1856. Hardness of metals. In *Reports on experiments on the strength and other properties of metals for cannon with a description of the machines for testing metals, and of the classification of cannon in service*, 259–75, 313–14. Philadelphia, PA: Henry Carey Baird.
12. Calvert, F. C., and R. Johnson. 1859. On the hardness of metals and alloys. *Philosophical Magazine* 17:114–21.
13. Martens, A. 1898. *Handbuch der Materialenkunde*. Berlin: Springer.
14. Shore, A. F. 1911. The property of hardness in metals and materials. *Proceedings of the American Society for Testing Materials* 11:733–43.
15. Lysaght, V. E. 1949. *Indentation hardness testing*. New York: Reinhold.
16. Turner, T. 1886. The hardness of metals. *Proceedings of Birmingham Philosophical Society* 5:282–312.
17. Turner, T. 1909. Notes on tests for hardness. *Journal of the Iron and Steel Institute* 79:426–43.
18. Brinell, J. A. 1900. Ein Verfahren zur Härtebestimmung nebst einigen Anwendungen desselben. *Baumaterialienkunde* 5:276–80, 294–97, 317–20, 364–67, 392–94, 412–16.
19. Springer, J. F. 1908. Methods of testing materials for hardness. *Cassier's Magazine* 34:387–401.
20. Wilde, H. R., and A. Wehrstedt. 2001. Introduction of Martens hardness HM. *Materialprufung* 42:468–70.
21. Meyer, E. 1908. Investigations of hardness testing and hardness. *Physikalische Zeitschrift* 9:66–74.
22. Shore, A. F. 1910. The scleroscope. *Proceedings of the American Society for Testing Materials* 10:490–517.
23. Ballentine, W. I. 1907. Processing of testing the hardness and density of metals and other materials. U.S. Patent 855,923.
24. Howe, H. M., and A. G. Levy. 1916. Notes on the Shore scleroscope test. *Proceedings of the American Society for Testing Materials* 16:36–52.
25. Shore, A. F., and R. Hadfield. 1918. Report on hardness testing: Relation between ball hardness and scleroscope hardness. *Journal of the Iron and Steel Institute* 98:59–88.
26. Lysaght, V. E. 1971. Hardness conversion tables for copper. *J. Mater.* 6:503–13.
27. Sakharova, N. A., J. V. Fernandes, J. M. Antunes, and M. C. Oliveira. 2009. Comparison between Berkovich, Vickers and conical indentation tests: A three-dimensional numerical simulation study. *International Journal of Solids and Structures* 46:1095–1104.
28. Scott, H., and T. H. Gray. 1939. Relation between the Rockwell "C" and diamond pyramid hardness scales. *Transactions of the American Society of Metals* 27:363–81.
29. Smith, R. L. 1916. Apparatus for determining the hardness of a body. U.K. Patent 11,936, 1916.
30. Smith, R., and G. Sandland. 1922. An accurate method of determining the hardness of metals, with particular reference to those of a high degree of hardness. *Proceedings of the Institution of Mechanical Engineers* 102:623–41.
31. Smith, R. L., and G. E. Sandland. 1925. Some notes on the use of a diamond pyramid for hardness testing. *Journal of the Iron and Steel Institute* 111:285–304.
32. Ludwik, P. 1908. *Die Kugeldruckprobe, ein neues Verfahren zur Härtebestimmung von Materialien*. Berlin: Springer.

33. Knoop, F., C. G. Peters, and W. B. Emerson. 1939. A sensitive pyramidal diamond tool for indentation measurements. *Journal of Research of the National Bureau of Standards* 23:39–61.
34. Berkovich, E. S. 1951. Three-faceted diamond pyramid for microhardness testing. *Industrial Diamond Review* 11:129–32.
35. Newey, D., M. A. Wilkins, and H. M. Pollock. 1982. An ultra-low-load penetration hardness tester. *Journal of Physics E: Scientific Instruments* 15:119–22.

4

Hardness and Elastic Modulus

Nilormi Biswas, Arjun Dey, and Anoop Kumar Mukhopadhyay

4.1 Introduction

In Chapter 3, we dealt with the brief history of indentation testing. It would sound silly if we were to ask you to distinguish between what is hard and what is elastic. You could easily tell me that the rubber band used to tie up a ponytail is elastic and that the wall on which you display your power point presentations is hard. So we kind of appear to know already about hardness and elasticity. But, really, do we know everything that we need to know? This is what we are going to examine and unfold briefly in this chapter.

4.2 Conceptual Issues

If you push your finger into dry sand on the beach, it goes into the sand. Even when the sand is wet, you can push your finger down to some extent. So, you can feel that it is not a hard material. On the other hand, if you try to push your finger through the wall beside your reading table, you will not be able to penetrate the wall. You will feel that it is hard. Similarly, to our sense of touch, dough made of ordinary flour will be soft, but a piece of marble floor tile will be hard. Thus it follows very easily that, conceptually, it is as if we know about hardness naturally, even though we may never have encountered a formal definition of the same. It would also be plausible to link the hardness to how strong a material is. For instance, we can make a big dent easily in sand and make sand piles on the beach. Thus the ease with which such a dent is made reflects the notion that the sand is not so strong. But can we repeat a similar success with the wall beside our reading table as easily as in the case of the sand? The answer is no. The reason is that the wall is strong!

Thus, it seems reasonable to surmise that hard things are strong. Soft things are not hard, as they are not so strong! Ceramics are classically strong, hard brittle solids. Why are ceramics strong and hard? Well, most ceramics have ionic or covalent bonding. These bonds themselves are strong. So, they are hard and strong by virtue of the strong bond strength that prevails in them. But then you may ask: if they are hard and strong, how come they are brittle? Well, they are brittle because they do not have the ability to plastically deform and adapt to the extra strain generated due to the applied stress! Then the question is: Why can't they deform plastically like metals at room temperature? The problem with ceramics is that—unless exposed to a state of hydrostatic stress within a very small confined volume—ceramics at room temperature generally do not have five or more independent slip systems actively operative. That is what is required to satisfy the von Mises criterion of plastic deformability. As the ceramics fail to satisfy this criterion in general, they do not show any appreciable plasticity at room temperature, so they remain unable to absorb the strain in the structure. Upon continued application of stress from the external system, the accumulated strain becomes high enough to cross the failure strain limit. When that happens, catastrophic failure occurs, especially so in classical brittle solids, like glass and ceramics. Widespread practical exploitation of the brittle solids like glass and ceramics is therefore somewhat limited due to the presence of flaws that are either intrinsic or are induced by nanoscale, microscale, or macroscale contact damage. These contact-induced flaws can act as a precursor to macroscopic defects. Therefore, it is essential to understand the nature and mechanism of contact-induced deformation in brittle solids at the macro-, micro-, and nanoscale levels.

All contact-related issues are tackled practically in terms of hardness. Therefore, the concept of hardness as it has developed through ages [1–9] is briefly discussed in Table 4.1.

TABLE 4.1

Development of the Concept of Hardness

Source	Evolution of Hardness Concepts
Aristotle [1]	Ability of a surface to remain "static" against external stress
Huygens [2]	Easy cleavage plane of calcite excited only for forward movement of knife
Haüy [3]	Mineral classification in accordance with ability to scratch other minerals
Mohs [4, 5]	Introduction of the 10-stage Mohs hardness scale
Seebeck [6]	The first scratch hardness tester with predefined load application
Martens [7]	New scratch hardness tester to measure the scratch width and the normal load
Calvert and Johnson [8]	Depth-controlled conical indentation
Hertz [9]	First quantitative definition of hardness

4.3 Beyond the Hertzian Era: Modern Contact Mechanics

Hertz [9, 10] gave us the first quantitative approach to define hardness as the almost constant contact pressure in a small circular area at the elasticity limit. This was the humble beginning of what we today know as contact mechanics [11–14]. On the basis of Hertz's definition, Auerbach [15–18] did a lot of research on the hardness of brittle solids.

4.4 The Experimental Issues

When the test load is below 2 N, the measured hardness is usually termed *microhardness* [19]. However, the distinction between micro- and macrohardness is yet to be unequivocally defined. Nevertheless, in a nanohardness test, the load can be lower than 10 mN and the depth may be even less than 10 μm. These values are much lower than those usually recorded in conventional micro- or macrohardness tests (force ≈ 0.001–100 N; depth ≈ 10–1000 μm), and that is why much better sensitivity and resolution are needed in nanoindentation machines [20]. According to Meyer [20], hardness (H) of the solid can be defined as the ratio

$$H = \frac{\text{Force, } P}{\text{Contact area, } A} \tag{4.1}$$

where P is the force perpendicular to the contact area A. This definition has become known as *Meyer hardness* (H_{meyer}) [21]. The physical meaning of Meyer hardness is the mean contact pressure P normal to the surface of the indent when the friction between the surface of the indenter and the sample is negligible. From the days of Meyer hardness, today we have nanohardness measurements. Li, Bhushan, and Takashima [22] have recently reviewed in detail the various aspects of nanoindentation test apparatuses, the associated data analysis, and the application of nanoindentation techniques for determination of mechanical properties, with special emphasis on thin films.

4.5 Elastic Modulus

Elastic modulus in a given direction is always the ratio of stress to strain. We have already defined stress and strain in Chapter 1. The rather simplistic definitions of the different elastic moduli of a solid are shown in Table 4.2.

TABLE 4.2

Definitions of Elastic Moduli

$$E_1 = \frac{\sigma}{\left(\frac{\Delta l}{l}\right)} \qquad \kappa = \frac{\pm\sigma}{\left(\frac{\Delta V}{V}\right)} \qquad \eta = \frac{\chi}{\left(\frac{\Delta\theta}{\theta}\right)} \qquad \nu = \frac{\left(\frac{\Delta D}{D}\right)}{\left(\frac{\Delta l}{l}\right)}$$

Here, E_1 is the Young's modulus, σ is the applied tensile stress, $\Delta l/l$ is the increase in length per unit length, κ is the bulk modulus, $\pm\sigma$ is tensile or compressive stress, $\Delta V/V$ is the change in volume per unit volume, η is the shear modulus, χ is the tangential stress, $\Delta\theta/\theta$ is angle of shear or shear strain, ν is the Poisson's ratio, and $\Delta D/D$ is the decrease in diameter per unit diameter.

4.6 Techniques to Determine Elastic Modulus

The more conventionally used main techniques to measure elastic modulus are the ultrasonic technique, the acoustic impedance measurement technique, the flexural resonance technique, and the beam bending technique [23].

In order to determine the elastic moduli via ultrasonic techniques, density (ρ), acoustic shear wave velocity (V_S), and acoustic compressional or longitudinal wave velocity (V_C) need to be measured [24]. For brittle solids, the density (ρ) is mostly measured by the Archimedes principle. To use the ultrasonic technique, an ultrasonic pulse is sent into the sample by means of a piezotransducer, and for the pulse-echo technique, the transducer that sends the pulses also receives them back. Thus, for a flat parallel sample of thickness (d) in the direction parallel to the direction of wave propagation, the velocities are calculated using:

$$V_S = 2d/\Delta t_S \tag{4.2}$$

$$V_C = 2d/\Delta t_C \tag{4.3}$$

Here, Δt_S and Δt_C are the corresponding times taken by the shear and the compressional or longitudinal acoustic waves to traverse across the thickness of the sample from the front end to the back end and then to traverse back from the back end to the front end, where these respective waves are picked up by the corresponding piezotransducers. Thus, once the Δt_S and Δt_C values are experimentally measured, the acoustic shear wave velocity (V_S) and acoustic compressional or longitudinal wave velocity (V_C) can be calculated using equations (4.2 and 4.3). Finally, the respective moduli are calculated using the following relationships:

$$\text{Shear modulus } (\mu) = \rho V_S^2 \tag{4.4}$$

$$\text{Bulk modulus } (\kappa) = \rho[(V_C^2 - 4V_S^2)/3] \tag{4.5}$$

$$\text{Young's modulus } (E) = [\rho V_S^2(3V_C^2 - 4V_S^2)]/(V_C^2 - V_S^2) \tag{4.6}$$

$$\text{Poisson's ratio } (\nu) = [(E/2\mu) - 1] \tag{4.7}$$

The elastic properties can be also characterized using the acoustic impedance technique. It involves the reflection and transmission of ultrasound at interfaces between two materials. Z_C, the compressional wave acoustic impedance, is defined as:

$$Z_C = \rho V_C \tag{4.8}$$

It then follows easily from equations (4.4–4.7) that we can also write:

$$[(\mu)] = V_S^2\,[(Z_C/V_C)] \tag{4.9}$$

$$[(\kappa)] = (V_C^2 - 4V_S^2/3)\,[(Z_C/V_C)] \tag{4.10}$$

$$[(E)] = [\{Z_C/V_C\}\,\{V_S^2(3V_C^2 - 4V_S^2)\}/\{(V_C^2 - V_S^2)\}] \tag{4.11}$$

$$[(\nu)] = [\{Z_C/V_C\}\,\{V_S^2(3V_C^2 - 4V_S^2)\}/\{(V_C^2 - V_S^2)\}/2\{V_S^2\,(Z_C/V_C)\}] - 1 \tag{4.12}$$

Thus, by measuring the acoustic impedance and the acoustic shear and compressional or longitudinal wave velocities, we can measure the elastic moduli of different solids.

The dynamic elastic moduli of isotropic, homogeneous ceramics are commonly determined by the flexural resonance methods [25]. In this technique, a prismatic beam specimen is vibrated in a flexural mode, and the resonant frequency is measured. The beam may have a square, rectangular, or circular cross section. Elastic modulus is determined from the resonant frequency, the mass or density of the prism, and the beam's physical dimensions.

The basic wave equation for the propagation of an elastic wave in an elastic medium is:

$$E = \rho v^2 \tag{4.13}$$

It can be shown [26] that elastic modulus (E) can be expressed as:

$$E = C_1 W f^2 \tag{4.14}$$

where W is weight of the prism, f is the flexural resonant frequency, and C_1 is given by:

$$C_1 = 4\pi^2 l^3 T_1/gI(4.73)^4 \qquad (4.15)$$

where l is the prism length, g is the gravitational constant, I is the second moment of inertia for the beam cross section, and T_1 is a dimensionless geometric constant that depends upon the radius of gyration of the prism cross section, the length of the prism, and the Poisson's ratio.

The three-point flexural test [23] can also be used for measurement of Young's modulus. In the three-point flexural test, a bar of mass m, width b, and depth d is put on two support points set apart a distance L and loaded at a point $L/2$ from the support points by a point force P. Under such condition, the flexural stress, σ_f of this bar is given by:

$$\sigma_f = \frac{3PL}{2bd^2} \qquad (4.16)$$

If the maximum deflection created just under the loading point in the bar is D, the corresponding flexural strain ε_f is given by:

$$\varepsilon_f = \frac{6Dd}{L^2} \qquad (4.17)$$

Thus, the flexural modulus, E_f, is given by:

$$E_f = \frac{L^3 m}{4bd^3} \qquad (4.18)$$

where m is the gradient of the initial straight-line portion of the load-deflection curve of the beam.

4.7 Conclusions

We have summarized what is meant by the hardness and modulus of solids and the various measurement approaches that may be employed to evaluate them at the macroscale. At this point, we need to understand why and how nanoindentation differs from the conventional indentation methods used to measure mechanical properties. We need to know that there are situations where nanoindentation can give us unique advantages that conventional indentation methods cannot possibly give—for instance, any evaluation of mechanical properties at the microstructural length scale! These aspects are discussed very briefly in Chapter 5.

References

1. Aristotle and I. Bekker. 1829. *Aristotelis Meteorologica*. Berolini Typis Academicis.
2. Huygens, C. H. R. 1690. *Traité de la lumière*. Leipzig: Gressner & Schramm.
3. Haüy, R. J. 1822. *Traité de mineralogie*. Paris: Chez Louis.
4. Mohs, F. 1812. *Versuch einer Elementar-Methode zur naturhistorischen Bestimmung und Erkennung der Foßilien*. Camesinische Buchhandlung.
5. Mohs, F. 1822. *Grund-Riß der Mineralogie*. Arnoldische Buchhandlung.
6. Seebeck, A. 1833. *Prüfungs-Programm des Cö lnischen Realgymnasiums*. Berlin.
7. Martens, A. 1898. *Handbuch der Materialienkunde für den Maschinenbau*. Springer.
8. Calvert, F. C., and R. Johnson. 1859. Ueber die Härte der Metalle und Legirungen. *Annalen der Physik* 108:575–82.
9. Hertz, H. 1882. Über die Berührung fester elastische Körper and liber die Harte. *Verhandlungen des Vereins zur Beförderung des Gewerbefleisses*. Leipzig.
10. Hertz, H. 1882. Ueber die Berührung fester elastischer Körper. *Journal für die reine und angewandte Mathematik* 92:156–71.
11. Johnson, K. L. 1996. *Contact mechanics*. Cambridge, U.K.: Cambridge University Press.
12. Fischer Cripps, A. C. 2000. *Introduction to contact mechanics*. New York: Springer.
13. Johnson, K. L. 1970. The correlation of indentation experiments. *Journal of Mechanics and Physics of Solids* 18:123–26.
14. Bhushan, B. 1998. Contact mechanics of tribology in rough surfaces. *Tribology Letters* 4:1–35.
15. Auerbach, F. 1890. Absolute Härtemessung. In *Nachrichten vonder Königl. Gesellschaft der Wissenschaften und der Georg-Augusts-Universität zu Göttingen*, 518–41.
16. Auerbach, F. 1891. Absolute Härtemessung. *Annalen der Physik* 43:61–100.
17. Auerbach, F. 1892. Plasticität und Sprödigkeit. *Annalen der Physik* 45:277–91.
18. Auerbach, F. 1896. Die Härtescala in absolutem Maasse. *Annalen der Physik* 58:357–80.
19. Lips, E. M. H. 1937. Härtemessungen an Gefügebestandteilen. *Zeitschrift für Metallkunde* 29:339–40.
20. Lips, E. M. H., and J. Sacks. 1936. A hardness tester for microscopical objects. *Nature* 138:328–29.
21. Meyer, E. 1908. Untersuchungen über Härteprüfung und Härte Brinell Methoden. *Zeitschrift des Vereins Deutscher Ingenieure, Berlin* 52:645–54.
22. Li, X., B. Bhushan, and K. Takashima. 2003. Mechanical characterization of micro/nanoscale structures for MEMS/NEMS applications using nanoindentation techniques. *Ultramicroscopy* 97:481–94.
23. Wachtman, J. B., R. Canon, and M. J. Matthewson. 2009. *Mechanical properties of ceramics*. New York: Wiley.
24. Canumalla, S., and M. G. Oravecz. 1998. In situ elastic property characterization of flip-chip underfills. Paper presented at Fourth International Symposium on Advanced Packaging Materials, Braselton, GA.

25. Quinn, G. D., and J. J. Swab. 2000. Elastic modulus by resonance of rectangular prisms: Corrections for edge treatments. *Journal of the American Ceramic Society* 83:317–20.

26. Picket, G. 1945. Equations for computing elastic constants from flexural and torsional resonant frequencies of vibration of prisms and cylinders. *Proceedings of the American Society for Testing and Materials* 45:846–65.

5

Nanoindentation: Why at All and Where?

Arjun Dey, Payel Bandyopadhyay, Nilormi Biswas,
Manjima Bhattacharya, Riya Chakraborty, I. Neelakanta Reddy,
and Anoop Kumar Mukhopadhyay

5.1 Introduction

In any new materials development, nanoindentation is often used when we need to know its mechanical integrity at the microstructural length scale. It can be applied for thin films and coatings, bulk materials, composites, alloys, hybrid nanocomposites, functionally graded materials, as well as living cells, etc. It can be used in three basic modes: (a) depth control, (b) location control, and (c) phase control, as shown schematically in Figure 5.1.

5.1.1 Depth-Control Mode

In depth-control mode, a nanoindentation technique is frequently used for multilayer thin films. The reason is that it has the ability to measure the stiffness at a precision level, e.g., a depth of a few nanometers. For example, if we consider a multilayered thin-film architecture (MLTFA) having four layers made of different stiffness values, the continuous stiffness measurement (CSM) procedure can tell us about the hardness and modulus at each depth level with a high spatial resolution. Thus, we can get the hardness and modulus properties of the MLTFA as a function of depth, as shown schematically in Figure 5.2. A recent review on nanoindentation by Li and Bhushan [1] showed how the CSM technique can be utilized to investigate the hardness and modulus of each layer (e.g., overcoat, magnetic layer, underlayer, and Ni-P layer) of a magnetic data-storage device. Further, a pop-in at a certain depth in a $P-h$ plot during CSM/progressive load measurement mode can tell us about the localized fracture strength of thin, brittle, amorphous carbon film on silicon [2].

5.1.2 Location-Control Mode

In location-control mode, we can measure the local mechanical properties at site-specific conditions as per our choice. For example, suppose someone wants

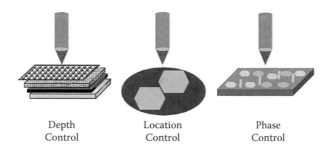

Depth   Location   Phase
Control   Control   Control

FIGURE 5.1 (See color insert.)
Different modes of nanoindentation technique to characterize materials.

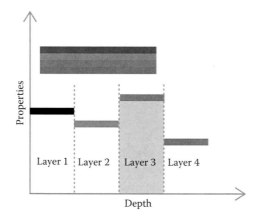

FIGURE 5.2 (See color insert.)
Schematic of variation of data at each layer in multilayer thin film utilizing CSM mode. (Adapted from Li and Bhushan [1].)

to measure the nanomechanical properties of a grain boundary; a coating–substrate interface far away from or very near to a pore in a porous structure; the interface of a weld zone; the interface of matrix and fiber, etc. In such situations, the nanoindentation technique is the only appropriate choice. Examples of location-specific nanoindentations are shown in Figures 5.3a–c. Figure 5.3a shows a footprint of a nanoindentation on plasma-sprayed HAp (hydroxyapatite) coating. Here, the location close to the natural defects has been deliberately avoided. A detailed study of this and related issues has been reported elsewhere [3]. In Figure 5.3b, we can see that the nanoindentations have been made on and very near to the grain boundary region of a polycrystalline coarse-grained sintered alumina ceramic. Further, the next example (Figure 5.3c) illustrates how nanoindentations can be performed very near to, near to, and far away from the dentin tubules. Thus, the point that we try to drive home here is that, using the nanoindentation technique, one can choose and actually measure the nanomechanical properties at the local microstructural length scale with or without the influence of the microstructural defects.

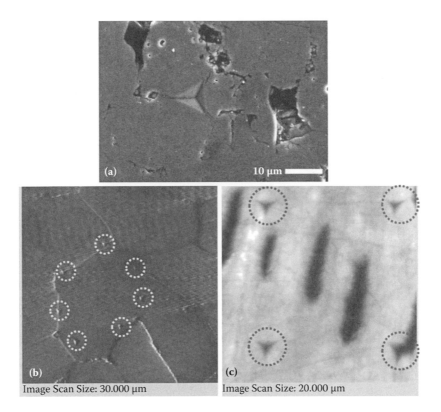

FIGURE 5.3 (See color insert.)
Location-specific nanoindentation: (a) on plasma-sprayed HAp coating showing footprint of indentation avoiding defects, (b) on a grain boundary and very near to a grain boundary region of dense alumina ceramics, and (c) very near, near, and far away from dentin tubules.

5.1.3 Phase-Control Mode

In phase-control mode, the nanoindentation technique can be used to evaluate the mechanical properties on individual phases in a multiphase material. However, in that case, we have to identify the phases by using an optical or scanning probe microscopy facility equipped with a nanoindenter. Figure 5.4 shows a micrograph of a biphasic (i.e., bright phase and dark phase) microstructure. Here, phase-specific nanoindentations on a bright phase have been made intentionally. In a recent work on nanopillars and nanotowers of silicon core with a silicon carbide shell [4], for instance, a nanoindentation technique was used to measure fracture toughness in the silicon core zone. Thus, what would have been otherwise impossible to evaluate becomes possible with the application of the nanoindentation technique on specific phases in multiphase materials at the ultrastructural length scale.

It is interesting to note that it is also possible to utilize the nanoindentation technique to evaluate the mechanical properties of biological materials as

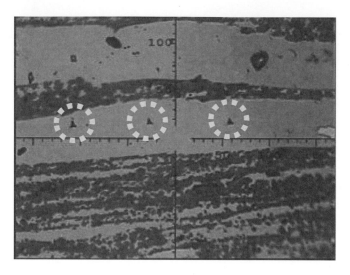

FIGURE 5.4
Phase-specific nanoindentation on a biphasic ceramic material.

well as living cells under atmosphere control and as well as with variation of temperature, although the maximum temperature is yet to exceed a couple hundred degrees Celsius, i.e., ≈400°C.

5.2 In Situ Nanoindentation

In the in situ nanoindentation technique, a very specially designed nanoindentation tip, along with the sensing devices for ultralow load and depth, are used under very high vacuum condition inside the chamber of a scanning (SEM) or transmission electron microscope (TEM), so that it becomes possible to observe, investigate, and measure the deformation at the nanoscale in situ in real time. The characterization of multifunctional material through in situ nanoindentation has recently been summarized [5]. Similarly, attempts have been made to correlate the occurrences of pop-ins, formation of shear bands, development of pile-ups, generation of cracking during loading, and delamination failure at higher loads with the in situ microstructural changes happening during the nanoindentation of various ceramic thin films [6–9]. On the other hand, in situ nanoindentations inside TEM have been conducted by several researchers on epitaxially grown titanium nitride thin films on single-crystal magnesium oxide (001) substrate [10], zirconia–magnesia spinel ceramic nanocomposites [11], and $YBa_2Cu_3O_{7-x}$ thin films [12]. The existence of various real-time deformation mechanisms operative at the nanostructural length scale could be noted by such novel attempts [10–12].

5.3 Conclusions

It is now well accepted beyond any doubt that in structural, functional, biological, biomedical, and/or nano-bio interface application frontiers on a wide variety of materials, including the brittle solids, nanoindentation is the only ex situ or in situ experimental technique that can be used to evaluate a system's mechanical integrity at the microstructural or submicrostructural length scale. In Chapter 6, we focus on data analysis methods that will be essential if we are to use the nanoindentation method as our experimental technique to probe zones that are not easily accessible to macrostructural testing formalities and the related scale formalisms.

References

1. Li, X., and B. Bhushan. 2002. A review of nanoindentation continuous stiffness measurement technique and its applications. *Materials Characterization* 48:11–36.
2. Borrero-Lopez, O., M. Hoffman, A. Bendavid, and P. J. Martin. 2009. Reverse size effect in the fracture strength of brittle thin films. *Scripta Materialia* 60:937–40.
3. Dey, A., and A. K. Mukhopadhyay. 2013. Fracture toughness of microplasma sprayed hydroxyapatite coating by nanoindentation. *International Journal of Applied Ceramic Technology* 8:572–90.
4. Beaber, A. R., S. Girschick, and W. W. Gerberich. 2011. Dislocation plasticity and phase transformation in Si–SiC core–shell nanotowers. *International Journal of Fracture* 71:177–83.
5. Nili, H., K. Kalantar-zadeh, M. Bhaskaran, and S. Sriram. 2013. In situ nanoindentation: Probing nanoscale multifunctionality. *Progress in Materials Science* 58:1–29.
6. Rabe, R., J. M. Breguet, P. Schwaller, S. Stauss, F. J. Haug, J. Patscheider, and J. Michler. 2004. Observation of fracture and plastic deformation during indentation and scratching inside the scanning electron microscope. *Thin Solid Films* 469/470:206–13.
7. Rzepiejewska-Malyska, K. A., M. Parlinska-Wojtan, K. Wasmer, K. Hejduk, and J. Michler. 2009. In situ SEM indentation studies of the deformation mechanisms in TiN, CrN and TiN/CrN. *Micron* 40:22–27.
8. Rzepiejewska-Malyska, K. A., M. Parlinska-Wojtan, K. Wasmer, K. Hejduk, and J. Michler. 2009. In situ scanning electron microscopy indentation studies on multilayer nitride films: Methodology and deformation mechanisms. *Journal of Materials Research* 24:1208–21.
9. Vecchione, N., K. Wasmer, D. S. Balint, and K. Nikbin. 2009. Characterization of EB-PVD yttrium-stabilised zirconia by nanoindentation. *Surface and Coatings Technology* 203:1743–47.
10. Minor, A., E. Stach, E. J. Morris, and I. Petrov. 2003. In situ nanoindentation of epitaxial TiN/MgO (001) in a transmission electron microscope. *Journal of Electronic Materials* 32:1023–27.

11. Lee, J. H., I. Kim, D. M. Hulbert, D. Jiang, A. K. Mukherjee, X. Zhang, and H. Wang. 2010. Grain and grain boundary activities observed in alumina zirconia-magnesia spinel nanocomposites by in situ nanoindentation using transmission electron microscopy. *Acta Materialia* 58:4891–99.
12. Lee, J. H., X. Zhang, and H. Wang. 2011. Direct observation of twin deformation in $YBa_2Cu_3O_{7-x}$ thin films by in situ nanoindentation in TEM. *Journal of Applied Physics* 109:83510–16.

6

Nanoindentation Data Analysis Methods

Manjima Bhattacharya, Arjun Dey, and Anoop Kumar Mukhopadhyay

6.1 Introduction

For decades, indentation has been perhaps the most commonly applied technique for measuring the mechanical properties of materials. Nanoindentation improves on the macro- and microindentation tests by (a) indenting at the nanoscale with a very precise tip shape, with high spatial resolutions to place the indents, and (b) providing real-time load-displacement (into the surface) data while the indentation is in progress. This has been practiced by many researchers on different materials [1–3]. To develop a better understanding about the methods typically utilized to analyze the nanoindentation data, here we shall again have a look at the schematic representation for load (P) versus depth of penetration (h) data plot (Figure 6.1) that is typically generated in a nanoindentation experiment. We assume that the loading is quasi-static in nature.

Now we will try to present a picture of the entire nanoindentation loading–unloading process. What happens is the following sequence of events: During the loading process, as the load starts to increase at and up to the stage B, when the load is really very low, the deformation remains elastic. Beyond this point, however, as the load is increased, the deformation process becomes more elastoplastic in nature (region BC) rather than remaining as purely elastic in nature (region AB). How does the indenter react to this process? Well, as the load is increased, the indenter sinks into the material due to both elastic and plastic deformation. Now, our next question is: How does the material react to such a process? We shall deal with this a little later in this book. However, we can generally expect that a metal will undergo plastic deformation; a brittle solid will undergo an elastoplastic deformation; and a polymeric material will undergo more of a viscoelastic deformation.

If the load is held constant, the indenter continues to sink into the material due to time-dependent deformation. This is called creep. It can happen even at room temperature in polymeric materials. However, in the case of brittle solids like glass and ceramics, the creep happens typically at a temperature more than half of its homologous temperature. In such cases, it can happen

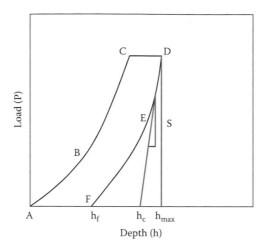

FIGURE 6.1
Schematic diagram showing different regions of a typical load–depth (*P–h*) plot obtained from a nanoindentation experiment.

when the applied stress is suddenly changed at a constant temperature. It can also happen when the temperature is suddenly changed when the material is held under constant applied stress. There are many other issues involved with creep deformation and the prevalent mechanisms, but that is beyond the scope of our present discussion. Because our main focus area in this book is nanoindentation of brittle solids, it suffices to know for the time being that there has not yet been any significant reported observation on room-temperature creep in brittle solids. After the predesignated hold time is over (region CD in Figure 6.1), the load is withdrawn through a predesignated period of time. How does the indenter feel about it? Well, the indenter starts to relax as the load continues to decrease at a given rate.

Our next query is then: What about the material that had been so long under the applied stress? How does it feel during the process of unloading? As soon as the load starts to decrease during the unloading cycle, the material starts to recover by a process that is primarily elastic (region DE in Figure 6.1). In most of the brittle solids, however, the unloading process does not ideally remain purely elastic. For example, in the region such as EF, it can still undergo elastoplastic deformation as it had undergone during the loading cycle. As will be shown later in this book, other modes of deformation, e.g., development of huge shear stress beneath the nanoindenter, are also possible. Such possibilities can and do translate to reality, depending on the material, the load, and the rate of loading and unloading. In fact, in situations when such shear stress exceeds (in highly localized regions inside and in the vicinity of the nanoindentation cavity) the theoretical shear strength of the material concerned, both shear-induced deformation and shear-induced grain boundary microcracking as well as microfracture can and really do happen.

6.2 Modeling of the Nanoindentation Process

There are numerous models available even for the nanoindentation process, which is a relatively new experimental technique. There are, in fact, excellent reviews as well as very detailed research papers available on the subject [1, 2], and it will not be our intention to describe all such models and efforts.

We would rather try to present a simplistic first-order picture of the whole scenario. Therefore, we shall now discuss the methods that have been developed to analyze all three parts (load, hold, and unload) of the load-displacement curves (Figure 6.1) typically generated during a nanoindentation experiment. We shall first describe the Oliver-Pharr model [3].

6.2.1 Oliver-Pharr Model

This is one of the most widely accepted models for nanoindentation behavior of brittle solids. Therefore, it deserves a special attention in our present discussion. This is the model that treats the unloading curve as the consequence of a fully elastic contact event. The basic assumptions of this approach are:

1. Deformation upon unloading is purely elastic.
2. The compliance of the sample and of the indenter tip can be combined as springs in series, e.g.,

$$\frac{1}{E_r} = \left(\frac{1 - v_s^2}{E_s}\right) + \left(\frac{1 - v_i^2}{E_i}\right) \tag{6.1}$$

   where, E_r is the reduced modulus, E is the Young's modulus, v is the Poisson's ratio, and subscripts i and s refer to the indenter and sample, respectively.
3. The contact can be modeled using an analytical model for contact between a rigid indenter of defined shape with a homogeneous isotropic elastic half-space using:

$$E_r = \left(\frac{\sqrt{\pi}}{2}\right) * \left(\frac{S}{\sqrt{A}}\right) \tag{6.2}$$

$$S = \frac{dP}{dh} = \left(\frac{2}{\sqrt{\pi}}\right) * \left(\frac{E_r}{\sqrt{A}}\right) \tag{6.3}$$

where S is the contact stiffness and A is the contact area. This relation was presented by Sneddon [4]. Now, it is already well known that the P–h data plot (Figure 6.1) obtained from the nanoindentation experiment gives the peak load (P_{max}), the maximum depth of penetration (h_{max}), the final depth of penetration (h_f), and the contact stiffness ($S = dP/dh$ at $h = h_{max}$). According to the Oliver-Pharr [3] model—the most well established, verified, and practiced model of the global research community—the nanohardness (H) is expressed as

$$H = \frac{P_{max}}{A_{cr}} \tag{6.4}$$

where the real contact area, A_{cr}, is given by [3]:

$$A_{cr} = 24.56h_c^2 + C_1h_c + C_2h_c^{\frac{1}{2}} + C_3h_c^{\frac{1}{4}} + \ldots + C_8h_c^{\frac{1}{128}} \tag{6.5}$$

Here, h_c is the true penetration depth if the unloading were purely elastic. Further, the constants C_1 to C_8 are determined by the standard calibration method [3]. Now, Sneddon's contact solution [4] predicts that the unloading data for an elastic contact for many simple indenter geometries (sphere, cone, flat punch, and paraboloids of revolution) follows a power law that can be written as follows:

$$P = \alpha h^m \tag{6.6}$$

In this equation, P is the instantaneous load on the indenter, h is the instantaneous elastic displacement of the indenter into the sample surface, and α, m are constants that are empirical in nature and may be actually determined by a data-fitting exercise.

Oliver and Pharr [3] were the first to apply this formulation to determine the contact area at maximum load, as it is valid even if the contact area changes during unloading. To do this, they derived the following relationship for the contact depth from Sneddon's [4] solutions:

$$h_c = h_{max} - \theta \frac{(P_{max})}{S} \tag{6.7}$$

where $\theta = 0.72$, 0.75, and 1, for cone-, sphere-, and flat-punch-geometries, respectively.

The relative stiffness (S_{max}/h_{max}) of the material in contact with the indenter is given by [4, 5]:

$$\frac{S_{max}}{h_{max}} = 5.52E_r\sqrt{\frac{A_{cr}}{A_{ca}}} \tag{6.8}$$

where

$$A_{ca} = 24.56\,(h_{max})^2 \tag{6.9}$$

6.2.2 Doerner-Nix Model

Although it is beyond the scope of the present book, it must be mentioned that the first complete nanoindentation data analysis was presented by Doerner and Nix [6]. These authors argued that the indenter can be treated as a flat punch [4] as long as the change in contact area is small during unloading. They made two simplifying assumptions. The first assumption is that as long as the loading process happens, all of the material in contact with the indenter is plastically deformed. The second assumption is that, outside the contact impressions of the indent, only elastic deformation occurs at the surface. Using these two assumptions, the model [4] could develop the connection between *P–h* data and the contact area.

6.2.3 Field-Swain Model

Further, Field and Swain [7] have extended the Hertzian approach [8–10] to incorporate plastic deformation in nanoindentation of an elastic half-space by a sphere. This model [7] treats the whole process as follows. These researchers imagine that there exists an impression of depth h_f and its shape is that of the residual impression as would have been created by an actual indentation. Then they proceed to consider that the whole process of the nanoindentation is nothing but a reloading of the preformed impression with depth h_f into reconformation with the indenter. From there on, using a loading and partially unloading technique, the model could determine hardness and reduced modulus from the appropriate nanoindentation data.

6.2.4 Mayo-Nix Model

Although most models assume a single monotonic relationship between stress and strain, it should better be admitted in reality that plastic deformation in all materials is time and temperature dependent at some scale. However, it should also be kept in mind that the time-dependent deformation assumes importance when the temperature is close to or greater than half the homologous temperature. If a constant uniaxial stress is applied to a material, the strain is often observed to increase with time, as shown in Figure 6.2.

In Figure 6.2, stage I represents the transient creep. It is a situation where the strain rate decreases with time. This situation continues until it reaches a steady state, shown as stage II in Figure 6.2. At this stage, the rate of change of strain rate with time is minimal. This stage is followed by the tertiary creep. It is shown in Figure 6.2 as stage III. This particular situation is characterized by an increasing strain rate that may precede fracture at a final time instant, t_f (Figure 6.2).

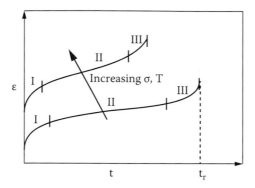

FIGURE 6.2
Schematic of the strain–time behavior of a crystalline sample in a creep test.

This time-dependent deformation is called *creep*. The steady-state creep rate (stage II in Figure 6.2) can often be described by an empirical equation of the form:

$$\dot{\varepsilon}_{II} = A\sigma^m e^{\left(\frac{-Q_c}{RT}\right)} \tag{6.10}$$

In this equation, A, m, and Q_c are material constants. Q_c is the creep activation energy, and m is the stress exponent. In a plot of $\log(\sigma)$ versus $\log(\dot{\varepsilon}_{II})$, the experimental data will fall on a straight line. The slope of this straight line is $n = 1/m$, where n is the strain rate sensitivity. What does it physically mean?

It gives us a measure of how the physical process of deformation of a given material in general, and a brittle solid in particular, will show its own responses when the applied strain rate changes. In other words, the process of dissipation of externally applied energy from the loading train first onto and then into the material will change. Now, we may ask ourselves: Well, then how do we measure this strain-rate sensitivity from a typical nanoindentation experiment? We shall describe that method now.

This method was actually proposed by Mayo and Nix [11]. What they did is as follows: They argued that stress can be calculated as the average pressure felt by the material that lies under the loaded indenter. But what is this quantity? We already know that this is nothing but the hardness of the material. The question that we now face then is how to measure the strain rate in an indentation experiment. Even before that, the question can be: What is the strain here? Well, then one needs to look at the physical process of the deformation from the end point.

Here the strain will be determined by the infinitesimal changes happening in the contact depth, considering that the final elastic contact depth is

h_c (Figure 6.1) when the instantaneous load P has reached the value of P_{max}. Thus, one can assume that the infinitesimal instantaneous strain ($\delta\varepsilon$) will be $\delta\varepsilon = (\delta h_c / h_c)$. Then, it follows that the infinitesimal instantaneous nanoindentation strain rate ($\delta\varepsilon/\delta t$) will be given by:

$$\left(\frac{\delta\varepsilon}{\delta t}\right) = \frac{\left(\dfrac{\delta h_c}{h_c}\right)}{\delta t}$$

In this entire ball game, an implicit assumption is involved. This implicit assumption is the innovative idea [11] that at each point under the indenter, the infinitesimal instantaneous nanoindentation strain rate felt by the loaded material must scale with the indenter descent rate divided by the current contact depth. Therefore, the overall grand average nanoindentation strain rate is given by

$$\left(\frac{d\varepsilon}{dt}\right) = \frac{\left(\dfrac{dh_c}{h_c}\right)}{dt}$$

To physically measure various stress and strain data, the nanoindentation experiments were conducted [11] at various loading rates. The strain rate sensitivity data were then obtained from a comparison of strain rates at a common depth at various loading rates and, hence, stresses. It was found that these strain-rate sensitivity data matched nicely with those measured from macroscopic uniaxial tests. Later, Mayo et al. [12] also showed that even with only one indent, it is possible to get the strain-rate sensitivity of a material, provided that the instantaneous nanoindentation stress and infinitesimal instantaneous nanoindentation strain rate can be experimentally measured during the holding segment of a typical nanoindentation test.

6.3 Conclusions

In this chapter we have learned some very basic ideas about the nanoindentation technique and the relevant models that are most commonly practiced. We have also tried to look at time-dependent deformation processes and the possibility of studying them using the nanoindentation technique. But we have not yet learned in detail about how to calculate the contact area from depth measurements and how to account for piling-up or sinking-in issues in such measurements. This is exactly what we address in Chapter 7.

References

1. Mukhopadhyay, N. K., and P. Paufler. 2006. Micro- and nanoindentation techniques for mechanical characterisation of materials. *International Materials Reviews* 51:209–45.
2. Marshall, D. B., and W. C. Oliver. 1990. An indentation method for measuring residual stresses in fiber-reinforced ceramics. *Materials Science and Engineering A* 126:95–103.
3. Oliver, W. C., and G. M. Pharr. 1992. An improved technique for determining hardness and elastic modulus using load and displacement sensing indentation experiments. *Journal of Materials Research* 7:1564–83.
4. Sneddon, I. N. 1965. The relation between load and penetration in the axisymmetric Boussinesq problem for a punch of arbitrary profile. *International Journal of Engineering Science* 3:47–57.
5. Sarkar, S., A. Dey, P. K. Das, and A. K. Mukhopadhyay. 2011. Evaluation of micromechanical properties of carbon/carbon and carbon/carbon–silicon carbide composites at ultralow load. *International Journal of Applied Ceramic Technology* 8:282–97.
6. Doerner, M. F., and W. D. Nix. 1986. A method for interpreting the data from depth sensing indentation instruments. *Journal of Materials Research* 1:601–9.
7. Field, J. S., and M. V. Swain. 1993. A simple predictive model for spherical indentation. *Journal of Materials Research* 8:297–306.
8. Lawn, B. R., and T. R. Wilshaw. 1975. Indentation fracture: Principles and applications. *Journal of Materials Science* 10:1049–81.
9. Lawn, B. R. 1983. Physics of fracture. *Journal of the American Ceramic Society* 66:83–90.
10. Fischer-Cripps, A. C. 2000. A review of analysis methods for sub-micro indentation testing. *Vacuum* 58:569–85.
11. Mayo, M. J., and W. D. Nix. 1988. A microindentation study of superplasticity in Pb, Sn, and Sn-28%Pb. *Acta Metallurgica* 36:2183–92.
12. Mayo, M. J., R. W. Siegel, A. Narayanasamy, and W. D. Nix. 1990. Mechanical properties of nanophase TiO_2 as determined by nanoindentation. *Journal of Materials Research* 5:1073–82.

7

Nanoindentation Techniques

Manjima Bhattacharya, Arjun Dey, and Anoop Kumar Mukhopadhyay

7.1 Introduction

7.1.1 Hardness Analysis

Based on the load versus depth-of-penetration data generated, the parameters involved in a nanoindentation experiment (Figures 7.1a and 7.1b) are given by h_f, the final depth of penetration; h_{max}, the maximum depth of penetration of the indenter when the load $P = P_{max}$; and h_c, the contact depth, i.e., the displacement where the indenter has maximum contact with the surface while unloading [1]. Any nanoindentation hardness calculation is also just like the conventional hardness calculation, i.e., load/load-bearing contact area. Now, there can be two area concepts used in nanoindentation evaluation. If the area is calculated from the contact depth, h_c, it is generally denoted by the contact area, A_s. There is another area called A_p, which is the projected contact area of an ideal nanoindenter. Since the nanoindenters have very small tip radius, typically about 40–200 nm, and the load may be ultralow, from a few microne-wtons to a few millinewtons, it is expected that the indentation imprints also will be very small. Therefore, to measure the contact area and the contact height directly experimentally is rather difficult because of the elastic recovery. Therefore, these are calculated based on the projected area, A_p. Assuming that a given nanoindenter (e.g., Vickers or Berkovich) has an ideal shape and the surfaces that it indents are perfectly flat, the contact depth (h_c) is given by

$$h_c = \sqrt{\frac{A_p}{24.56}}$$

and when h_c is known, the contact area A_c is given by, e.g., $A_c = 26.44\, h_c^2$, while the corresponding projected contact area A_p is given by, e.g., $A_p(h_c) = 24.56\, h_c^2$. The quantity A_p can be measured with sufficient accuracy from the images of the nanoindent, provided that the tangent height is set at typically about 90% below the surface height. Once A_p is known, h_c can be calculated. When h_c is calculated, the value of A_s can be easily predicted,

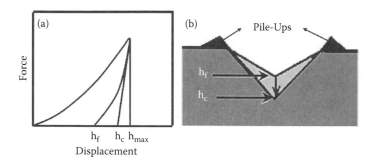

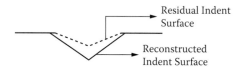

FIGURE 7.1
Definition of typical physical parameters in nanoindentation: (a) load (*P*) versus depth (*h*) plot; (b) concept of contact depth (h_c) and final depth of penetration (h_f). (Adapted from Fischer-Cripps [4].)

Residual Indent
Surface

Reconstructed
Indent Surface

FIGURE 7.2
Reconstructed nanoindent cavity surface and the residual nanoindent cavity surface.

as mentioned previously. Because of the repeated contact events and friction with the surface being indented, the tip of the nanoindenter gets worn out and, as a result, it becomes rounded. Thereby, it physically shifts from its ideal shape. Hence, the calibration of the indenter area function is done to account for the deviation from the ideal shape.

$$\text{Projected area: } A_p\left(h_c\right) = a_0 h_c^2 + a_1 h_c + a_2 h_c^{\frac{1}{2}} + a_3 h_c^{\frac{1}{4}} + \ldots + a_8 h_c^{\frac{1}{128}} \qquad (7.1)$$

$$\text{Surface area: } A_s\left(h_c\right) = b_0 h_c^2 + b_1 h_c + b_2 h_c^{\frac{1}{2}} + b_3 h_c^{\frac{1}{4}} + \ldots + b_8 h_c^{\frac{1}{128}} \qquad (7.2)$$

The quantity a_0 in equation (7.1) is approximately 24.56, whereas the quantity b_0 in equation (7.2) is approximately 26.44 [2]. The calculations of the projected area functions are done through dedicated software packages (which may vary from machine to machine) [3, 4] by analyzing the load–unload curves generated from a series of nanoindents made over a range of prespecified depths. Further, h_c is related to h_f by $h_c = h_f\,(1 + e_c)$, where e_c physically is about 10%–30% for most metals and reflects the amount of elastic recovery following the unloading of the nanoindenter. Thus, from the experimentally measured value of h_f using standard software packages, h_c is derived to digitally reconstruct the fully loaded nanoindentation cavity surface from the residual nanoindentation cavity surface (Figure 7.2). To improve the accuracy

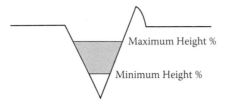

FIGURE 7.3
The minimum height (%) and the maximum height (%) that can be used to improve the accuracy of reconstruction.

of the reconstruction results, the minimum and maximum height of the indent depth to be used for indenter area function calculation (Figure 7.3) can be judiciously chosen in such a way as to avoid the influences of the surface roughness and sharp-tip area in the data.

7.2 Conclusions

The main point of this chapter has been to learn in detail about how to calculate the contact area from depth measurements. We have also attempted to clarify how to account for piling-up or sinking-in issues in such measurements. But how do we do the actual measurements in practice? How do the real machines function? How is the load measured? How are the depths measured? How accurate can we claim such measurements to be? What would be the typical resolutions linked to such measurements? There are still many questions unanswered than resolved in this chapter. Thus, in Chapter 8 we shall attempt to find answers to these queries.

References

1. Oliver, W. C., and G. M. Pharr. 1992. An improved technique for determining the hardness and elastic modulus using load and depth sensing indentation experiments. *Journal of Materials Research* 7:1564–83.
2. Oliver, W. C., and G. M. Pharr. 2004. Measurement of hardness and elastic modulus by instrumented indentation: Advances in understanding and refinements to methodology. *Journal of Materials Research* 19:3–20.
3. Mukhopadhyay, N. K., and P. Paufler. 2006. Micro- and nanoindentation techniques for mechanical characterisation of materials. *International Materials Reviews* 51:209–51.
4. Fischer-Cripps, A. C. 2007. *Introduction to contact mechanics*. New York: Springer.

8

Instrumental Details

Payel Bandyopadhyay, Arjun Dey, and Anoop Kumar Mukhopadhyay

8.1 Introduction

By now, you are quite familiar with the terms *contact deformation* and *nanoindentation* from the preceding Chapters 1–7. In this chapter, we discuss the instrumental details of the different varieties of the commercially available nanoindentation machines that are typically used for performing the nanoindentation experiments. There are some essential units in a nanoindenter machine. First of all, there is an indenter, which is generally made of hard materials, mostly synthetic diamond. Then there is a load cell that converts the electrical energy into mechanical energy and vice versa. Finally, there are some units whose job it is to measure the displacement.

8.2 Nanoindenters: Tip Details and Tip Geometries

There are various types of commercially available nanoindenters with different tip geometries. The accepted tolerance is defined by the standard called ISO 14577-22. The dimensions and the angles of any geometry of any nanoindenter or microindenter are acceptable if and only if they are within the tolerance limit specified by the ISO 14577-22 standard. There are various geometries available for nanoindenters in market. The three-sided pyramids, four-sided pyramids, wedges, cones, cylinders, and spheres are the common shapes of the nanoindenters. The tip end of the nanoindenter can be made sharp, flat, or rounded to a cylindrical or spherical shape. Diamond and sapphire are the primary materials of nanoindenters, but other hard materials can also be used, such as quartz, silicon, tungsten, steel, tungsten carbide, and almost any other hard metal or ceramic. Nanoindenters are mounted on holders, which could be the standard design from the manufacturer of a given piece of nanoindentation equipment or, if required, a customized design is also possible. The holder material is usually steel, titanium, etc. In most cases

FIGURE 8.1
An example of a nanoindenter with its three parts: the diamond, the holder, and the bond.
(Modified from http://eng.thesaurus.rusnano.com/wiki/article873.)

TABLE 8.1

The Magnitudes of the Semi-Apex Angles of Different
Tip Geometries

Tip Geometry	α (semi apex angle)
Berkovich	65.27°
Cube corner	35.26°
Customized, three-sided	20°–80°
Vickers	68°
Knoop	δ = 172.5°, γ = 130°
Customized, four-sided	20°–80°

the indenter is attached to the holder using a rigid metal-bonding process. Figure 8.1 shows a typical sketch of a nanoindenter in a holder. The typical values of the apex angles [1, 2] of different geometries are shown in Table 8.1.

Berkovich, Vickers, Knoop, and the conical nanoindenters are shown schematically in Figures 8.2a–d, respectively. The defining angle α is the angle between the axis and any of the faces. The tip of the Berkovich nanoindenter has a sharp three-sided geometry. The three faces are symmetrically placed 120° apart from each other, all placed around the axis. Similarly, the tips of the Vickers and Knoop nanoindenters have four-sided geometry. Most of the nanoindenters that have four-sided tips have their corresponding pyramidal faces placed symmetrically around the axis, 90° apart from each other. The standard Knoop indenter has a special geometry, as shown in Figure 8.2c. A sharp four-sided pyramid tip always ends in a small line called the *line of conjunction*.

Now we shall look into the other instrumental details of the nanoindentation machine. There are three major activities that need to be performed by a nanoindenter. It should be able to make the nanoindents on the surface of the sample at precisely predefined locations while operating in load-control mode or depth-control mode. Therefore, it must have the ability to apply load and measure the displacement, or it must have the ability to go up to a

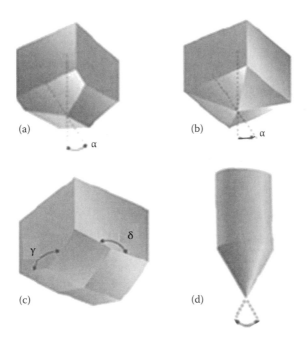

FIGURE 8.2
Different type of indenters: (a) Berkovich, (b) Vickers, (c) Knoop, and (d) conical. (Modified from http://eng.thesaurus.rusnano.com/wiki/article873.)

predesignated displacement and then apply the load [3, 4]. So, first of all, let us discuss the force actuation and measurement parts. The nanoindenters are designed in a smart manner such that a small change in displacement should not change the applied force significantly. There can be electromagnetic, electrostatic, and mechanical force actuations in a nanoindentation machine.

The most common and earliest means of applying force is to insert a coil inside a permanent magnet, as shown in Figure 8.3a. The magnetic field is generated by varying the current inside the coil. This generated field interacts with the field of the permanent magnet, which generates the force. For such electromagnetic actuation, the current versus force generated plot remains linear, even up to the millimeter range of displacements, corresponding to a force magnitude of several newtons. Because of the higher force actuation capability, it can pick up the indenter and push it down onto the sample. However, the greatest disadvantage of electromagnetic actuation, apart from being large (about 10 cm in size and heavy, about a kilogram), is that the current that passes through the load coil for force actuation also generates joule heating, which leads to almost unmanageable thermal drift in the machine's position and depth controls.

In the case of the electrostatic force actuation (Figure 8.3b), the voltage is applied between the middle and upper/lower plate, and the generated force is proportional to the square of the voltage. The vertical displacement is measured using a capacitance technique, i.e., by measuring the downward

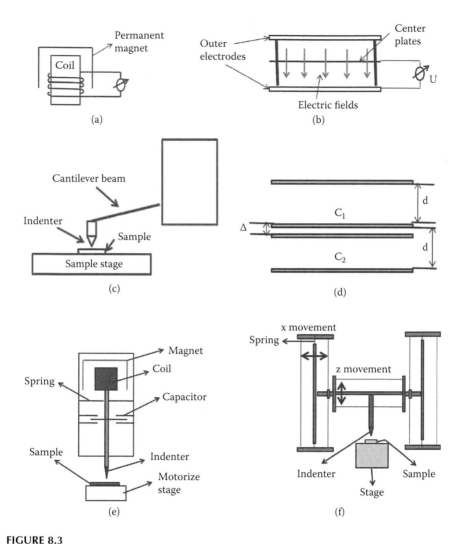

FIGURE 8.3
Schematic illustrations of (a) magnet/coil force generation, (b) force-generating transducer device, (c) cantilever beam approach of spring-based force actuation, (d) differential capacitor, (e) Nano Indenter (MTS/Agilent), and (f) TriboScope (Hysitron). (Modified from www.nanoindentation.cornell.edu.)

displacement of the plate at the center with respect to those of the two outer plates. The greatest advantage of the electrostatic force actuation lies in its temperature stability and hence, a good control over thermal drift. As a result of lower thermal drift, better control over the machine's position and depth controls is achieved. The other big advantage is that the size is so tiny that the whole system can be compacted within a few tens of millimeters. However, the force magnitude will typically be below ≈0.1 mN and, hence, the displacement range becomes limited to ≈0.1 μm [3, 4].

In both electromagnetic and electrostatic force actuation systems, the force calibration is done in a simple fashion. A series of small weights are hung from the indenter shaft. Then, the load current needed to lift the shaft to the same position for each weight is recorded. Once these data are known, the machine control software and the corresponding hardware interface can be guided to pick up exactly those load currents that will give the desired low forces.

The spring-based force actuation system is very common in nanoindenters. In this case, accurate knowledge of the spring constant and the relative displacements determine the accuracy of the force calibration. However, it must be kept in mind that the accuracy of force calibration is also a strongly sensitive function of many factors, e.g., temperature-dependent geometric changes, misalignments, etc. The cantilever beam approach is the most common method of spring-based actuation (Figure 8.3c). This method is very common for a scanning probe microscope (SPM) or atomic force microscope (AFM). Here, a silicon nanoindenter tip is typically attached to the free end of a cantilever beam that also works as the SPM/AFM tip. The sample is attached to a piezoelectric displacement actuator. The surface undulations created due to the nanoindentation process itself displace the tip attached to the cantilever beam, and the free end of the cantilever beam gets deflected. These displacements are picked up by using a laser beam reflected from the free end of the beam.

There is, in general, a conflict between the range and resolution of force and displacement sensing. At the lower range, one can have very high resolution of sensing and measurements. For instance, in conventional AFMs, force resolution can be in the piconewton range. In typical nanoindenters, that goes up to 10,000 μN or so, and the force resolution is about 1 nN. However, at the higher range, one has to compromise on resolution issues. It is obvious that at higher force resolution, greater displacements are needed to effect the corresponding changes in force. Now the next question is how we measure the displacement!

But for the dimensions in the range of pico-, nano- and micrometers, you could have blamed us—what a silly question it is! Well, in such cases we can use a capacitive displacement gauge. What is a capacitive displacement gauge? It is nothing but a parallel plate capacitor. The voltage and the area of the plates are held constant. Thus, all changes in current are made to be due only to changes in the distance between the plates. To manage issues related to nonlinearity of response, a differential capacitor (Figure 8.3d) may be used. In this case, the difference between the two capacitances C_1 and C_2 due to the displacement Δ is measured. The unique advantage that is gained here is that the first nonlinearity appears as a cubic Δ^3/δ^3 term in a differential capacitor, as opposed to the first nonlinearity appearance as a Δ^2/δ^2 term in a simple capacitor. As a result, even <1 Å resolution can be attained in a tiny size. However, one then has to accept that the total displacement range will be relatively very small indeed.

Some commercial nanoindentation machines may use an inductive force generation system (Figure 8.3e), wherein the displacement is continuously measured by a capacitive displacement gauge. Electromagnetic force actuation coupled with a sensitive capacitive transducer is also used by some manufacturers. However, when nanotribology experiments are to be conducted, the instrument consists of two perpendicular transducer systems (Figure 8.3f). The z system is responsible for indentation measurements. The x system allows scratching along with lateral force detection. The displacement of the center plate with respect to the outer electrodes is continuously measured by the corresponding capacitive displacement gauge.

8.3 Conclusions

In this chapter we have tried to explain how a nanoindenter actually works in practice, what the essential components of a nanoindenter are, and how these different components work in coordination with each other. We have also touched upon the issues of varieties in the types of load and displacement sensors and their respective working principles, as well as their relative limitations in terms of respective ranges and resolutions. However, we have not yet discussed the really huge variety of materials that we have used in our research and the machines that we have used for conducting our research. These details will be briefly discussed in Chapter 9.

References

1. Mukhopadhyay, N. K., and P. Paufler 2006. Micro- and nanoindentation techniques for mechanical characterisation of materials. *International Materials Reviews* 51:209–45.
2. Palacio, M. L. B., and B. Bhushan. 2013. Depth-sensing indentation of nanomaterials and nanostructures. *Materials Characterization* 78:1–20.
3. Fischer-Cripps, A. C. 2006. Critical review of analysis and interpretation of nanoindentation test data. *Surface and Coatings Technology* 200:4153–65.
4. Li, X., and B. Bhushan 2002. A review of nanoindentation continuous stiffness measurement technique and its applications. *Materials Characterization* 48:11–36.

9

Materials and Measurement Issues

Arjun Dey, Riya Chakraborty, Payel Bandyopadhyay,
Nilormi Biswas, Manjima Bhattacharya, Saikat Acharya,
and Anoop Kumar Mukhopadhyay

9.1 Introduction

In this chapter, we shall briefly discuss the huge variety of materials and the typical nanoindentation techniques used to characterize their mechanical properties, mainly at the micro- and nanoscale of the microstructure. All samples were mirror-polished down to 0.25 μm grit size diamond paste.

9.2 Materials Studied

For our studies on both the static and dynamic contact deformation behavior of glass (Chapters 10–15), we used commercial soda-lime-silica (SLS) glass with the principal components of ≈70 wt% of SiO_2, 14 wt% of Na_2O, and 8 wt% of CaO. The samples were either thin (≈330 μm) coverslips or about 1.4-mm-thick glass slides [1, 2].

To study the nanoscale static contact-damage evolution behavior of monolithic coarse-grain ceramics in as-received and as-gas-gun-shocked conditions (Chapters 16–22), 10 μm and ≈20 μm grain size alumina samples were used [3, 4]. The samples were pressureless sintered at 1650°C–1700°C in air. The 10 and 20 μm grain size alumina ceramics had respective densities of about 99.9% and 95% of theoretical. The high-pressure gas-gun-shock experiments were conducted at 6.5 and 12 GPa pressures, which were more than about two to seven times the Hugoniot elastic limit of the 10-μm grain size alumina [3]. The fragmented pieces of the gas-gun-shocked samples were recovered by an appropriately designed momentum trap [3, 5].

To study the nanoscale static contact deformation behavior of composite materials, carbon/carbon (C/C) and carbon/carbon-silicon carbide (C/C-SiC) composites (Chapter 23), multilayered ceramic matrix composites (MLCCs) of alumina and lanthanum phosphate (Chapter 24), and hydroxyapatite-mullite (HAp-mullite) as well as hydroxyapatite-calcium titanate (HAp-CT) composites (Chapter 25) were used. The C/C composites were prepared by vacuum infiltration with coal-tar pitch followed by carbonization and graphitization in inert atmosphere. Further, liquid silicon infiltration was done in a controlled atmosphere at 1600°C to convert the matrix carbon of the C/C composites into silicon carbide [6]. The multilayered ceramic matrix composites (MLCC) of alumina and lanthanum phosphate tapes were produced by tape-casting lamination followed by pressureless sintering at 1650°C in air [7]. On the other hand, the HAp-mullite composite was prepared by a ball-milling technique using varying amounts of mullite (10, 20, and 30 wt%), subsequently pressed and sintered at 1250°C–1350°C in normal atmospheric condition [8]. In contrast, the HAp-CT composites were prepared by a critical spark-plasma sintering process that was highly optimized for time and temperature as well as hold time [9].

Silicon wafer samples were provided by Fischer, Switzerland (Chapter 26). Further, the zirconia-toughened alumina (ZTA) samples were pressureless sintered at 1650°C with 10, 20, and 40 vol% of zirconia in alumina matrix (Chapter 27). For actuator materials, PZN-BT ($[Pb_{0.88}Ba_{0.12}]$ $[(Zn_{1/3}Nb_{2/3})_{0.88}Ti_{0.12})]O_3$) and PZN-BT-PT ($[Pb_{0.8}Ba_{0.2}][(Zn_{1/3}Nb_{2/3})_{0.8}Ti_{0.2})]O_3$) were used. These samples (Chapter 28) were synthesized by a three-step method to minimize secondary-phase formation. Finally, the pellets were sintered at 950°C in air [10]. In addition, a nano bismuth ferrite ($BiFeO_3$) was employed to study the nanoindentation behavior of magnetoelectric multiferroic materials (Chapter 29). The nano $BiFeO_3$ powder was synthesized by a sol–gel technique starting with the nitrates of the respective elements. Finally, pellets were annealed at a very low temperature of only 300°C to avoid grain coarsening [11].

The solid oxide fuel cells (SOFCs) were fabricated by tape casting followed by pressureless sintering (Chapter 30) [12] of all the layers, i.e., the Ni-YSZ anode, the dense YSZ (yttria-stabilized zirconia) electrolyte, and the screen-printed LSM (lanthanum-strontium-manganite) cathode. Further, for the sealing purpose with electrolyte, special-grade glass ceramics have been developed (Chapter 31). Barium, lanthanum, aluminum, and magnesium oxide were incorporated in the glass structure to achieve glass ceramics with better thermomechanical properties [13].

To study the static contact damage evolution behavior of ceramic coatings at the nanoscale, phase-pure and porous HAp coatings were mainly utilized (Chapters 32–36). The coatings were deposited on SS316L and Ti6Al4V substrates by a microplasma spraying (MIPS) technique with a very low plasmatron power of ≈1.5 kW. The thickness was around 200 μm, and volume percent open porosity was ≈20% for the plan section and ≈11% for the cross section. The degree of crystallinity of the coating was ≈80%–90%. To understand its in vitro behavior, the HAp coating was further immersed in simulated body fluid (SBF) solution prepared according to Kokubo's formulation [14–19] for

1, 4, 7, and 14 days. In another related study, an oxide/ceramic coating was grown on AZ31B magnesium alloy by employing the micro arc oxidation (MAO) technique (Chapter 37). In this case, dual electrolytes (sodium silicate and sodium fluoride) were used. To further understand its corrosion resistance behavior, DC polarization tests were also conducted. The nanoindentation studies were also conducted in pre- and post-corrosion situations on these MAO coatings [20].

Several ceramic thin films have been developed to understand their nanoscale contact deformation behavior. For example, a magnesium hydroxide thin film of ≈1500 nm thickness was developed (Chapter 38). These films were deposited on SLS glass slides by a simple, inexpensive chemical dipping method utilizing aqueous solutions of magnesium nitrate and sodium hydroxide precursors [21]. In other studies, titanium nitride films of ≈1200 nm thickness and alumina films of ≈1000 nm thickness were deposited by a magnetron-sputtering technique on SS304 and quartz (Chapters 39 and 40). Further, pure DLC (diamond-like carbon) and nanocrystalline gold-doped DLC films of 70–120 nm thickness were deposited [22] on the SLS glass by a capacitatively coupled RF (radio frequency) plasma CVD (chemical vapor deposition) technique using premixed methane and argon gas mixtures with 50%, 60%, 70%, and 80% of argon in methane (Chapter 41).

In Chapters 42–44, natural hybrid composites, e.g., human tooth enamel as well as bone and fish scale, have been utilized for nanoindentation study. Both tooth enamel and bone samples were extracted freshly from broken body parts, and fish scales of *Catla catla* were collected from the Ganges River (India). All samples were properly cleaned and dried prior to the nanoindentation study. In particular, the premolar tooth was freshly extracted from a 65-year-old male human being. The tooth sample was sectioned along the longitudinal direction by a slow-speed diamond saw. This step was followed by ultrasonic cleaning of the longitudinally sectioned tooth samples by sequentially using analytical reagent grade acetone, ethanol, and deionized water. The cleaned tooth samples were embedded in epoxy resin and mirror-polished with diamond pastes down to 0.25 μm grit size prior to microstructural and nanomechanical characterizations [23–25]. Similar treatment was followed in the case of the bone samples.

Just as a means of illustrative examples, some typical microstructural photographs are presented in Figures 9.1–9.3. Figure 9.1a shows an SEM image of nanoindentation on the same SLS glass. An optical photomicrograph of the scratch groove made at 10 N load on the SLS glass is shown in Figure 9.1b. Similarly, Figures 9.1c and 9.1d show typical FESEM photomicrographs of, respectively, 10 μm and 20 μm grain size alumina microstructures.

Figure 9.2a shows an optical photomicrograph of the plan section of the C/C composite, while Figure 9.2b depicts a FESEM photomicrograph of the C/C composite in cross section. Figure 9.2c shows the microstructure of the sintered, polished, thermally etched surface of pure HAp (sintered

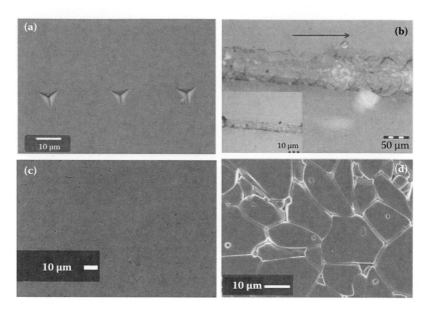

FIGURE 9.1 (See color insert.)
(a) SEM image of nanoindentation on glass. (b) Optical photomicrograph of the scratch groove made at 10 N load on glass. FESEM photomicrographs of alumina microstructure of (c) 10 µm and (d) 20 µm grain size. ([a]: Reprinted with permission of Chakraborty, R., A. Dey, and A. K. Mukhopadhyay. 2010. Role of the energy of plastic deformation and the effect of loading rate on nanohardness of soda–lime–silica glass. *Physics and Chemistry of Glasses: European Journal of Glass Science and Technology B* 51:293–303, from Society of Glass Technology and Deutsche Glastechnische Gesellschaft.)

at 1250°C for 2 hours), while Figure 9.2d depicts the microstructure of the HAp-30 wt% mullite (sintered at 1350°C for 2 hours). The microstructure of the pure mullite (sintered at 1700°C for 4 hours) showed elongated grains, as expected (Figure 9.2e). An optical micrograph of an AlN-SiC composite is shown in Figure 9.2f, and Figure 9.2g shows an optical micrograph of the cross section of a human cortical bone. Optical micrographs showing typical indents on human dentin and pulp are shown respectively in Figures 9.2h and 9.2i. Finally, FESEM images of a typical nanoindentation cavity on human dental enamel and the microstructure of a human dentin enamel junction (DEJ) are shown respectively in Figures 9.2j and 9.2k.

SEM photomicrographs of MgO and $Mg(OH)_2$ thin films are shown in Figures 9.3a and 9.3b, respectively. SEM photomicrographs of as-sprayed and polished plan sections of MIPS–HAp coatings are shown respectively in Figures 9.3c and 9.3d. Similarly, SEM photomicrographs of SU-MAO (i.e., MAO coatings by silicate-based electrolyte and unsealed condition) coating and SS-MAO (i.e., MAO coatings by silicate-based electrolyte and sealed condition) coatings are shown in Figures 9.3e and 9.3f, respectively.

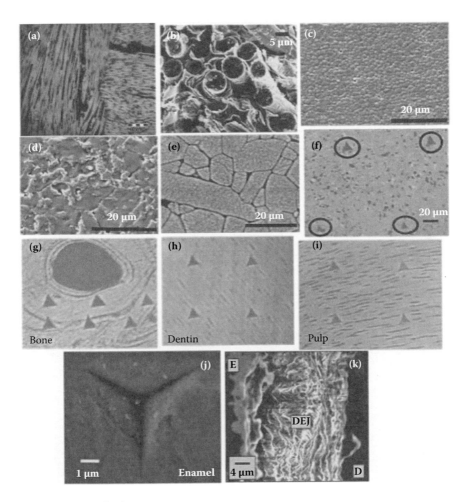

FIGURE 9.2 (See color insert.)
(a) Optical photomicrograph of plan section of carbon/carbon composite. (b) FESEM photomicrograph of cross section of carbon/carbon composite. (c) Microstructure of the sintered, polished/thermally etched surface of pure HAp (sintered at 1250°C for 2 h). (d) HAp-30wt% mullite (sintered at 1350°C for 2 h). (e) Pure mullite (sintered at 1700°C for 4 h). (f) Optical micrograph of AlN-SiC composite. (g) Optical micrograph of cross section of a human cortical bone. (h) Optical micrograph showing typical indents on human dentin. (i) Optical micrograph showing typical indents on human pulp. (j) FESEM image of a typical nanoindentation cavity on human enamel. (k) FESEM photomicrograph of human dentin–enamel junction (DEJ). ([c–e]: Reprinted with permission of Nath et al. [8] from Elsevier. [k]: Reprinted with permission from Biswas, N., A. Dey, and A. K. Mukhopadhyay. 2013. Mechanical properties of enamel nanocomposite. *ISRN Biomaterials* 2013: Article ID 253761, from Hindawi and ISRN.)

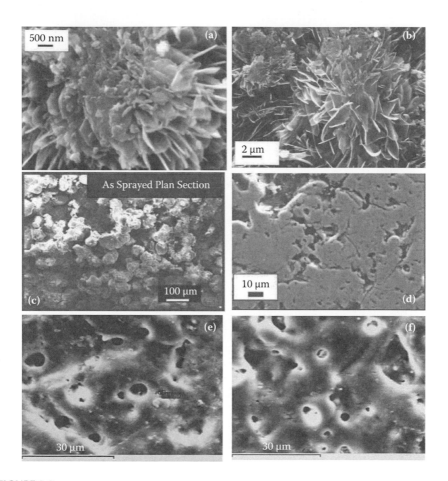

FIGURE 9.3
(a) SEM photomicrograph of MgO coating. (b) SEM photomicrograph of Mg(OH)$_2$ thin film.
(c) SEM photomicrograph showing the surface morphology of as-sprayed plan section of
microplasma-sprayed HAp coating. (d) SEM photomicrograph showing the surface morphology of polished top surface of microplasma-sprayed HAp coating. (e) SEM photomicrograph
of SU-MAO coating. (f) SEM photomicrographs of SS-MAO coating. ([c, d]: Reprinted with
permission of Dey et al. [14], from Elsevier. [e, f]: Reprinted/modified with permission of Dey
et al. [20] from Elsevier.)

9.3 Nanoindentation Studies

The nanoindentation studies were carried out employing three nanoindenters:

1. Fischerscope H100-XYp, Fischer, Switzerland
2. Tribo Indenter UBI 700, Hysitron, USA
3. Nano Indenter G200, Agilent-MTS, USA

The nanohardness and Young's modulus data of the present studies were calculated using the Oliver-Pharr model [26]. The nanoindentation data presented in Chapters 10, 12, 15–28, 30–37, and 42–45 were measured with the Fischerscope H100-XYp machine. The rest of the nanoindentation data were evaluated mostly with the Tribo Indenter, UBI 700 (Hysitron, USA), and in some occasions (see Chapters 40, 46) with the Nano Indenter G200 (Agilent-MTS, USA). We briefly discuss these machines in the following subsections.

9.3.1 Fischerscope H100-XYp

The Fischerscope machine was operated according to the DIN 50359-1 standard. The depth- and force-sensing resolutions of the machine were 1 nm and 0.2 μN, respectively. The machine was calibrated with nanoindentation-based independent evaluation of $H \approx 4.14 \pm 0.1$ GPa and $E \approx 84.6 \pm 3.5$ GPa of a standard reference block, Schott BK7 Glass (Schott, Germany), provided by the supplier. The machine had a load range of 0.4–1000 mN and was equipped with both Berkovich and Vickers indenters that had tip radii of 150 nm. In connection with the acquisition of data by the Fischerscope system, no major variation was observed in the calibration data. Load–depth plots of the BK7 standard reference calibration block provided by the supplier of the machine are shown as additional evidence in Figure 9.4. The smooth load–depth curves are almost overlapped with each other, which proved the repeatability factor. In addition, Figures 9.5a and 9.5b show, respectively, the hardness and Young's modulus data of the BK7 standard reference calibration block as a function of the number of measurements. The average values of the hardness and the Young's modulus were measured as $H_{avg} = 4.14 \pm 0.05$ GPa and $E_{avg} = 84.09 \pm 1.09$ GPa, respectively. This data matched quite well with the calibration reference data ($\approx H = 4.14 \pm 0.1$ GPa and $E = 84.6 \pm 3.5$ GPa) provided by the supplier.

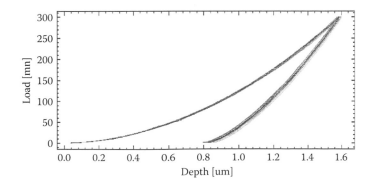

FIGURE 9.4 (See color insert.)
Load–depth (*P–h*) plots of the Schott BK-7 calibration glass block.

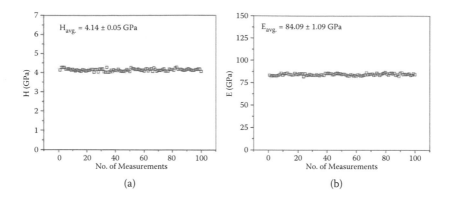

FIGURE 9.5
(a) Hardness and (b) Young's modulus data of the Schott BK-7 calibration glass block as a function of the number of measurements.

The calibration was repeated before each and every experiment. The reason for this exercise was twofold. The first was to ensure that the calibration values of hardness and Young's modulus for the standard calibration reference glass block, Schott BK7, a standard material, was within the prescribed limits of experimental error. The second was to ensure that the load–depth plot was free from any temporal effects such as thermal drifts, etc. This procedure of calibration check was deliberately done to check also the extent of reproducibility of the experimental data, which was found to be satisfactory. We have used this machine for the studies reported in Sections 3–9 and 11 (except for Chapters 29 and 30) of this book. A Vickers tip was utilized for studies reported in Chapters 24, 25 (only for HAp-mullite system), 27, and 31, while a Berkovich tip was used for the others. The amount of energy spent in both elastic and plastic deformations was also provided by the machine's dedicated software.

9.3.2 Tribo Indenter UBI 700

The Tribo Indenter UBI 700 had load and depth resolutions of 1 nN and 0.04 nm along the z axis, respectively, while the thermal drift was <0.05 nm·s^{-1}. The machine provided a surface topography of constant force in scanning probe microscopy (SPM) mode and a load versus depth of penetration plot in nanoindentation mode, using a Berkovich tip of $\approx$150 nm tip radius and 65.3° semi-apex angle. The sample was mounted on a motorized table that allowed for a movement in the plane normal to the axial motion of the tip. The transducer that measured both load and depth consisted of a three-plate capacitor, with the same Berkovich tip attached to the central plate. The instrument was calibrated by performing indents of increasing depth in a standard fused-quartz sample that had a known nanohardness of 9.25 ± 0.93 GPa and reduced modulus (E_r) of 69.6 ± 3.48 GPa. Figure 9.6a

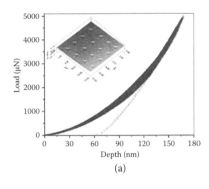
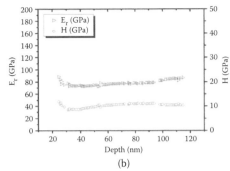

(a) (b)

FIGURE 9.6 (See color insert.)
(a) Partial-unloading–depth plot of the standard fused-quartz sample and (b) hardness (H) and reduced modulus (E_r) as a function of the depth of penetration for the standard fused-quartz sample (inset: SPM image of the indentation array.)

shows a typical partial-unloading–depth plot of a standard fused-quartz sample. The inset of Figure 9.6a shows an SPM image of the impression of the indentation array. Figure 9.6b shows the hardness and reduced modulus data as a function of depth for the standard fused-quartz sample. The data were also in the range, as mentioned previously. The diamond tip had a Poisson's ratio of 0.07 and Young's modulus of 1140 GPa. The machine had a load range of 0.01–12,000 µN. We have used this machine for the studies reported in Chapters 25 (only for the HAp-CT system) and 29 and Section 10 (except Chapter 40) of this book using only a Berkovich tip.

9.3.3 Nano Indenter G200

The Nano Indenter G200 (Agilent-MTS) is compliant with the ISO 14577 standard. It had a load range of 0.4–500 mN, and it was equipped with a Berkovich indenter of 120 nm tip radius. The displacement resolution was less than 0.01 nm, while load resolution was 50 nN. The maximum indentation depth was less than 500 µm. We have used this machine for the study reported in Chapter 40 using a Berkovich tip.

9.3.4 The Typical Protocol

Typically, the nanoindentation experiments were conducted mostly in the Fischerscope machine with a Berkovich indenter at various loads in the range of 10–1000 mN. By keeping the load constant, the loading as well as the unloading rates were varied in the range of 1 mN·s⁻¹ to 1000 mN·s⁻¹. In some cases, a Vickers nanoindenter tip was also used with this machine. However, on some occasions, i.e., for the studies reported in Chapter 11, the Tribo Indenter UBI 700 was utilized for our investigation on the 330 µm thin soda-lime-silica glass sample. The nanoindentation was done with a Berkovich indenter as described previously. In this work, the peak load was kept constant at 10,000 µN, while both the loading and the unloading times were varied from 0.5 to 1000 s to change the loading rates in the range of 10–20,000 µN·s⁻¹. For all the experiments, at least five nanoindents were made at each of five randomly chosen locations on the sample at a given loading rate for a given peak load. Thus, each reported value of nanohardness was an average of at least 25 or more individual data points. However, for some studies reported in Section 12 of this book, all three of the machines mentioned here were used.

9.4 The Scratch Tests

A commercially available scratch tester (Model TR-102-M3, Ducom, Bangalore, India) equipped with a Rockwell C diamond indenter with 200 µm tip radius was used to perform the scratch tests on the SLS glass at four different constantly applied normal loads (P) of 2, 5, 10, and 15 N with scratching speeds (v) of 100, 200, 500, and 1000 µm·s⁻¹. At least three or more scratch experiments were conducted for each load–speed combination, and the average data were taken. The error bars for all data reported here signify ±1 standard deviation. A compound force transducer with a resolution of ±0.01 N and attached to the scratch tester measured both normal load (P) and lateral (F) forces up to 20 N. An acoustic emission sensor attached separately with the compound force transducer measured the acoustic emission data in all the scratch tests. For all scratch experiments, the stroke length and scratch offset were kept at 3 and 0.5 mm, respectively. The depth and width data of the scratches were measured with a profilometer (Form Talysurf 120, Taylor Hobson, U.K., diamond stylus radius = 2 µm). The nanoindentation tests were conducted through the scratch grooves at 100 mN load and the time kept fixed at 30 s. In the case of the 20 µm grain size alumina, the loading rate was deliberately varied from ≈10³ to 10⁶ µN⁻¹ at three different peak loads of 10⁵, 5 × 10⁵, and 10⁶ µN. A 5 × 5 array matrix was utilized for this purpose. All of these aforesaid experiments were done using the Fischerscope machine, as mentioned previously.

9.5 Microstructural Characterizations

Microstructural characterizations and damage evolution studies were carried out initially by optical microscopy (GX51, Olympus, USA) technique. Further characterizations were done by either environmental scanning electron microscopy (E-SEM, Hitachi S-3400N, Japan) technique or normal scanning electron microscopy (SEM, s430i, LEO, UK) technique. Still further details were revealed by field emission scanning electron microscopy (FESEM, Supra VP35, Carl Zeiss, Germany) technique. Shock debris of alumina and some other samples were also studied by transmission electron microscopy (TEM: Tecnai G2 30, S-Twin, 300 KV, FEI, Netherlands, LaB_6 filament, line resolution 0.14 nm, and point resolution 0.2 nm) technique. In addition, in some cases, scanning probe microscopy (SPM, Tribo Indenter UBI 700, Hysitron, USA) technique was used for characterization of contact damage as well as nanoindentation cavities. Further compositional analysis was done with energy dispersive X-ray (EDX) technique by the same SEM machines as described previously. Prior to insertion in the sample chamber for electron microscopy, a 50–70 nm gold coating was deposited on the tooth samples by the arc-deposition technique to avoid charging.

9.6 Conclusions

In this chapter, we have tried to present a very brief overview of the huge variety of brittle materials that we shall come across as we go through the subsequent chapters. We discuss their preparation methods, followed by a brief description of the relevant nanoindentation machines and the particular techniques associated with the corresponding research works that are described later in this book. In Chapter 10 we shall try to understand what will happen to the nanoindentation response of a typical brittle, solid-like glass if the rate of contact is deliberately varied with time.

References

1. Chakraborty, R., A. Dey, and A. K. Mukhopadhyay. 2010. Loading rate effect on nanohardness of soda-lime-silica glass. *Metallurgical and Materials Transactions A* 41A:1301–12.
2. Dey, A., R. Chakraborty, and A. K. Mukhopadhyay. 2011. Enhancement in nanohardness of soda–lime–silica glass. *Journal of Non-Crystalline Solids* 357:2934–40.
3. Mukhopadhyay, A. K., K. D. Joshi, A. Dey, R. Chakraborty, A. Rav, S. K. Biswas, and S. C. Gupta. 2010. Shock deformation of coarse grain alumina above Hugoniot elastic limit. *Journal of Material Science* 45:3635–51.
4. Bhattacharya, M., R. Chakraborty, A. Dey, A. K. Mandal, and A. K. Mukhopadhyay. 2012. Improvement in nanoscale contact resistance of alumina. *Applied Physics A* 107:783–88.

5. Chakraborty, R., A. Dey, A. K. Mukhopadhyay, K. D. Joshi, A. Rav, A. K. Mandal, S. Bysakh, S. K. Biswas, and S. C. Gupta. 2012. Nanohardness of as-sintered and shock-deformed alumina. *Metallurgical and Materials Transaction A* 43:459–70.

6. Sarkar, S., A. Dey, P. K. Das, A. Kumar, and A. K. Mukhopadhyay. 2011. Evaluation of micromechanical properties of carbon/carbon and carbon/carbon-silicon carbide composites at ultra-low load. *International Journal of Applied Ceramic Technology* 8:282–97.

7. Dey, A. 2007. Toughening of ceramic matrix composites. M. Tech. diss. Bengal Engineering and Science University, Shibpur, India.

8. Nath, S., A. Dey, A. K. Mukhopadhyay, and B. Basu. 2009. Nanoindentation response of novel hydroxyapatite–mullite composites. *Materials Science and Engineering A* 513/514:197–201.

9. Dubey, A. K., G. Tripathi, and B. Basu. 2010. Characterization of hydroxyapatite-perovskite ($CaTiO_3$) composites: Phase evaluation and cellular response. *Journal of Biomedical Materials Research Part B: Applied Biomaterials* 95:320–29.

10. Himanshu, A. K., S. K. Bandyopadhyay, P. Sen, D. C. Gupta, R. Chakraborty, A. K. Mukhopadhyay, B. K. Choudary, and T. P. Sinha. 2010. Dielectric and micromechanical studies of barium titanate substituted (1-y)Pb ($Zn_{1/3}Nb_{2/3}$) O_3-yPT ferroelectric ceramics. *Indian Journal of Pure and Applied Physics* 48:349–56.

11. Sen, P., A. Dey, A. K. Mukhopadhyay, S. K. Bandyopadhyay, and A. K. Himanshu. 2012. Nanoindentation behaviour of nano $BiFeO_3$. *Ceramics International* 37:1347–52.

12. Dey, T., A. Dey, P. C. Ghosh, M. Bose, A. K Mukhopadhyay, and R. N. Basu. Forthcoming. Influence of microstructure on nano-mechanical properties of single planar solid oxide fuel cell in pre- and post-reduced conditions. *Materials and Design* 53:182–91. http://www.sciencedirect.com/science/article/pii/S0261306913005864.

13. Ghosh, S., A. D. Sharma, A. K. Mukhopadhyay, P. Kundu, and R. N. Basu. 2010. Effect of BaO addition on magnesium lanthanum alumino borosilicate-based glass-ceramic sealant for anode-supported solid oxide fuel cell. *International Journal of Hydrogen Energy* 35:272–83.

14. Dey, A., A. K. Mukhopadhyay, S. Gangadharan, M. K. Sinha, D. Basu, and N. R. Bandyopadhyay. 2009. Nanoindentation study of microplasma sprayed hydroxyapatite coating. *Ceramics International* 35:2295–2304.

15. Dey, A., A. K. Mukhopadhyay, S. Gangadharan, M. K. Sinha, and D. Basu. 2009. Development of hydroxyapatite coating by microplasma spraying. *Materials and Manufacturing Processes* 24:1249–58.

16. Dey, A., A. K. Mukhopadhyay, S. Gangadharan, M. K. Sinha, and D. Basu. 2009. Weibull modulus of nano-hardness and elastic modulus of hydroxyapatite coating. *Journal of Materials Science* 44:4911–18.

17. Dey, A., A. K. Mukhopadhyay, S. Gangadharan, M. K. Sinha, and D. Basu. 2009. Characterization of microplasma sprayed hydroxyapatite coating. *Journal of Thermal Spray Technology* 18:578–92.

18. Dey, A., S. K. Nandi, B. Kundu, C. Kumar, P. Mukherjee, S. Roy, A. K. Mukhopadhyay, M. K. Sinha, and D. Basu. 2011. Evaluation of hydroxyapatite and β-tri calcium phosphate microplasma spray coated pin intra-medullarly for bone repair in a rabbit model. *Ceramics International* 8:1377–91.

19. Dey, A., and A. K. Mukhopadhyay. Forthcoming. In vitro dissolution, micro-structural and mechanical characterizations of microplasma sprayed hydroxy-apatite coating. *International Journal of Applied Ceramic Technology*. http://onlinelibrary.wiley.com/doi/10.1111/ijac.12057/abstract.
20. Dey, A., R. U. Rani, H. K. Thota, A. K. Sharma, P. Bandyopadhyay, and A. K. Mukhopadhyay. 2013. Microstructural, corrosion and nanomechanical behaviour of ceramic coatings developed on magnesium AZ31 alloy by micro arc oxidation. *Ceramics International* 39:3313–20.
21. Das, P. S., A. Dey, M. R. Chaudhuri, S. Roy, A. K. Mandal, N. Dey, and A. K. Mukhopadhyay. 2012. Chemically deposited magnesium hydroxide thin film. *Surface Engineering* 28:731–36.
22. Paul, R., S. Hussain, S. Majumder, S. Varma, and A. K. Pal. 2009. Surface plasmon characteristics of nanocrystalline gold/DLC composite films prepared by plasma CVD technique. *Materials Science and Engineering B* 164:156–64.
23. Biswas, N., A. Dey, S. Kundu, H. Chakraborty, and A. K. Mukhopadhyay. 2012. Orientational effect in nanohardness of functionally graded microstructure in enamel. *Journal of the Institute of Engineers India: Series D* 93:87–95.
24. Biswas, N., A. Dey, and A. K. Mukhopadhyay. 2012. Loading rate effect on nano-hardness of human enamel. *Indian Journal of Physics* 86:569–74.
25. Biswas, N., A. Dey, and A. K. Mukhopadhyay. 2013. Micro-pop-in issues in nanoscale contact deformation resistance of tooth enamel. *ISRN Biomaterials*. http://www.hindawi.com/isrn/biomaterials/2013/545791/.
26. Oliver, W. C., and G. M. Pharr. 1992. An improved technique for determining the hardness and elastic modulus using load and depth sensing indentation experiments. *Journal of Materials Research* 7:1564–83.

Section 3

Static Contact
Behavior of Glass

10

What If the Contact is Too Quick in Glass?

Riya Chakraborty, Arjun Dey, and Anoop Kumar Mukhopadhyay

10.1 Introduction

In this and the next chapter, we shall discuss the effect of loading rate on the nanoscale Contact resistance of glass. Glass is a classical example of a brittle solid and, hence, has been used in many research studies over centuries. The loading rate is nothing but the rate at which the contact happens. Thus, basically it relates to the process of energy transfer between the loading train and the material in question. Glass as an amorphous solid offers a simple means to look at different scales of deformation when the contact loads are varied, for instance from a few newtons of force to a few millinewtons of force to a few micronewtons of force applied on the indenter. In other words, the indentation technique in general, and the nanoindentation technique in particular, provides an opportunity to study the scale issues in deformation and fracture in a controlled manner. It also follows, then, that by proper choice of appropriate load scales, the corresponding microstructural length scale issues or issues involving more than one microstructural length scale can be addressed.

The importance of such studies emerges from the fact that today glass plays a unique role in our daily life. It also plays a unique aesthetic role in the social and architectural texture of society. Further, across centuries, it has borne the historical footprints of the evolution of human civilization. The simple household applications of glass can span from simple pots to window glasses. But the truly advanced applications today can cover a spectacular range that is both fascinating and highly divergent in nature. For instance, consider the case of the microwave-safe crockery glass or the advanced car cover glass with hydrophobic coating. The advanced super-tough glasses decorate the electronic gadgets like touchscreen mobiles, tablet computers, I-pads, and ultra-slick cameras, which today have become items of almost daily utility. Further, glass with anti-reflecting, hydrophilic/hydrophobic, and heat-reflecting coatings has huge demand in the building industries sectors. In all modern-day shopping malls, the startling applications of such glasses stand a testimony to their ever-increasing value as a high-end technology product. Similarly, glass has huge technological applications in advanced radio telescopes in particular and in the space technology in general.

In this context, however, it must be borne in mind that glass is a brittle solid and, hence, is prone to suffer from contact-damage-induced brittle fracture. In this connection, the factor of utmost importance is static contact damage. The property that governs the intrinsic resistance against static-contact-induced damage is the hardness of glass. When we look at such a scenario at the microstructural length scale, we call this intrinsic resistance against contact-induced permanent deformation at the scale of the local microstructure the material's *nanohardness*.

Understanding the details of such mechanical response at the microstructural length scale therefore becomes important because it is at such a scale that the issue of structural integrity gets ultimately determined for a brittle solid like glass. One way to simulate such low-dimensional contact events is to use the nanoindentation technique [1]. Thus, in the present chapter we present the results of nanoindentation experiments conducted at three peak loads (100, 500, and 1000 mN) at loading rates (LR) in the range of 1–1000 mN·s⁻¹ on the same thin (≈330 μm) SLS (soda-lime-silica) glass cover slip as described in Chapter 9.

10.2 Effect of Loading Rate on Nanohardness

The data presented in Figures 10.1a–d provide the first-ever evidence that, depending on peak load and loading rate combination, the nanohardness of the present SLS glass can increase by about 6%–9% [2]. Further, there was a threshold loading rate (TLR) up to which there was a sharp increase of nanohardness with loading rate (LR). The values of TLR were 10 (Figure 10.2a), 50 (Figure 10.2b), and 50 mN·s⁻¹ (Figure 10.2c) for P_{max} of 100, 500, and 1000 mN, respectively. In addition, for a given peak load of 1000 mN as shown in Figures 10.2a–c, pronounced serration appeared in the *P–h* plots up to the TLR, but above the TLR, the serration was not so frequently observed in the *P–h* plots (see Figure 10.2d). These data would suggest that the deformation mechanisms below and above the TLR must be different. That this was indeed the case could be confirmed from the results of the SEM studies presented in Figures 10.3a–e, corresponding to peak load of 1000 mN. A large number of shear bands were present in and around the nanoindent made at the low loading rate of 1 mN·s⁻¹ (Figures 10.3a and 10.3b). This LR was below the TLR of 50 mN·s⁻¹. But their number slightly decreased at a higher loading rate of 77 mN·s⁻¹ (Figures 10.3c and 10.3d). However, at a much higher loading rate of 770 mN·s⁻¹, only a couple of shear bands were found (Figure 10.3e) around the nanoindent. Both of these LRs were higher than the TLR.

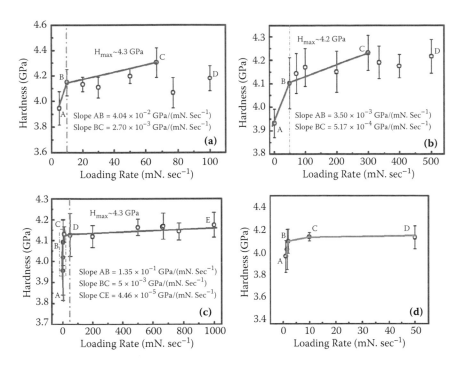

FIGURE 10.1

Nanohardness as a function of loading rates at various peak loads, P_{max} (mN): (a) 100, (b) 500, (c) 1000, and (d) blown-up view of portion marked as AC in Figure 10.1c. (Reprinted with permission of Chakraborty, Dey, and Mukhopadhyay [2] from Springer.)

10.3 Damage Evolution Mechanism

Thus, from the SEM-based evidence presented here, it is proved that the deformation mechanisms below and above the TLR were different and thereby contributed to enhancement of the nanohardness of the present SLS glass. Note that plastic flow in glass occurs by an inhomogeneous mechanism, i.e., large shear displacements occur on certain planes, but only elastic strains occur between the planes [3]. The favored planes can arise by nucleation at weak points in the network for glasses, such as soda-lime-silica glass that contains a large amount of network modifiers.

The present SLS glass had about 13 and 8 wt% of Na_2O and CaO, respectively (Chapter 9), which were designated to act as network modifiers. What might have happened here is that the large amount of modifiers (Na_2O and CaO) included in the present SLS glass led to the formation of nonbridging oxygen and weaker bonds such as Na^+-O^-. It is at the localized spatial positions of such weaker bonds that the SLS glass can suffer a relatively larger amount of shear-flow-induced plastic deformation. Further, from

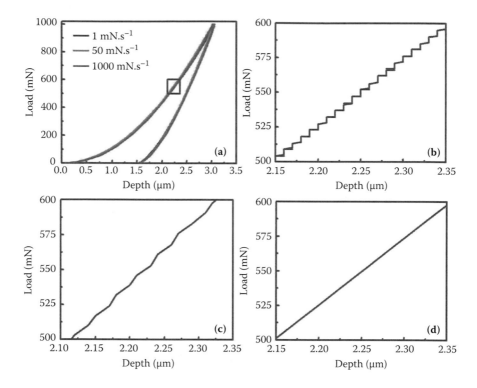

FIGURE 10.2
(a) Typical load (P) versus depth (h) plots for P_{max} (1000 mN) as a function of loading rate for 1, 50, and 1000 mN·s^{-1}. (b) Enlarged view of the marked portion of Figure 10.2a for 1 mN·s^{-1} loading rate. (c) Enlarged view of the marked portion of Figure 10.2a for 50 mN·s^{-1} loading rate. (d) Enlarged view of the marked portion of Figure 10.2a for 1000 mN·s^{-1} loading rate.

the classical Hertzian analysis [4], the maximum shear (τ_{max}) stress below the indenter was predicted as ≈8.09, 7.19, and 6.88 GPa for the given peak loads of 100, 500, and 1000 mN, respectively. But the theoretical shear strength (τ_{theor}) for the SLS glass can be estimated as ≈6 GPa [5, 6]. Thus, the maximum shear stress generated just below the tip of the Berkovich indenter used in the present work was on the order of or slightly greater than the theoretical shear strength of the SLS glass. Therefore, it is plausible to argue that the applied loading provided enough shear stress to cause shear-induced flow and/or deformation in the indented soda-lime-silica glass. Thus, it may be suggested that the first displacement burst observed in the present experiments corresponded to the condition when the maximum shear stress generated just below the tip of the indenter was on the order of or slightly higher than the theoretical shear strength of the material. Therefore, it was likely that the first burst corresponded to the initiation of plastic deformation in the material through the nucleation of shear bands, as was also experimentally observed (Figures 10.3a and 10.3b).

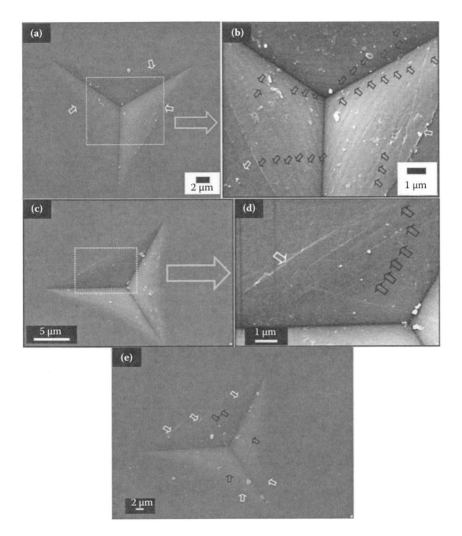

FIGURE 10.3
Typical SEM photomicrographs of shear bands in nanoindents made with 1000 mN applied load at different loading rates: (a, b) 1 mN·s⁻¹ at low and high magnification; (c, d) 77 mN·s⁻¹ at low and high magnification; and (e) 770 mN·s⁻¹ (black and white hollow arrows indicate the shear bands inside the nanoindentation cavity and around the nanoindentation cavity, respectively). (Reprinted with permission of Chakraborty, Dey, and Mukhopadhyay [2] from Springer.)

It is suggested further that, for a given lower loading rate, because of the long time available during the indentation process, the action of the next and all such consequent shear bands formed could have come into play nearly sequentially, one after another. When this happens, due to the sequential, discrete action of multiple shear bands, multiple serrations should appear in the *P–h* plot, as was experimentally observed (Figure 10.2a). It is also evident that such a generic process most likely was only locally active. This localization would possibly make the plastic deformation field and its effect highly inhomogeneous in and around the nanoindents made at lower loading rates [4].

Thus, it is proposed that for a given peak load as the loading rate was increased, initially there was a moderate increment in nanohardness, possibly due to the relatively larger number of slip bands formed and the consequent shear flow process, as suggested by other researchers [7]. It is expected that the greater the number of slip bands formed, the larger was the amount of strain accommodated. The genesis of the permanent strain is linked to the atomic environment in the SLS glass. The range of atomic environments in SLS glass is such that some atoms can always reside in regions where the local topology is unstable, e.g., at positions of network modifiers. In such localized regions, the response to shear stress provided by the nanoindentation process could be twofold: (a) atomic displacements and (b) an anelastic reshuffling of the atomic near-neighbors. The actual fraction of atoms involved in such a process may be a small number. However, the local strains caused by their displacements and anelastic reshuffling will be large enough. It is the cumulative effect of such localized strains that finally contributes to locally generate a significant amount of macroscopic strain.

Similarly, at higher loading rates, the loading as well as the unloading times were extremely small, at about 1 s. It is suggested that during this extremely short duration of quasi-static contact, only the overall average elastoplastic and plastic deformation processes can come into play. At such high rates ($1000 \ \mathrm{mN \cdot s^{-1}}$), it has been suggested further [4] that multiple shear bands can operate in a simultaneous fashion (rather than in a sequential discrete fashion) within the extremely small time span of contact. The simultaneous operation of these multiple shear bands possibly forced a homogenization of the plastic deformation response of the glass such that only a couple of shear bands had occurred around the nanoindentation cavity (Figure 10.3e). The response of the glass to this homogenized deformation resulted in a *P–h* plot where the signature of serrations was not as prominent as that of the serrations obtained at lower loading rates. As the number of shear bands formed was reduced, so possibly was the extent of shear flow, as explained previously. It follows then that the more such reduction occurs, the lesser will be the resultant enhancement in nanohardness. This picture provided a basis to explain why the rate of increase of the nanohardness value was slowed down at loading rates higher than the TLR (Figures 10.1a–d). It is expected that, at a point very close to the indenter's tip, the stress would be maximum. As a result of this intense local stress distribution, the strain (about 8%) would be very high, and its spatial gradient would be very steep near the base of the nanoindent in the cavity [8]. It is proposed that this acute high strain and its steep spatial gradient in the small deformation volume below the indenter possibly initiated the formation of one first shear band to quickly accommodate the strain locally, without initiating a crack.

10.4 Conclusions

Depending on the peak load (P_{max}) in the range of 100–1000 mN, the nano-hardness of very thin (≈330 μm) soda-lime-silica glass increased by 6%–9% with an increase in loading rate (1–1000 mN·s⁻¹). For a given P_{max}, the nano-hardness of the glass sharply increased initially with loading rate up to the TLR, which was different for different peak loads. For a given P_{max}, the SEM photographs showed a profusion of shear bands in and around the nanoindentation cavity made at loading rates lower than the TLR. However, the number of shear bands was comparatively much lower in and around the nanoindentation cavity made at loading rates higher than the TLR. It is suggested that the position of the shear-flow-induced shear-band formation would be governed by the local, short-range atomic arrangements in the glass, i.e., the positions of local weakness provided by the network modifiers. Although we now have gotten some insight into the deformation mechanisms, we still do not know what will happen if the load is reduced to the micronewton load range while the loading rate is varied over a large range, i.e., much larger than what we have experienced in this chapter. This is exactly what we shall explore in Chapter 11. We shall also explore whether it is possible to have a sizeable enhancement in the nanohardness of SLS glass.

References

1. Mukhopadhyay, N. K., and P. Paufler. 2006. Micro- and nanoindentation techniques for mechanical characterisation of materials. *International Materials Reviews* 51:209–45.
2. Chakraborty, R., A. Dey, and A. K. Mukhopadhyay. 2010. Loading rate effect on nanohardness of soda-lime-silica glass. *Metallurgical and Materials Transactions A* 41A:1301–12.
3. Peter, K. W. 1970. Densification and flow phenomena of glass in indentation experiments. *Journal of Non-Crystalline Solids* 5:103–15.
4. Packard, C. E., and C. A. Schuh. 2007. Initiation of a shear band near a stress concentration in metallic glass. *Acta Materialia* 55:5348–58.
5. Honeycombe, R. W. K. 1984. *Plastic deformation of metals*. 2nd ed. London: Edward Arnold Ltd.
6. Puthucode, A., R. Banerjee, S. Vadlakonda, R. Mirshams, and M. J. Kaufman. 2008. Incipient plasticity and shear band formation in bulk metallic glass studied using indentation. *Metallurgical and Materials Transactions A* 39A:1552–59.
7. Arora, A., D. B. Marshall, and B. R. Lawn. 1979. Indentation deformation/fracture of normal and anomalous glasses. *Journal of Non-Crystalline Solids* 31:415–28.
8. Shikimaka, O., and D. Grabco. 2008. Deformation created by Berkovich and Vickers indenters and its influence on surface morphology of indentations for LiF and CaF₂ single crystals. *Journal of Physics D: Applied Physics* 41:074012-1-6.

11

Enhancement in Nanohardness of Glass: Possible?

Riya Chakraborty, Arjun Dey, and Anoop Kumar Mukhopadhyay

11.1 Introduction

In Chapter 10, we presented the results of nanoindentation experiments conducted at three peak loads of 100, 500, and 1000 mN at loading rates (LR) in the range of 1–1000 mN·s^{-1} on the same thin ($\approx$330 µm) SLS glass cover slip, as described in Chapter 9. In this chapter, we shall examine the effect of loading rate variation (10–20,000 µN·s^{-1}) during the nanoindentation experiments conducted at a constant peak load of 10,000 µN on the nanohardness of the same thin ($\approx$330 µm) SLS (soda-lime-silica) glass cover slip. The results provide more evidence that the loading rate variation at the range of (10–20,000 µN·s^{-1}) can significantly enhance the nanohardness of the present SLS glass. The details of the experimental methods are discussed in Chapter 9.

11.2 Nanomechanical Behavior

With an increase in loading rate, the nanohardness and Young's modulus of the SLS glass sample increased respectively by $\approx$74% (Figure 11.1a) and $\approx$15% (Figure 11.1b), while the average projected area of contact (A_p) decreased in a corresponding manner (Figure 11.1a) [1, 2]. The load versus depth (P–h) plots for nanoindentations on the SLS glass sample at loading rates of 10, 100, 2,000, and 20,000 µN·s^{-1} are shown respectively in Figures 11.2a–d. From the data presented in Figure 11.2, it is observed that pronounced serrations appeared in the P–h plots, often more so at lower loading rates of 10 µN·s^{-1} (Figure 11.2a) than at higher loading rates of 20,000 µN·s^{-1} (Figure 11.2d).

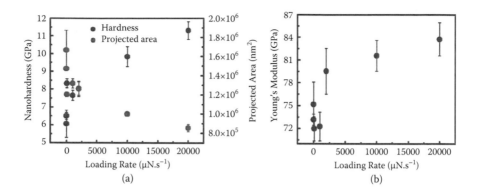

FIGURE 11.1 (See color insert.)
(a) Nanohardness and projected area of contact. (b) Young's modulus of the SLS glass as a function of loading rate. (Reprinted with permission of Dey, Chakraborty, and Mukhopadhyay [2] from The American Ceramic Society and Wiley).

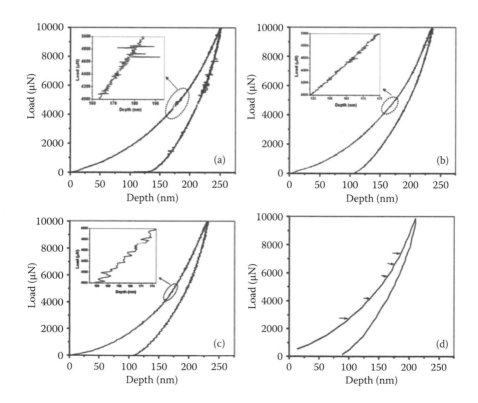

FIGURE 11.2
Load–depth plots for nanoindentations with a constant load of 10,000 μN on SLS glass at loading rates of (a) 10, (b) 100, (c) 2,000, and (d) 20,000 μN·s⁻¹. (Reprinted with permission of Dey, Chakraborty, and Mukhopadhyay [1] from Elsevier.)

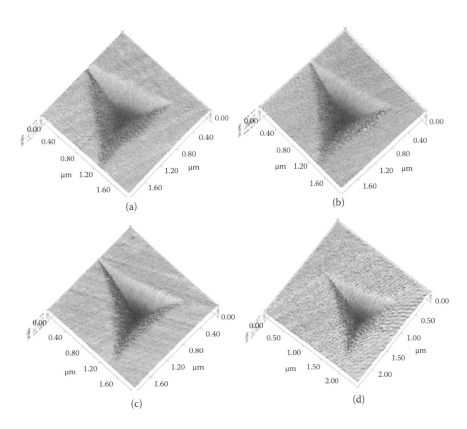

FIGURE 11.3 (See color insert.)
SPM images of nanoindentations with a constant load of 10,000 µN on SLS glass at loading rates of (a) 10, (b) 100, (c) 2,000, and (d) 20,000 µN·s⁻¹. (Reprinted with permission of Dey, Chakraborty, and Mukhopadhyay [1] from Elsevier.)

Similarly, the corresponding SPM images (Figures 11.3a–d) confirm the presence of a large number of shear bands in the nanoindents at the higher loading rate of 20,000 µN·s⁻¹ (Figure 11.3d) compared to those made at the lower loading rate of 10 µN·s⁻¹ (Figure 11.3a). The occurrence of these shear bands and the enhancements of nanohardness can be explained by the same argument posited in Chapter 10, i.e., that the position of the shear-flow-induced shear-band formation is governed by the local weakness positions of the network modifiers. Further, at respective loading rates of 10, 100, 2,000, and 20,000 µN·s⁻¹, the shear stresses are predicted to be ≈1.3, 1.4, 7.1, and 10.6 GPa [3, 4], which was on the order of or greater than the theoretical shear strength (τ_{theor} ≈ 2–6 GPa) for the SLS glass [5, 6]. Therefore, just beneath the tip of the nanoindent in the SLS glass, the applied loading provided enough shear stress to cause formation of shear-induced deformation bands (Figures 11.3a–d). Hence, it was likely that the first burst corresponded to the initiation of plastic deformation in the material (Figures 11.3a–d) through the nucleation of shear bands [7–9]. At loading rates greater than 2,000 $\mu N \cdot s^{-1}$ (Figure 11.2d), the increase in nanohardness of the soda-lime-glass was much less compared to that at loading rates lower than 2,000 $\mu N \cdot s^{-1}$ (Figures 11.2a and 11.2b), possibly because the stored elastic energy had been exhausted [9].

11.3 Conclusions

The significant improvement ($\approx$74%) in nanohardness of a thin ($\approx$330 μm) commercial SLS glass cover slip with loading rates of 10–20,000 $\mu N \cdot s^{-1}$ was commensurate with the significant presence of more shear-induced serrations in load–depth plots and the formation of more shear deformation bands at lower loading rates, as revealed by SPM images of the nanoindentation cavity. The applied loading provided enough shear stress just beneath the tip of the nanoindent to cause shear-induced flow and/or deformation in the SLS glass at positions of local weakness provided by the network modifiers. At loading rates higher than 2,000 $\mu N \cdot s^{-1}$, the forced homogenization of the plastic-deformation response of the glass caused a reduction in the extent of shear flow that limited the resultant enhancement in nanohardness, presumably because most of the stored elastic energy was already exhausted. Thus, the role of energy partition appears to play an important role, and this is what we are going to probe in Chapter 12.

References

1. Dey, A., R. Chakraborty, and A. K. Mukhopadhyay. 2011. Enhancement in nanohardness of soda-lime-silica glass. *Journal of Non-Crystalline Solids* 357:2934–40.
2. Dey, A., R. Chakraborty, and A. K. Mukhopadhyay. 2011. Nanoindentation on soda-lime-silica glass: An effect of loading rate. *International Journal of Applied Glass Science* 2:144–55.
3. Packard, C. E., and C. A. Schuh. 2007. Initiation of a shear band near a stress concentration in metallic glass. *Acta Materialia* 55:5348–58.
4. Shang, H., T. Rouxel, M. Buckley, and C. Bernard. 2006. Viscoelastic behavior of soda-lime-silica glass determined by high temperature indentation test. *Journal of Materials Research* 21:632–38.
5. Honeycombe, R. W. K. 1984. *Plastic deformation of metals.* 2nd ed. London: Edward Arnold Ltd.

6. Puthucode, A., R. Banerjee, S. Vadlakonda, R. Mirshams, and M. J. Kaufman. 2008. Incipient plasticity and shear band formation in bulk metallic glass studied using indentation. *Metallurgical and Materials Transactions A* 39A:1552–59.
7. Shim, S., H. Bei, E. P. Georgea, and G. M. Pharr. 2008. A different type of indentation size effect. *Scripta Materialia* 59:1095–98.
8. Ji, H., V. Keryvin, T. Rouxel, and T. Hammouda. 2006. Densification of window glass under very high pressure and its relevance to Vickers indentation. *Scripta Materialia* 55:1159–62.
9. Li, J., K. J. Van Vliet, T. Zhu, S. Yip, and S. Suresh. 2002. Atomistic mechanisms governing elastic limit and incipient plasticity in crystals. *Nature* 418:307–10.

12

Energy Issues in Nanoindentation

Riya Chakraborty, Arjun Dey, and Anoop Kumar Mukhopadhyay

12.1 Introduction

The previous two chapters (Chapters 10 and 11) gave us some idea about how the loading rate affects the hardness of SLS glass at nano- and micro-scales. However, there is another important viewpoint involved. We now ask: What will the role of plastic deformation energy be with regard to the effect of loading rate on the nanohardness of SLS glass? What do we mean by plastic deformation energy? During the nanoindentation, the machine continuously monitors the load (P) and depth (h) of penetration into the sample for a given indenter, which in the present investigation was a Berkovich tip. These data are utilized to get the P–h data plot. The important physical quantities obtained from the P–h plot (Figure 12.1) are: the peak load (P_{max}), maximum penetration depth (h_{max}), final penetration depth (h_f), and the contact stiffness (S). It is well known that, except for fully elastic material, the load versus depth of penetration plot will consist of two separate parts (Figure 12.1). The area specifically under the *unloading* curve is the amount of reverse deformation energy released (W_e) when the test load is released. However, the area encompassed by the *loading–unloading* curve is the amount of energy (W_p) dissipated in plastic deformation processes during the indentation test (Figure 12.1). Thus, for a material characterized mostly by elastic deformation, e.g., a carbon fiber, we have $W_e \gg W_p$. However, when a material, e.g., a polymer, is characterized mostly by plastic deformation, we have $W_p \gg W_e$. The situation will be intermediate when elastoplastic deformation, e.g., as in a glass or a ceramic, is concerned. The sum of these two is the total mechanical work of indentation, W_t. Thus,

$$W_t = W_p + W_e \tag{12.1}$$

or

$$W_t = \int_0^{h_{max}} P\,dh = \int_0^{h_f} P\,dh + \int_{h_f}^{h_{max}} P\,dh = W_p + W_e \tag{12.2}$$

where P and h are indentation load and depth of penetration, respectively.

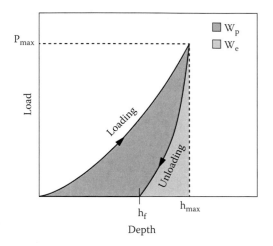

FIGURE 12.1
Schematic of a typical load (*P*) versus depth (*h*) plot obtained during nanoindentation in a brittle material (glass). (Reprinted with permission of Chakraborty, Dey, and Mukhopadhyay [8] from Society of Glass Technology and Deutsche Glastechnische Gesellschaft.)

12.2 Energy Models

Researchers have proposed different energy models [1–5] to link the W_p/W_t ratio to the ratio of hardness (*H*) to elastic modulus (*E*), i.e., *H/E*. It is important to recognize that both the hardness and elastic modulus are macroscopic parameters and can be easily evaluated without precise knowledge in micromechanics [2]. However, the models are often framed from different perspectives and supported through the use of data on a very wide variety of materials [1–5].

12.2.1 Lawn-Howes Model

For instance, the model proposed by Lawn and Howes [1] is based on (a) the premise that the loading is elastic–plastic while the unloading is purely elastic and (b) the recognition that a residual field term must be taken into consideration to accommodate the postindentation permanent deformation, which makes a significant departure from the initial, flat test surface. They predicted that the relationship is:

$$\frac{W_p}{W_t} = \left[1 - 2\tan\alpha \frac{H\left(1-v^2\right)}{E}\right]^{3/2} \tag{12.3}$$

where α is the semi-apex angle of the indenter. This relationship was successfully applied to Vickers microindentation hardness of soda lime glass, pyroceram, hot-pressed silicon nitride, tungsten carbide, steel, etc.

12.2.2 Sakai Model

On the other hand, the Sakai model [2] assumes that elastoplastic surface deformation is the result of a process in which the purely elastic deformation component is connected *in series* to a purely plastic deformation component, such that the ratio of W_p/W_t is given by:

$$\frac{W_p}{W_t} = \frac{1}{\left[1 + \sqrt{2}\tan(\alpha)\sqrt{\frac{H(1-v^2)}{E}}\right]^3} \tag{12.4}$$

This relation was successfully applied to Vickers microindentation hardness data of metals, e.g., aluminum and copper with $h_f \approx h_{max}$; brittle ceramics, e.g., silicon nitride and silicon carbide (SiC) with $0 < h_f < h_{max}$; and highly elastic glassy carbons with $h_f \approx 0$.

12.2.3 Cheng-Cheng Model

Unlike the aforesaid approaches, the Cheng and Cheng model [3] utilizes a scaling approach combined with dimensional analysis and finite element calculations for stress fields created due to indentations by pyramidal and conical indenters in an elastoplastic solid with *work hardening*. These researchers [6] predict a linear relationship between W_p/W_t and H/E:

$$\frac{W_p}{W_t} = \left[1 - a\frac{H(1-v^2)}{E}\right] \tag{12.5}$$

where the factor a is 5 for Vickers indenter and 4.7 for Berkovich indenter [4], and v is the Poisson's ratio of the indenter. This relation was successfully applied to experimental data on fused silica and sapphire as well as on aluminum, copper, etc., metals.

12.2.4 Malzbender-With Model

However, when applied to the nanoindentation data obtained with a Berkovich indenter on a hybrid organic-inorganic coating, it was reported [5] that different models [1–3] grossly underpredicted the ratio W_p/W_t. This situation led to another modified relationship, i.e.,

$$\frac{W_p}{W_t} = \left(1 - 5.59\frac{H}{E}\right)^3 \tag{12.6}$$

which was claimed to give the best fit to the experimental data [5] on hybrid organic–inorganic coatings.

These discrepancies highlight that the energy dissipation processes during an indentation contact event are yet to be fully understood. Further, none of the aforesaid models takes the loading-rate aspect directly into account. This is why the importance of the energy dissipation process in nanohardness of glass as a function of loading rate becomes an issue of significant concern.

12.3 Energy Calculation

As discussed previously, to identify whether there is any interconnection between the relative amount of energy spent in creating plastic deformation and the loading rates, the corresponding data on the relative amount of energy spent in plastic deformation for the three peak loads of 100, 500, and 1000 mN are shown, respectively, in Figures 12.2a–c. These data confirmed that the maximum enhancements in the relative amount of energy spent in plastic deformation $(W_p/W_t)\%$ occurred at the same threshold loading rates of 10, 50, and 50 mN·s^{-1} below and up to which the maximum rate of change of nanohardness occurred for the three corresponding peak loads of 100, 500, and 1000 mN (see Chapter 10, Figures 10.1a–d). This observation suggested that there was an interrelation between the maximum rate of change in nanohardness that occurred at the TLR for the SLS glass sample and the relative amount of energy spent in plastic deformation $(W_p/W_t)\%$, thereby identifying it as a key energy-related parameter [7].

In this connection, it is worth mentioning that several workers [1–3, 5] attempted to predict the relative amount of energy spent in plastic deformation $(W_p/W_t)\%$ in terms of the quantity H/E. Therefore, in our work, an attempt was made to utilize these models [1–3, 5] to predict the experimentally measured data of $(W_p/W_t)\%$ in terms of the quantity H/E, which were all measured in the present experiments for the given peak loads of 100, 500, and 1000 mN. The results of this effort are shown in Figures 12.3a–c.

The experimental data (Figures 12.3a–c) are shown in the gray band, while the model predictions are depicted with hollow symbols without any such band. These results show that none of the models could predict the experimental data correctly. It may be noted from the results shown in Figures 12.3a–c that, for any given peak load, the Lawn and Howes model [1] and the Cheng and Cheng model [3] overpredicted the experimental data, while both the Sakai model [2] and the Malzbender and With model [5] had underpredicted the experimental data. This information suggested that a new approach is indeed very much needed to explain the present experimental data.

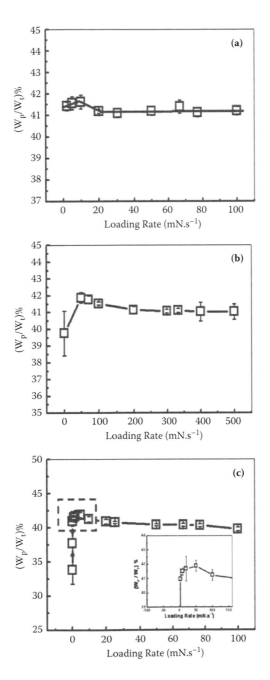

FIGURE 12.2
Variation of the relative percentage of plastic energy $(W_p/W_t)\%$ as a function of the loading rate for different peak loads: (a) 100 mN, (b) 500 mN, and (c) 1000 mN. (Reprinted with permission of Chakraborty, Dey, and Mukhopadhyay [8] from Society of Glass Technology and Deutsche Glastechnische Gesellschaft.)

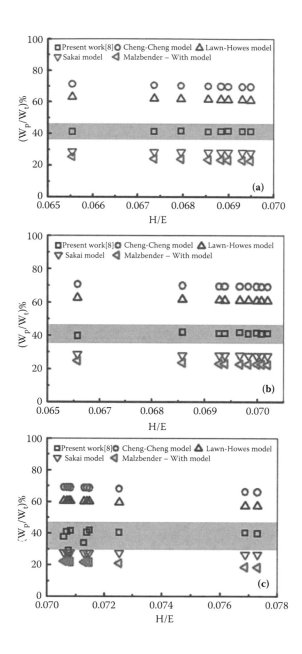

FIGURE 12.3
Variation of the relative amount of energy spent in plastic deformation ($[W_p/W_t]\%$) as a function of the hardness to Young's modulus ratio $[H/E]$, based on experimental data shown in gray bands for various peak loads: (a) 100 mN, (b) 500 mN, and (c) 1000 mN. The data predicted according to the Lawn and Howes [1], Cheng and Cheng [3], Sakai [2], and Malzbender and With [5] models are included for the purpose of comparison. (Reprinted with permission of Chakraborty, Dey, and Mukhopadhyay [8] from Society of Glass Technology and Deutsche Glastechnische Gesellschaft.)

12.3.1 Inelastic Deformation (IED) Parameter

The data on the inelastic deformation (IED) parameter, plotted as a function of loading rate, are shown in Figures 12.4.a–c. These data plots clearly established that 10 mN·s⁻¹ (for a peak load of 100 mN) and 50 mN·s⁻¹ (for peak loads of 500 and 1000 mN) were indeed the threshold loading rates (TLR).

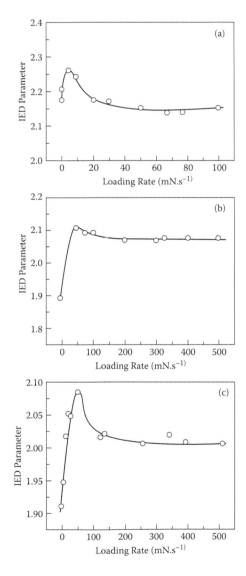

FIGURE 12.4

Variation of inelastic deformation (IED) parameter as a function of loading rate at different peak loads: (a) 100 mN, (b) 500 mN, and (c) 1000 mN. (Reprinted with permission of Chakraborty, Dey, and Mukhopadhyay [8] from Society of Glass Technology and Deutsche Glastechnische Gesellschaft.)

The IED parameter had the maximum values at these TLRs and decreased thereafter slowly with further increase in loading rate [7].

From the deformation physics point of view, the IED parameter, proposed for the first time by the present authors [8], represents the ratio of maximum depth (h_{max}) to recovered depth ($h_{max}-h_f$). It is interesting to note that the maximum increase in the IED parameter had also occurred exactly at the loading rates of 10 mN·s^{-1} (for 100 mN load) and 50 mN·s^{-1} (for peak loads of 500 and 1000 mN), which are exactly the threshold loading rates as mentioned in Chapter 10 (Figures 10.1a–d). It slowly decreased thereafter with further increase in loading rate. This explains why the rate of increase of nanohardness with loading rate was much higher, up to a loading rate of 10 mN·s^{-1} (for 100 mN peak load) and 50 mN·s^{-1} (for 500 and 1000 mN·s^{-1} peak loads) but became much lower beyond it (e.g., see Chapter 10, Figures 10.1a–d). Further, it logically follows that, to facilitate the maximum amount of inelastic deformation to occur, the relative amount of energy spent in plastic deformation must also maximize at these very TLR values, as was also observed, from our experimental data (see Chapter 10, Figures 10.1a–d), thereby supporting the hypothesis framed here.

12.4 Conclusions

For soda-lime-silica (SLS) glass, about 5%–10% enhancement of nanohardness is observed with an increase in the loading rate (1–1000 mN·s^{-1}). A threshold loading rate (TLR) of 10 and 50 mN·s^{-1} was found, above which a change in the energy dissipation process happens, thereby effecting a change in the deformation mechanism. The relative amount of energy spent in plastic deformation (W_p/W_t)% reached its maximum at the same TLR values. Further, the inelastic deformation (IED) parameter had the maximum values at these same TLR values and thereafter decreased slowly with further increase in loading rate. These results suggested that the maximum magnitude of the IED parameter occurred when the relative amount of energy spent in plastic deformation during the nanoindentation process was also the maximum, thereby contributing to the maximum rate of change in enhancement of nanohardness of the present glass with loading rate at the TLR values. But so far, we have dealt with only static contact situations in the SLS glass and, as such, have not examined the change of scenario if the contact situation becomes a dynamic one, e.g., as during the scratching of glass by an airborne hard particle, say, a tiny particle of sand. This is what we are going to look at in Chapter 13.

References

1. Lawn, B. R., and V. R. Howes. 1981. Elastic recovery at hardness indentations. *Journal of Material Science* 16:2745–752.
2. Sakai, M. 1993. Energy principle of the indentation-induced inelastic surface deformation and hardness of brittle materials. *Acta Metallurgical Materials* 41:1751–58.
3. Cheng, Y. T., and C. M. Cheng. 1998. Relationships between hardness, elastic modulus, and the work of indentation. *Applied Physics Letter* 73:614–16.
4. Giannakopoulos, A. E., and S. Suresh. 1999. Determination of elastoplastic properties by instrumented sharp indentation. *Scripta Materialia* 40:1191–98.
5. Malzbender, J., and G. D. With. 2000. Energy dissipation, fracture toughness and the indentation load-displacement curve of coated materials. *Surface and Coating Technology* 135:60–68.
6. Lonnroth, N., L. M. Christopher, C. Pantano, and Y. Yue. 2008. Nanoindentation of glass wool fibers. *Journal of Non-Crystalline Solids* 354:3887–95.
7. Chakraborty, R., A. Dey, and A. K. Mukhopadhyay. 2010. Loading rate effect on nanohardness of soda-lime-silica glass. *Metallurgical and Materials Transactions A* 41A:1301–12.
8. Chakraborty, R., A. Dey, and A. K. Mukhopadhyay. 2010. Role of the energy of plastic deformation and the effect of loading rate on nanohardness of soda-lime-silica glass. *Physics and Chemistry of Glasses: European Journal of Glass Science and Technology B* 51:293–303.

Section 4

Dynamic Contact Behavior of Glass

13

Dynamic Contact Damage in Glass

Payel Bandyopadhyay, Arjun Dey, and Anoop Kumar Mukhopadhyay

13.1 Introduction

By this time, we can hope that all of you have become at least somewhat familiar with the static-contact-induced damage in glass. But what happens when two contacting bodies are in relative motion with respect to each other? Well, then what you get is a dynamic contact situation that is totally different from static contact situations. As such, any two surfaces in contact basically pose an intimate contact between the corresponding asperities of the two surfaces, even under self load or applied load during the static contact. However, the situation of relative motion between two such contacts can always be equivalently made to happen between one moving body and another body that is at relative rest with respect to the moving body. In such a situation, the asperities of the harder surface try to plow through the asperities of the comparatively softer surface. It is due to this interaction between the asperities that the scratches are created, and genesis of the coefficient of friction in dynamic conditions happens between the contacting surfaces. This is the fundamental simplistic picture of any machining operation that is undertaken to get a desired surface finish in a given surface.

OK, so far it appears fine. But you may still wonder what could be behind the scientific interest in such dynamic contact deformation mechanism studies. Here comes the importance of the science behind the deformations and damages induced by dynamic contact. You all know that the advanced applications of many materials demand very smooth surface finishes that can only be achieved by perfect grinding and polishing. Grinding and polishing processes are nothing but material removal by multiple single-point scratches, all happening simultaneously. In this case, however, the damages created by neighboring very closely spaced scratches interact between themselves and effect material removal. Nevertheless, the science of material-removal mechanisms, even for an apparently simple material like glass, is yet to be fully understood.

Therefore, we can start with a very simple and easily available material, for instance, glass. From time almost as close to immortal the material glass has found many applications in our daily life. The art of glassmaking is generally known to be about 5000 years old, but the ancient form of glass, i.e., obsidian, has been used since the dawn of civilization. Nowadays, glass has many applications spanning the ranges from common kitchen utensils (microwave-safe crockery and utensils) to transparent armors to telescope lenses and mirrors to biomedical instruments and biosensors. The scientific reasons behind the material removal mechanisms during the grinding and polishing of materials in general, and brittle materials in particular, are far from well understood. The class of brittle materials may include, but is not necessarily limited to, glasses, ceramics, glass-ceramics, ceramic coatings, glass and ceramic matrix composites, thin films, and various elastoplastic materials. It is for such brittle materials that the material removal mechanisms and the optimization of the machining parameters constitute a huge area of active research today [1–9]. Despite the wealth of information in the literature, however, even for a commercially important material like the soda-lime-silica (SLS) glass, there is not enough information about how the applied normal load affects the growth of both surface and subsurface damage under dynamic contact situations. You should be able to appreciate that such knowledgebase is absolutely essential for optimization of the submicro- or nano-scale machining parameters, without which the extraordinarily stringent surface-finish conditions demanded by the advanced high-end applications cannot possibly be met. So, in this chapter we have made a humble effort to understand the basic scientific processes working behind the material removal mechanisms. Here, controlled single pass scratch experiments were conducted at applied normal loads of 2, 5, 10, and 15 N on polished SLS glasses at a constant speed of 1000 μm$\cdot$s^{-1}. The scratches were made with a Rockwell C diamond indenter of 200 μm tip radius.

13.2 Damage Due to Dynamic Contact

The results (Figures 13.1a–f) show that lateral force, coefficient of friction, acoustic emission, width, depth, and wear volume of scratch grooves increased with normal load. The changes in slopes with load range reflect the change in the correspondingly active mechanisms of material removal [8]. The microfractures and the microcracks create chipping and wear debris in the scratch grooves. This wear debris gets entrapped between the indenter and the scratch grooves and, hence, may increase the coefficient of friction. The width (Figure 13.1d), depth (Figure 13.1e), and wear volume (Figure 13.1f) of the scratch grooves increased with the normal load, as the propensity of the occurrences of both microcracks and the microfractures also increased with the normal load (Figures 13.2a–d). This happened

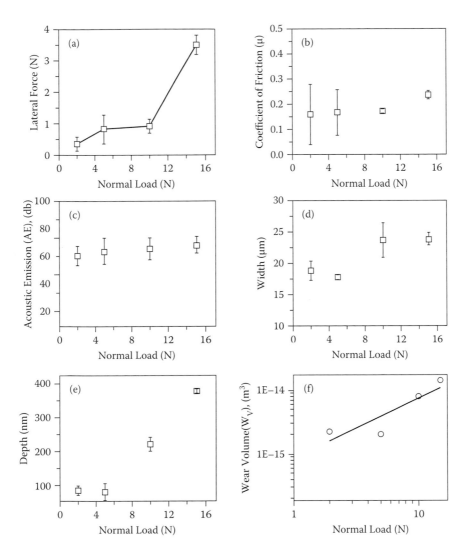

FIGURE 13.1
Variations of (a) lateral force, (b) coefficient of friction, (c) acoustic emission, (d) width of scratch groove, (e) depth of scratch groove, and (f) wear volume as a function of normal load (*P*). (Reprinted/modified with permission of Bandyopadhyay, Dey, and Mukhopadhyay [8] from American Ceramic Society and Wiley.)

because both the tensile contact stresses (Figure 13.3a) and the critical loads (Figure 13.3b) necessary to initiate the microcracks under dynamic contacts increased with the applied normal loads [4–9]. Initially, there were both partial and complete ring cracks (2N, Figure 13.4a). Their own interactions increased along with the appearance of edge cracks (5N, Figure 13.4b), and the interaction among the ring and edge cracks created a possible zone for

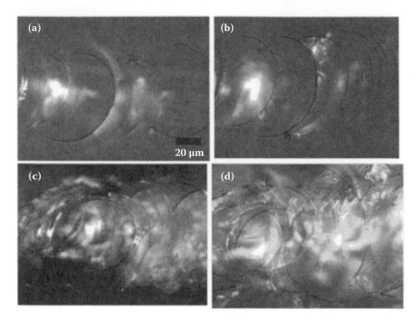

FIGURE 13.2
Optical photomicrographs of the scratch grooves at applied normal loads of (a) 2 N, (b) 5 N, (c) 10 N, and (d) 15 N.

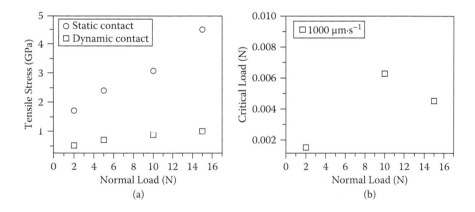

FIGURE 13.3
Variations in (a) maximum tensile stress as a function of normal load (*P*) for static and dynamic contact situations and (b) critical loads for dynamic condition. (Reprinted/modified with permission of Bandyopadhyay, Dey, and Mukhopadhyay [8] from American Ceramic Society and Wiley.)

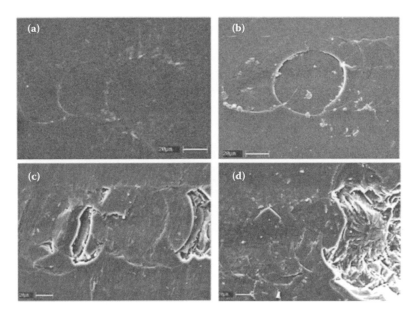

FIGURE 13.4
SEM photomicrographs of scratch grooves at applied normal loads of (a) 2 N, (b) 5 N, (c) 10 N, and (d) 15 N. (Reprinted/modified with permission of Bandyopadhyay, Dey, and Mukhopadhyay [8] from American Ceramic Society and Wiley; Bandyopadhyay et al. [7] from Elsevier; and Bandyopadhyay et al. [4] from Springer.)

microchipping. At applied normal load of 10 N (Figure 13.4c), some portions were already or were about to be chipped out. The severity of damage was the highest for the highest applied normal load of 15 N (Figure 13.4d) apart from the appearance of cracks nearly parallel to the scratch grooves. Microchipping led to localized formation of microwear debris inside the scratch grooves (Figure 13.4).

The detailed FESEM photomicrographs (Figures 13.5–13.8) of the subsurface damages provide evidence for the very first time that the severity of damage, as well as the degree of interactions among the various microcracks, increased with the applied normal load. This was the cause that enhanced further the probability of localized microchipping. The presence of shear deformation bands (Figures 13.5c, 13.6d, 13.7d, 13.8d), microcracks (Figures 13.5b, 13.6b, 13.7c, 13.8b), microwear debris (Figures 13.5f, 13.6e, 13.7e, 13.8c) and rodlike structures (Figure 13.5e) were prominent in the subsurface regions of all the scratch grooves.

Thus, to summarize, only a few Hertzian tensile cracks were present at low load (2N, Figures 13.9a, 13.2a, and 13.4a). The Hertzian tensile cracks began to interact among themselves, and the number of edge cracks, which further interact with Hertzian tensile cracks, had increased with load (5 N, Figures 13.9b, 13.2b, and 13.4b). These two types of interactions

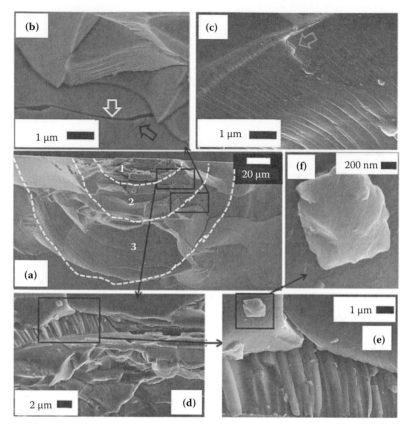

FIGURE 13.5
Subsurface deformation zone at applied normal load of 2 N. (Reprinted with permission of Bandyopadhyay and Mukhopadhyay [9] from Elsevier.)

enhanced the chance of microchipping. The spatial density of both Hertzian tensile cracks and edge cracks increased for a relatively higher load (10 N, Figures 13.9c, 13.2c, and 13.4c), leading to more interaction and microchip formations, which enhanced the amount of microwear debris. An additional set of cracks generated at an angle ($0° < θ < 90°$) with the scratch direction and spatially located nearly parallel to each other in the scratch track were observed clearly at still higher magnitude of the applied normal load (15 N, Figures 13.9d, 13.2d, and 13.4d). Similarly, in the subsurface, the presence of three distinct damage zones could be marked as 1, 2, and 3 in Figure 13.10 [9]. The zone marked as 1 is nearest to the moving indenter and is severely damaged (Figures 13.10a–d). The main features of this region were microcracks (indicated with black lines in Figure 13.10), shear bands (indicated with comblike lines in Figure 13.10), as well as less crushing. The huge numbers of shear bands were present in

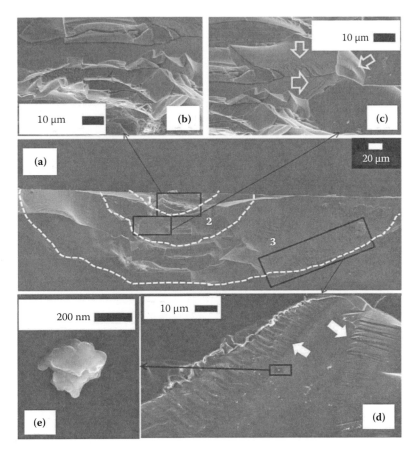

FIGURE 13.6
Subsurface deformation zone in the SLS glass, scratched at an applied normal load of 5N. (Reprinted with permission of Bandyopadhyay and Mukhopadhyay [9] from Elsevier.)

zone 3, though the numbers of microcracks were significantly less in this region (Figure 13.10). Secondary cracks (indicated with fishbonelike blue lines in Figure 13.10a), fishbonelike shear bands (indicated with comblike green lines in Figure 13.10b), and comblike bands (indicated with red lines in Figures 13.10b–d) were very common characteristic features for the subsurface damage zones. The crack propagation through different planes in the damage zone (the brown line in Figure 13.10c) was noticed for applied normal load of 10 N. At applied normal load of 15 N, the non-linear (i.e., curved) nature of propagation of the microcracks (as indicated by the line green in Figure 13.10d) was also frequently observed in the damage zones. The microwear debris (as marked by solid gray polygons in Figures 13.10a–d) were also present in zones 2 and 3 of the scratch grooves, as mentioned previously.

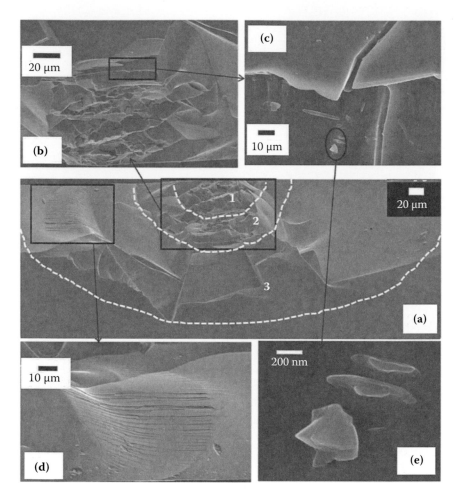

FIGURE 13.7
Subsurface deformation zone at applied normal load of 10 N. (Reprinted with permission of Bandyopadhyay and Mukhopadhyay [9] from Elsevier.)

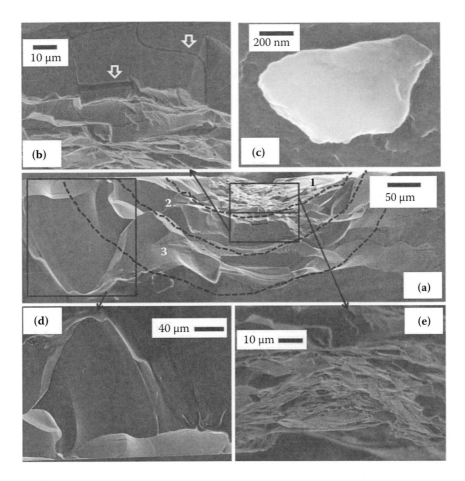

FIGURE 13.8
Subsurface deformation zone of the SLS glass scratched at an applied normal load of 15 N. (Reprinted with permission of Bandyopadhyay and Mukhopadhyay [9] from Elsevier.)

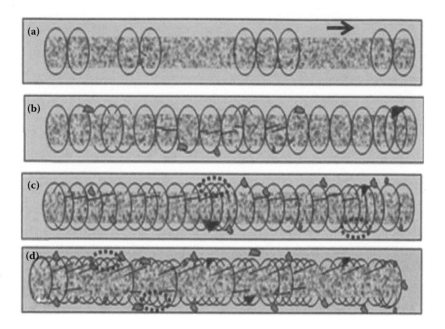

FIGURE 13.9
Schematic diagram of the surface damage zones for (a) 2 N, (b) 5 N, (c) 10 N, and (d) 15 N. The
arrow shows the direction of the scratch. (Reprinted with permission of Bandyopadhyay et al.
[6] from Elsevier.)

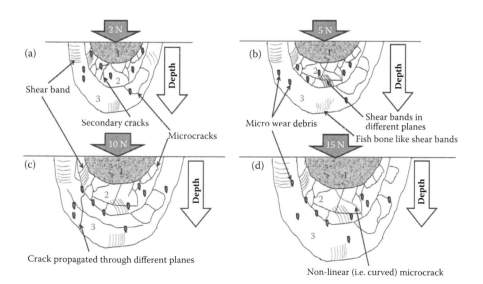

FIGURE 13.10 (See color insert.)
Schematic diagrams of the surface damage zones for (a) 2 N, (b) 5 N, (c) 10 N, and (d) 15 N.
(Reprinted with permission of Bandyopadhyay and Mukhopadhyay [9] from Elsevier.)

13.3 Conclusions

The severity of damages and their interactions at both surface and subsurface regions of the present SLS glass increased with the applied normal load and led to microchipping. This process gave genesis to localized microwear debris formation in the scratch grooves. There were three distinct, roughly semicircular subsurface damage zones with different damage characteristics present in the current in the SLS glass. But, so far, we have no idea about how the rate or the speed of dynamic contact changes the damage evolution scenario in the SLS glass. Therefore, this will be our focus in Chapter 14.

References

1. Schneider, J., S. Schula, and W. P. Weinhold. 2012. Characterisation of the scratch resistance of annealed and tempered architectural glass. *Thin Solid Films* 520:4190–98.
2. Larsson, P. L. 2013. On the correlation of scratch testing using separated elasto-plastic and rigid plastic descriptions of the representative stress. *Materials and Design* 43:153–60.
3. Lawn, B. R. 1967. Partial cone crack formation in a brittle material loaded with a sliding spherical indenter. *Proceedings of the Royal Society of London A* 299:307–16.
4. Bandyopadhyay, P., A. Dey, A. K. Mandal, S. Roy, N. Dey, and A. K. Mukhopadhyay. 2012. Effect of scratching speed on deformation of soda-lime-silica glass. *Applied Physics A* 107:685–90.
5. Bandyopadhyay, P., A. Dey, S. Roy, N. Dey, and A. K. Mukhopadhyay. 2012. Nanomechanical properties inside the scratch grooves of soda-lime-silica glass. *Applied Physics A* 107:943–48.
6. Bandyopadhyay, P., A. Dey, S. Roy, and A. K. Mukhopadhyay. 2012. Effect of load in scratch experiments on soda-lime-silica glass. *Journal of Non-Crystalline Solids* 358:1091–1103.
7. Bandyopadhyay, P., A. Dey, A. K. Mandal, N. Dey, and A. K. Mukhopadhyay. 2012. New observations on scratch deformations of soda-lime-silica glass. *Journal of Non-Crystalline Solids* 358:1897–1907.
8. Bandyopadhyay, P., A. Dey, and A. K. Mukhopadhyay. 2012. Novel combined scratch and nanoindentation experiments on soda-lime-silica glass. *International Journal of Applied Glass Science* 3:163–79.
9. Bandyopadhyay, P., and A. K. Mukhopadhyay. 2013. Role of shear stress in scratch deformation of soda-lime-silica glass. *Journal of Non-Crystalline Solids* 362:101–13.

14

Does the Speed of Dynamic Contact Matter?

Payel Bandyopadhyay, Arjun Dey, and Anoop Kumar Mukhopadhyay

14.1 Introduction

Chapter 13 gave us some idea about how the normal load controls damage evolution and the subsequent interactions during dynamic contact in an SLS glass. When a vehicle runs across the road, the road gets damaged due to the dynamic contact between the wheel and the road. Our common understanding is that the type of the vehicle, i.e., if it is a lightweight private car or a heavyweight truck, determines the kind of damage inflicted on the road. This picture means that the effective applied load on the wheel, i.e., the weight of the vehicle, is an important parameter in the creation of damage. The previous chapter gave us a scenario of damage in SLS glass under various contact loads that matches with this picture. But now imagine this story. You are running behind your friend toward a finish line consisting of a fixed sharp glass plate with sharp edges, and both of you just mildly touch this winning post as you pass by, say within few seconds of each other. Will either of you suffer any physical damage? The answer is no. Now imagine that both of you run superfast; you run much faster than your friend and, failing to control, bump into the fixed sharp glass plate at the finish line. Maybe your friend does almost the same thing but bumps into the same fixed sharp glass plate at a slightly slower speed than yours. Common sense tells us that you are the person who gets hurt the most; your friend also gets hurt, but a little less than you. In other words, the speed of contact can and really does matter. Therefore, especially in the case of dynamic contact-induced damage initiation and growth in a brittle material like SLS glass, we need to understand to what extent the speed of contact matters.

We can draw an inference immediately if we bear in mind the fact that, in a similar fashion, the speed of the grinding wheel may also control the extent of damage initiated on the surface of the glass material to be polished. The glass is a brittle material, as you know. So, if the surface gets damaged during polishing, it can never achieve a smooth surface finish. Therefore, besides the applied normal load, the speed of the grinding wheel should be optimized to achieve perfect polishing. The easiest way to simulate the grinding and polishing process is to conduct scratch testing at different scratching speeds. Despite the wealth of literature on this topic [1–4], the scientific knowledge about the mechanisms of surface and subsurface damage initiation and their subsequent growth due to variations in scratching speed of SLS glass is far from complete. To experimentally verify these issues, single-pass scratch experiments were conducted at four different scratching speeds of 100, 200, 500, and 1000 μm.s^{-1} at each of the applied normal loads of 2, 5, 10, and 15 N on the same commercial SLS glass described in Chapter 9. The other details of the scratch experiments have already been given in Chapters 9 and 10.

14.2 Effect of Speed of Dynamic Contacts and Damage Evolution

The results of the scratch experiments (Figures 14.1a–c) show that the lateral force, coefficient of friction, and acoustic emission all had increased with the scratching speeds. This occurred because the predicted [5–10] values of tensile stress (Figure 14.1d) active at the wake of the indenter also increased with the scratching speed. However, the width, w (Figure 14.2a), depth, d (Figure 14.2b), wear volume, W_v (Figure 14.2c), and wear rate, W_R (Figure 14.2d) decreased with the scratching speeds, following inverse power law dependencies as predicted by a model developed by us [8] for scratching speeds of 100, 500, and 1000 μm.s^{-1} and applied normal loads of 5, 10, and 15 N.

The SEM photomicrographs of the scratch grooves (Figure 14.3) show that the damage had, in fact, decreased with the increase in the scratching speed. Hertzian tensile crack, plastic grooving, microcracking, formation of microwear debris, and additional crack formations in the groove are the main damage features at the lowest scratching speed of 100 μm.s^{-1}. The interactions among the Hertzian cracks are so intense for the lowest scratching speed that it creates some portions that were either already chipped out or were about to chip out from the rest of the material. The density of the Hertzian tensile crack and the additional crack formations in the groove gradually decreased with the scratching speeds. These photomicrographs confirmed beyond doubt that the number of damage features in the SLS glass at lower scratching speeds were indeed greater than those that happened at higher scratching speeds.

The higher magnification FESEM photomicrographs (Figure 14.4) show a general feature that the areas entrapped between the Hertzian tensile ring cracks were heavily plastically deformed, showing signatures of extensive shear-induced deformations and microfractures. Further, for

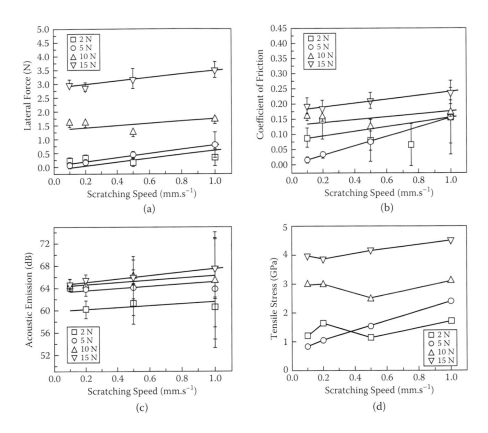

FIGURE 14.1 (See color insert.)
Variations of (a) lateral force, (b) coefficient of friction, (c) acoustic emission, and (d) tensile stress with the scratching speeds. (Reprinted with permission of Bandyopadhyay, Dey, and Mukhopadhyay [9] from the American Ceramic Society and Wiley.)

a given applied normal load of 5 N, the initiation of edge cracks from the existing tensile ring cracks were apparently more evident in the case of scratches made at a scratching speed of 100 µm.s⁻¹ (Figure 14.4a) than in those made at a scratching speed of 1000 µm.s⁻¹ (Figure 14.4b). Also, at higher scratching speeds more intratensile ring cracks were evident (Figure 14.4b).

At an applied normal load of slightly higher magnitude (10 N), an additional feature of a large number of shear bands was present in the surface of the scratch grooves, especially in the region between the Hertzian contact-stress-induced tensile ring cracks (Figure 14.4c). Interestingly, at lower scratching speed of 100 µm.s⁻¹, the shear bands were mutually perpendicular to each other. However, at a higher scratching speed of 1000 µm.s⁻¹ (Figure 14.4d), the shear bands were oriented at small angles with respect to the direction of scratching and also with respect to each other. The extent

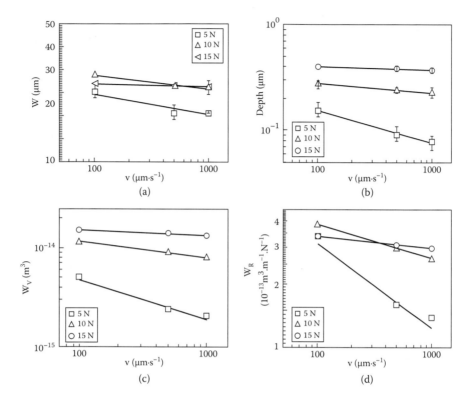

FIGURE 14.2
Variations of (a) width, (b) depth, (c) wear volume, and (d) wear rate with the scratching speeds. (Reprinted/modified with permission of Bandyopadhyay et al. [8] from Elsevier.)

of microcracking was more pronounced in the scratch grooves produced at a lower scratching speed of 100 $\mu m.s^{-1}$ (Figure 14.4c) than the extent of microcracking that occurred in the scratch grooves produced at a relatively higher scratching speed of 1000 $\mu m.s^{-1}$ (Figure 14.4d). Generally, the shear-deformation band formation occurred within the space bounded by the two consecutive Hertzian tensile ring cracks but did not cross the boundary of the ring cracks (Figure 14.4d).

At the applied normal load of still higher magnitude (15 N), the spatial presence of the microfracture events was more than those of the shear-deformation events (Figures 14.4e and 14.4f). At a lower scratching speed of 100 $\mu m.s^{-1}$ (Figure 14.4e), there were several pieces of thin glass slabs in an almost detached state lying on the surface of the scratch groove. These thin layers of the SLS glass appeared to be almost sheared out, but still faintly attached to the bulk material.

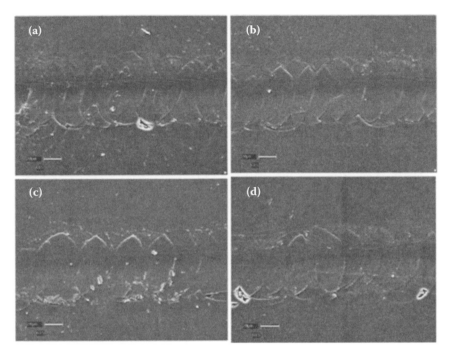

FIGURE 14.3
SEM photomicrographs of scratch grooves made under a constant applied normal load of 15 N at (a) 100 µm·s^{-1}, (b) 200 µm·s^{-1}, (c) 500 µm·s^{-1}, and (d) 1000 µm·s^{-1}. (Reprinted/modified with permission of Bandyopadhyay, Dey, and Mukhopadhyay [9] from the American Ceramic Society and Wiley.)

In these sheared-out portions, there was the clear presence of huge shear-deformation lines nearly parallel to the boundary of the Hertzian tensile crack. A large number of nearly parallel shear deformation bands had also intersected them nearly orthogonally. When any one of two such heavily sheared portions tried to come over the other, the registry was not perfect, and such regions apparently became the favorite zones for shear-induced microcrack formation, probably during the unloading cycle.

Many microfractured regions were evident near the outer periphery of the Hertzian tensile ring crack (Figure 14.4e), which appeared to be a little lifted out. Further, microwear chip formations had happened due to the thin broken-out pieces from regions in the most immediate vicinity of the surface. The localized comminution process of these thin broken-out pieces led to the genesis of the microwear debris. At a higher scratching speed of 1000 µm·s^{-1}, the additional feature present was that the region near the boundary of the Hertzian tensile crack had suffered more extensive microfracture and more localized comminution of the thin broken-out glass layers (Figure 14.4f).

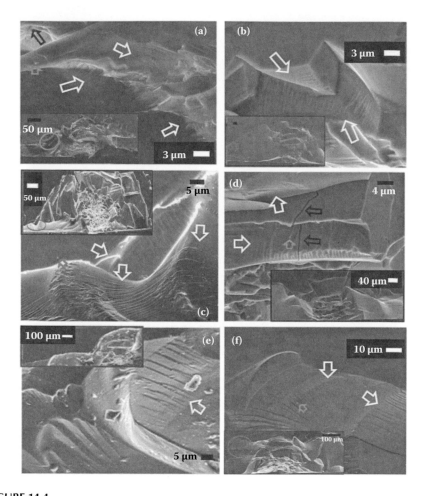

FIGURE 14.4
FESEM photomicrographs of scratch grooves made at 5 N [(a) 100 μm·s⁻¹ and (b) 1000 μm·s⁻¹],
10 N [(c) 100 μm·s⁻¹ and (d) 1000 μm·s⁻¹], and 15 N [(e) 100 μm·s⁻¹ and (f) 1000 μm·s⁻¹]. (Reprinted/
modified with permission of Bandyopadhyay et al. [8] from Elsevier.)

14.3 Conclusions

In a commercial SLS glass, the width and depth of scratch grooves, the wear
volume, and the wear rate decreased with scratching speeds following
inverse power law dependencies, which can easily be explained by a mathematical model proposed by the authors. Photomicrographs (optical, SEM,
and FESEM) showed that the damage in the scratch grooves actually
decreased with an increase in the scratching speed, but the shear stress
component played a major role in damage initiation. It is fine to know

about the individual influences of load and speed on the characteristics of dynamic-contact-induced damage evolution, but this knowledge does not help us to know the extent to which nanomechanical properties are degraded inside the scratch groove of the SLS glass. You may be wondering whether it is possible to measure this. Our response is Yes, and this is exactly what we will discuss in Chapter 15.

References

1. Schneider, J., S. Schula, and W. P. Weinhold. 2012. Characterisation of the scratch resistance of annealed and tempered architectural glass. *Thin Solid Films* 520:4190–98.
2. Larsson, P.-L. 2013. On the correlation of scratch testing using separated elasto-plastic and rigid plastic descriptions of the representative stress. *Materials and Design* 43:153–60.
3. Lawn, B. R. 1967. Partial cone crack formation in a brittle material loaded with a sliding spherical indenter. *Proceedings of the Royal Society of London A* 299:307–16.
4. Li, K., Y. Shapiro, and J. C. M. Li. 1998. Scratch test of soda-lime glass. *Acta Materialia* 46:5569–78.
5. Bandyopadhyay, P., A. Dey, A. K. Mandal, S. Roy, N. Dey, and A. K. Mukhopadhyay. 2012. Effect of scratching speed on deformation of soda-lime-silica glass. *Applied Physics A* 107:685–90.
6. Bandyopadhyay, P., A. Dey, S. Roy, N. Dey, and A. K. Mukhopadhyay. 2012. Nanomechanical properties inside the scratch grooves of soda-lime-silica glass. *Applied Physics A* 107: 943–48.
7. Bandyopadhyay, P., A. Dey, S. Roy, and A. K. Mukhopadhyay. 2012. Effect of load in scratch experiments on soda-lime-silica glass. *Journal of Non-Crystalline Solids* 358:1091–1103.
8. Bandyopadhyay, P., A. Dey, A. K. Mandal, N. Dey, and A. K. Mukhopadhyay. 2012. New observations on scratch deformations of soda-lime-silica glass. *Journal of Non-Crystalline Solids* 358:1897–1907.
9. Bandyopadhyay, P., A. Dey, and A. K. Mukhopadhyay. 2012. Novel combined scratch and nanoindentation experiments on soda-lime-silica glass. *International Journal of Applied Glass Science* 3:163–79.
10. Bandyopadhyay, P., and A. K. Mukhopadhyay. 2013. Role of shear stress in scratch deformation of soda-lime-silica glass. *Journal of Non-Crystalline Solids* 362:101–13.

15

Nanoindentation Inside the Scratch: What Happens?

Payel Bandyopadhyay, Arjun Dey, and Anoop Kumar Mukhopadhyay

15.1 Introduction

Chapters 13 and 14 have given us some idea about how the surface of the brittle glass samples gets damaged due to the effect of both the applied normal load and scratching speed. We also observed the scratch grooves by using the optical microscope, SEM, and FESEM. The qualitative measurements were done. But how do we quantify the damage inside the scratch grooves? Though we also measure the scratch parameters, they do not convey much information about the quantitative measurement of the extent of damage inside the scratch groove. A literature survey [1–13] also did not convey much information in this regard.

Here, we use the unique advantages of nanoindentation techniques to measure the Young's modulus and nanohardness at the microstructural length scale in and outside of the scratch groove. The nanoindentation experiments were conducted at 100 mN load using a Fischerscope H100-XYp machine equipped with a Berkovich tip. Thus, the nanohardness (H) and Young's modulus (E) measurements were deliberately made at the vicinity of, inside, and across the scratch grooves. Both the loading and unloading times were kept fixed at 30 seconds. The measurements of H and E were done from the experimentally obtained load versus depth of penetration plots using the well-known Oliver-Pharr model [14].

15.2 Nanoindentation Inside a Scratch Groove

The first experimental observation results are described in Figure 15.1 [12]. The arrowhead in Figure 15.1 indicates the direction of scratching. Further, the insets in Figures 15.1a–f represent the typical corresponding

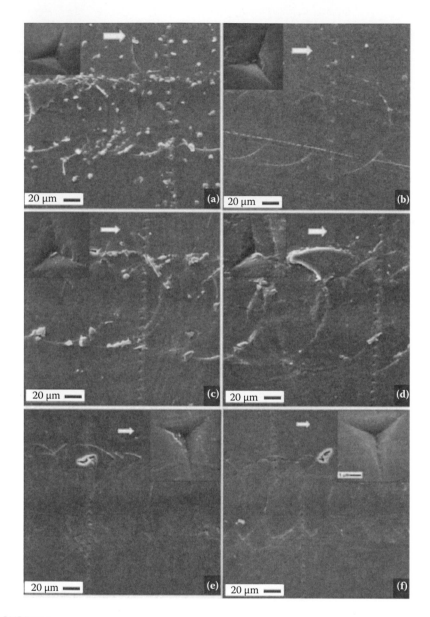

FIGURE 15.1
The nanoindentation experiments through the scratch grooves at (a) 5 N, 100 µm·s⁻¹; (b) 5 N, 1000 µm·s⁻¹; (c) 10 N, 100 µm·s⁻¹; (d) 10 N, 1000 µm·s⁻¹; (e) 15 N, 100 µm·s⁻¹; and (f) 15 N, 1000 µm·s⁻¹. (Reprinted with permission of Bandyopadhyay, Dey, and Mukhopadhyay [12] from the American Ceramic Society and Wiley.)

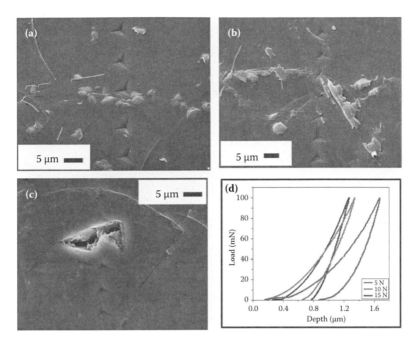

FIGURE 15.2 (See color insert.)
The nanoindentation conducted through the edge of the scratch grooves for (a) 5N, (b) 10 N, and (c) 15 N; (d) load–depth plots of the indent made at the middle of the damage zone. (Reprinted with permission of Bandyopadhyay et al. [9] from Springer.)

high magnification SEM photomicrographs of the indents made inside the scratch grooves of the soda-lime-silica glass. The scale bar shown in the inset of Figure 15.1f is valid for all the insets in Figures 15.1a–e.

The typical illustrative SEM photomicrographs of the indents through the edges of the scratch grooves made at 5, 10, and 15 N applied normal loads are shown in Figures 15.2a–c, while the representative load (P) versus depth (h) plots are shown in Figure 15.2d [9]. The maximum depth of penetration (h_{max}) was the highest for the indent made at the scratch groove corresponding to the highest applied normal load and was lowest corresponding to the lowest applied normal load (Figure 15.2d) [9, 12].

The nanohardness as well as the elastic recovery (Figures 15.3a, 15.3c, and 15.3e) and Young's moduli (Figures 15.3b, 15.3d, and 15.3f) were minimum at the deepest point of the scratch grooves, then gradually increased on both sides of the center of the groove, to finally attain the values of the undamaged SLS glass at locations far away from the scratch grooves [9]. As the elastic recovery was minimum at the center of the scratch grooves (Figures 15.3a, 15.3c, and 15.3e), the amounts of energy dissipated to create a plastic impression were maximum at those points (Figures. 15.3b, 15.3d, and 15.3f). The rates of changes of both H and E with the distance through

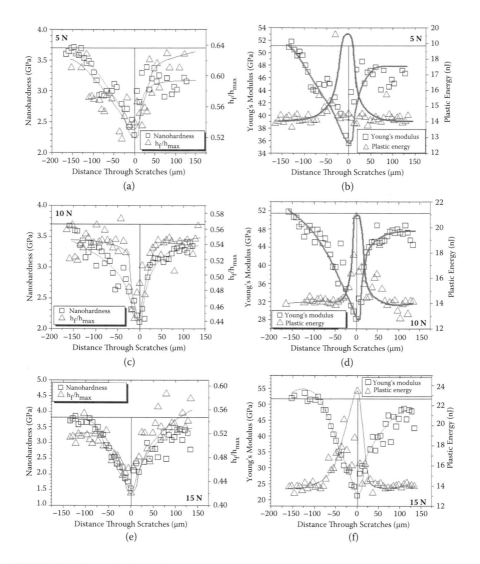

FIGURE 15.3 (See color insert.)
Plots of nanohardness and elastic recovery (h_f/h_{max}) (a, c, e) and Young's modulus and plastic energy with distance through scratches (b, d, f) for applied loads of 5 N (a, b), 10 N (c, d), and 15 N (e, f). (Reprinted with permission of Bandyopadhyay et al. [9] from Springer.)

the scratches were maximum at the middle of the scratch grooves and then fell with a steep gradient within the first 25 μm on both sides of the center of the scratch grooves; thereafter, the rates of changes reduced further in the zones between 25 and 50 μm on both sides before becoming nearly constant at still further distances. Thus, depending on the applied normal load, the decreases in *H* (Figure 15.4a) and *E* (Figure 15.4b) values in the

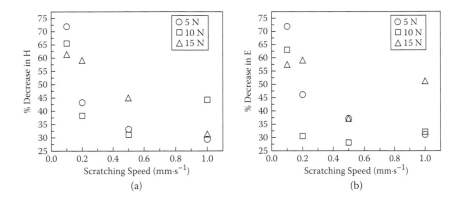

FIGURE 15.4 (See color insert.)
Percentage decrease in (a) nanohardness and (b) Young's modulus. (Reprinted with permission of Bandyopadhyay, Dey, and Mukhopadhyay [12] from the American Ceramic Society and Wiley.)

scratch-damaged regions were as high as ≈30%–60%. The value of $[d\{\Delta H(\%)\}/dP]$ was ≈1 for the applied normal load range of 5–10 N but jumped to ≈3 for the load range of 10–15 N. However, $[d\{\Delta E(\%)\}/dP]$ was ≈3 for the applied normal load range of 5–15 N. The data on the percentage of decrease in nanohardness and Young's modulus values inside the groove of the soda-lime-silica glass as a function of the scratching speed are shown respectively in Figures 15.4a and 15.4b. It may be seen from the data presented in Figure 15.4a that, depending on the scratching speed for the applied normal loads of 5, 10, and 15 N, respectively, the nanohardness values decreased by about 30%–65%, 29%–71%, and 30%–60%.

Similarly, the data of Figure 15.4b illustrate that, depending on the scratching speed for the applied normal loads of 5, 10, and 15 N, respectively, the Young's modulus values decreased by about 27%–63%, 30%–71%, and 36%–58% [9, 12]. Thus, the extent of local degradation in the nanomechanical properties, i.e., nanohardness and Young's modulus inside the scratch groove of the present soda-lime-silica glass, was quite significant and, hence, deserves an explanation.

15.3 The Model of Microcracked Solids

The only way that the hardness and Young's modulus values of the soda-lime-silica glass could decrease is due to the presence of microcracks in the scratch groove. A closer inspection of the scratch tracks (Figures 15.1a–f) indeed reveals that there were microcracks present in the grooves. It has been reported [15] that the Young's modulus of a material can decrease due to the presence of the microcracks, which make it weaker and, hence, more compliant under a given condition of applied normal load. If we assume that a material contains N number of microcracks of equal size l per unit volume, then the Young's modulus (E) of the microcracked material can be related to the Young's modulus of the material without microcracks (E_0) by the following relationship [9, 12, 15]:

$$E = E_0 \left(1 + \frac{16(1 - v_s^2)Nl^3}{9(1 - 2v_s)} \right)^{-1} \tag{15.1}$$

In equation (15.1), v_s is the Poisson's ratio of the SLS glass material. The quantity Nl^3 basically denotes the total number of microcracks.

Thus, assuming $v_s = 0.25$ [9, 12] for the present SLS glass and using the experimentally measured values of the Young's modulus (E) of the microcracked material and the Young's modulus (E_0) of the material without microcracks, the quantities Nl^3 were calculated as a function of the applied normal loads and the scratching speeds. The lowest values of E at the *center of the zone*, with highest magnitudes of damage, were utilized as the Young's modulus (E) of the microcracked SLS glass samples under different conditions of the applied normal load and scratching speed, pertaining to the present experiments.

The results of such an exercise are depicted in Figures 15.5a and 15.5b, which display the total number of microcracks as a function of the applied normal loads and the scratching speeds. The data presented in Figure 15.5a

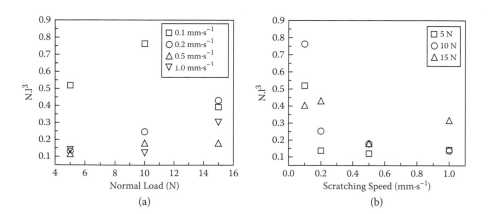

FIGURE 15.5 (See color insert.)
Variations of microcrack population with (a) applied normal load (b) scratching speed. (Reprinted with permission of Bandyopadhyay, Dey, and Mukhopadhyay [12] from the American Ceramic Society and Wiley.)

clearly prove that the total number of microcracks increased with normal loads for a given scratching speed, as was indeed experimentally observed (Figures 15.1a–f). Similarly, it is evident from the data presented in Figure 15.5b that, for a given normal load, the total number of damage features, i.e., microcracks, decreased with scratching speed.

15.4 Conclusions

The nanomechanical properties decreased inside the scratch grooves due to the presence of microcracks inside the scratch grooves. In this and in previous chapters we have gained some unique insight into static and dynamic deformation of a typical amorphous brittle solid such as SLS glass. Now, we might ask ourselves whether the situation will remain same if we had taken a polycrystalline brittle ceramic such as alumina. Unlike glass, it is highly crystalline in nature, and there are innumerable grains separated by many grain boundaries present in this brittle solid. Therefore, we shall have a look at the nanomechanical properties of ceramics in Chapter 16.

References

1. Schneider, J., S. Schula, and W. P. Weinhold. 2012. Characterisation of the scratch resistance of annealed and tempered architectural glass. *Thin Solid Films* 520:4190–98.
2. Larsson, P.-L. 2013. On the correlation of scratch testing using separated elasto-plastic and rigid plastic descriptions of the representative stress. *Materials and Design* 43:153–60.
3. Lawn, B. R., and M. V. Swain. 1975. Microfracture beneath point indentations in brittle solids. *Journal of Materials Science* 10:113–22.
4. Li, K., Y. Shapiro, and J. C. M. Li. 1998. Scratch test of soda-lime glass. *Acta Materialia* 46:5569–78.
5. Gu, W., Z. Yao, and X. Liang. 2011. Material removal of optical glass BK7 during single and double scratch tests. *Wear* 270:241–46.
6. Abdelounis, H. B., K. Elleuchb, R. Vargiolu, H. Zahouani, and A. L. Bot. 2009. On the behaviour of obsidian under scratch test. *Wear* 266:621–26
7. Houerou, V. L., J. C. Sangleboeuf, S. Deriano, T. Rouxel, and G. Duisit. 2003. Surface damage of soda-lime-silica glasses: Indentation scratch behaviour. *Journal of Non-Crystalline Solids* 316:54–63.
8. Bandyopadhyay, P., A. Dey, A. K. Mandal, S. Roy, N. Dey, and A. K. Mukhopadhyay. 2012. Effect of scratching speed on deformation of soda-lime-silica glass. *Applied Physics A* 107:685–90.

9. Bandyopadhyay, P., A. Dey, S. Roy, N. Dey, and A. K. Mukhopadhyay. 2012. Nanomechanical properties inside the scratch grooves of soda-lime-silica glass. *Applied Physics A* 107:943–48.

10. Bandyopadhyay, P., A. Dey, S. Roy, and A. K. Mukhopadhyay. 2012. Effect of load in scratch experiments on soda-lime-silica glass. *Journal of Non-Crystalline Solids* 358:1091–1103.

11. Bandyopadhyay, P., A. Dey, A. K. Mandal, N. Dey, and A. K. Mukhopadhyay. 2012. New observations on scratch deformations of soda-lime-silica glass. *Journal of Non-Crystalline Solids* 358:1897–1907.

12. Bandyopadhyay, P., A. Dey, and A. K. Mukhopadhyay. 2012. Novel combined scratch and nanoindentation experiments on soda-lime-silica glass. *International Journal of Applied Glass Science* 3:163–79.

13. P. Bandyopadhyay, and A. K. Mukhopadhyay. 2013. Role of shear stress in scratch deformation of soda-lime-silica glass. *Journal of Non-Crystalline Solids* 362:101–13.

14. Oliver, W. C., and G. M. Pharr. 1992. An improved technique for determining hardness and elastic modulus using load and displacement sensing indentation experiments. *Journal of Materials Research* 7:1564–83.

15. Bueno, S., and C. Baudina. 2006. Instrumented Vickers microindentation of alumina-based materials. *Journal of Materials Research* 21:161–73.

Section 5

Static Contact Behavior of Ceramics

16

Nanomechanical Properties of Ceramics

Riya Chakraborty, Manjima Bhattacharya, Arjun Dey,
and Anoop Kumar Mukhopadhyay

16.1 Introduction

Structural ceramics like alumina (Al_2O_3) comprise more than 70% of the global advanced ceramics market. The recent forecast is that the global advanced ceramics market is to reach US$56.4 billion (INR 5640 crore) by 2015 according to a new report by Global Industry Analysts, Inc., USA. However, ceramics are inherently brittle in nature. Hence, it is urgently needed to develop better microstructurally engineered alumina for various applications ranging from agricultural sector equipment, to cutting tool inserts, to biomedical prostheses, to advanced spacecraft windows, to futuristic electronics, to other strategic sectors, especially for high strain rate resistant usage. These needs therefore demand a thorough understanding of basic contact-induced deformation mechanisms and damage initiation as well as growth mechanisms at macro-, micro-, and nano-structural length scale of the alumina microstructure. Despite the wealth of existing literature, these mechanisms are yet to be comprehensively and unequivocally understood. Moreover, besides examining the static contact deformation mechanisms of alumina at various length scales, the dynamic contact properties such as coefficient of friction, etc., are also important for in-service applications of a brittle ceramic like alumina. The grain size dependencies of both macro- as well as micro-hardness and scratch resistance have been extensively studied mainly for fine and intermediate grain-size alumina [1, 2], but information about the resistance against contact-induced permanent deformation, i.e., nanohardness of alumina, is quite rare [3].

It is well established today that the origin of any macroscopic failure process, especially in the coarse-grain alumina, must have had its genesis linked to the nano- and submicro-scale deformation followed by crack initiation at the microstructural length scale. Therefore, in this chapter we shall try to answer two simple questions. Our first query will be about the typical nanomechanical properties (nanohardness and Young's modulus) of a dense (99.9% of theoretical), coarse-grain (10 µm) alumina sample (Figures 16.1a and 1.6b). Our second query would be how the applied load

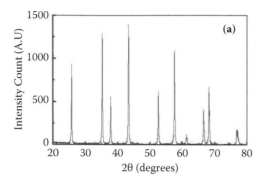
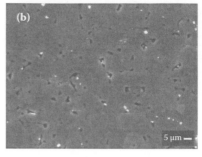

FIGURE 16.1
(a) XRD pattern and (b) SEM photomicrograph of polished and thermally etched alumina. (Reprinted with permission of Chakraborty et al. [3] from The Institute of Engineers, India.)

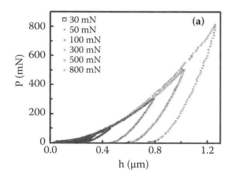
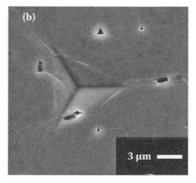

FIGURE 16.2 (See color insert.)
(a) Load–depth (P–h) plots for various nanoindentation loads and (b) SEM photomicrograph of the residual impression of a typical nanoindent at 10^6 µN load in alumina. (Reprinted with permission of Chakraborty et al. [3] from The Institute of Engineers, India.)

affects the nanomechanical properties of the same dense, coarse-grain alumina sample.

16.2 Nanoindentation Study

Typical P–h plots at different loads of the present sample are given in Figure 16.2a. At higher nanoindentation loads, there was an expected large residual depth and, consequently, a larger area encompassed by the P–h plot, implying dissipation of a higher amount of energy as compared to those

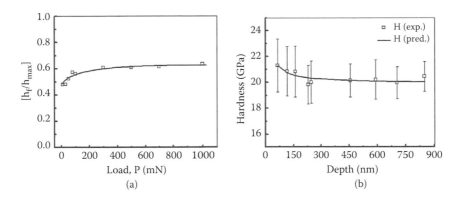

FIGURE 16.3
(a) Variation of the ratio of final depth of penetration (h_f) to the maximum depth of penetration (h_{max}) as a function of load (P) and (b) variation of nanohardness (H) as a function of depth. The solid line represents the predicted trends following the Nix and Gao model [6]. (Reprinted with permission of Chakraborty et al. [3] from The Institute of Engineers, India.)

involved at lower or intermediate loads. The FESEM photomicrograph of a typical nanoindent made at 1000 mN load is shown in Figure 16.2b. The indentation area, however, looked smooth without any imminent sign of severe damage growth or accumulation.

Now, to verify the applicability of the Oliver-Pharr model [4], it is always necessary to calculate the (h_f/h_{max}) ratio from the P–h plot. According to the theory, the value of this ratio should lie below 0.7 to obtain reliable results [4]. The average values of (h_f/h_{max}) ratio in the present study were almost constant at ≈0.6 (Figure 16.3a). This information justified the basic applicability of the Oliver-Pharr model to analyze the nanoindentation data of the present sintered alumina ceramic sample. The Young's modulus value measured at any given load (i.e., 10–1000 mN) was ≈400 GPa, as expected for a dense alumina sample [4]. Further, the variation of nanohardness (H) data as a function of the indentation depth is shown in Figure 16.3b. At a low load of 10 mN, the alumina sample showed H value ≈21.3 GPa at a depth of about 65 nm, which slightly dropped by about 4.5% to ≈20.4 GPa at a depth of ≈850 nm for a higher load of 1000 mN (Figure 16.3b). The decrease in nanohardness with the increase in indentation depth as well as load revealed the presence of an indentation size effect (ISE) in the present data.

16.3 Indentation Size Effect (ISE) in Alumina

Indentation size effect (ISE) as encountered in both metallic and brittle materials as well as nanocrystalline and quasi-crystalline materials is a significant phenomenon in micro- and nanoindentation tests describing the increase in hardness values with decrease of the indentation loads and/or depth [5]. The large number of explanations developed for ISE has been reviewed recently [5]. However, the most critically appraised and theoretically well-founded explanation of ISE is believed to be due to the organization and reorganization of the dislocation network under the indent during the indentation process, as described by the Nix and Gao model [6]. The model is based on the dislocation theory of Taylor [7, 8]. The parameters of the model are connected to the dislocation density and a scale factor in depth in the following fashion [6]:

$$\left(\frac{H}{H_0}\right)^2 = 1 + \frac{h^*}{h} \tag{16.1}$$

where H is the experimentally measured nanohardness at a depth h, h^* is a characteristic length that depends on the properties of the indented material and the indenter angle, and H_0 is the indentation hardness at infinite depth. Accordingly, it was decided to fit the whole data of the present work to equation (16.1).

The result of this exercise is shown by the solid line indicating the predicted trend of the nanohardness data (Figure 16.3b). Clearly, the model could rightly predict the basic trend of increase in nanohardness with decrease in depth as observed in the experimental data. The fitting of the present data to the Nix and Gao model [6] predicts $H_0 = 19.93$ GPa and $h^* = 9$ nm. The predicted value of H_0 matched well with the reported hardness value of alumina [4] as well as the grand average of the experimental nanohardness value of ≈20.3 GPa. Huang et al. [9] suggested that the characteristic length scale along depth up to the point where the gradient in local strain would affect the nanohardness data at maximum would be about 100 nm for typical oxide ceramics, e.g., MgO. In the present case, for the alumina ceramics, the characteristic length scale was much below (i.e., 9 nm) that of the limiting value of 100 nm [9]. Therefore, the applicability of the Nix and Gao model [6] in the present work to explain the ISE was justified.

16.4 Conclusions

The nanohardness ($H ≈ 20.4$–21.3 GPa) and Young's modulus ($E ≈ 400$ GPa) of a dense (≈99.9%), coarse grain (≈10 μm) alumina were measured as a function of load (10–1000 mN) by means of the nanoindentation technique. The presence of a slight indentation size effect in the present data was explained successfully through the application of the well-established Nix and Gao model that predicted $H_0 = 19.93$ GPa and $h^* = 9$ nm. Thus we have got some valuable information in this chapter about the typical nanoindentation response of a typical coarse-grain alumina ceramic, but we have no idea yet how the response will change, if it does at all, if the rate of energy input from the loading train into the system exposed to nanoindentation load is made fast or slow. We shall try to look at this aspect in Chapter 17.

References

1. Rice, R. W., C. C. Wu, and F. Borchelt. 1994. Hardness-grain-size relations in ceramics. *Journal of the American Ceramic Society* 77:2539–53.
2. Xu, H. H. K., and S. Jahanmir. 1995. Effect of grain size on scratch damage and hardness of alumina. *Journal of Materials Science Letters* 14:736–39.
3. Chakraborty, R., A. K. Mukhopadhyay, A. Dey, K. D. Joshi, A. Rav, S. K. Biswas, and S. C. Gupta. 2010. Nanomechanical properties of coarse grain alumina ceramic. *Journal of the Institution of Engineers (India), Part MM: Metallurgy and Material Science Division* 91:9–14.
4. Oliver, W. C., and G. M. Pharr. 1992. An improved technique for determining hardness and elastic modulus using load and displacement sensing indentation experiments. *Journal of Materials Research* 7:1564–83.
5. Mukhopadhyay, N. K., and P. Paufler. 2006. Micro and nanoindentation techniques for mechanical characterisation of materials. *International Materials Reviews* 51:209–45.
6. Nix, W. D., and H. Gao. 1998. Indentation size effects in crystalline materials: A law for strain gradient plasticity. *Journal of the Mechanics and Physics of Solids* 46:411–25.
7. Horstemeyer, M. F., M. I. Baskes, and S. J. Plimpton. 2001. Length scale and time scale effects on the plastic flow of FCC metals. *Acta Materialia* 49:4363–74.
8. Iost, A., and R. Bigot. 1996. Indentation size effect: Reality or artifact? *Journal of Materials Science* 31:3573–77.
9. Huang, Y., F. Zhang, K. C. Hwang, W. D. Nix, G. M. Pharr, and G. Feng. 2006. A model of size effects in nano-indentation. *Journal of the Mechanics and Physics of Solids* 54:1668–86.

17

Does the Contact Rate Matter for Ceramics?

Manjima Bhattacharya, Riya Chakraborty, Arjun Dey,
and Anoop Kumar Mukhopadhyay

17.1 Introduction

Among all well-known structural and functional ceramics, alumina deserves special mention owing to its large-scale applications as wear-resistant inserts, biomedical implants, high strain rate impact-resistant plates, etc. Now the question is what is the major noticeable property of alumina that makes it suitable for such large-scale applications? What makes it different from other structural ceramics? The hardness of alumina is the single most important mechanical property that plays a major role in all contact-related applications of alumina. You may wonder why the focus is on hardness but not the other mechanical properties. Actually, materials science has been linked with all the mechanical properties of a material that ultimately describe the characteristics of that material, but hardness is important in materials science because it describes the intrinsic contact deformation resistance of the material. In particular, the hardness evaluated at the nanoscale of the microstructure assumes significant importance because it is at this scale that the mechanical integrity of a structural ceramic in-service gets determined. The defects that ultimately define the mechanical integrity of a structural ceramic originate at the ultralow length scale (nano- or microscale) of the microstructure during its service lifetime. Thus, evaluation of the nanohardness of alumina is a very important checkpoint in order to make it suitable for global applications.

17.2 Effect of Loading Rate and "Multiple Micro Pop-in" and "Multiple Micro Pop-out"

The hardness of a dense ($\approx$99.9%), coarse grain ($\approx$10 μm) alumina possesses moderately high value ($\approx$20 GPa), as we have already discussed in Chapter 16. In this chapter we shall try to understand the influence of loading rates

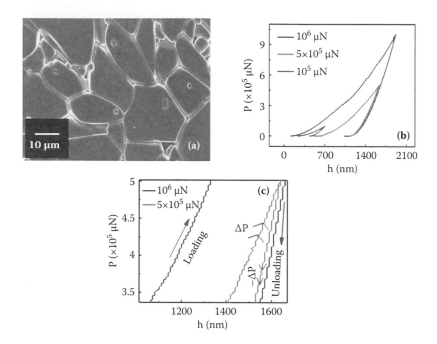

FIGURE 17.1 (See color insert.)
(a) Surface microstructure of dense, coarse-grain alumina, (b) *P–h* plots at three different loads of 10^5, 5×10^5, and 10^6 μN, and (c) exploded view of *P–h* plots. (Reprinted with permission of Bhattacharya et al. [1] from Elsevier.)

(10^3 to 10^6 μN·s^{-1}) on the nanoindentation response and the nanoscale contact deformation resistance of another high-density ($\approx$95% of theoretical) and even coarser ($\approx$20 μm) grain sized alumina ceramic (Figure 17.1a). The reason for varying the loading rates was to vary the rate of contact and see how it affected the ceramic's response. The load-versus-depth (*P–h*) plots obtained from the nanoindentation experiments at various loading rates and their exploded views are shown in Figures 17.1b and 17.1c, respectively. A large number of serrations as could be seen from the *P–h* plots occurred during the loading and unloading cycles (Figures 17.1b and 17.1c). These serrations are nothing but the signatures of "multiple micro pop-in" and "multiple micro pop-out" events occurring in the alumina during the loading and unloading cycles, respectively. These observations actually suggested the initiation of nanoscale plasticity events [2] during the nanoindentation experiments.

It was interesting to note that for any given loading rate, the magnitude of increase in load (ΔP) at which two consecutive nanoscale plasticity events, i.e., micro pop-ins, occurred during the loading cycle was different from the magnitude of decrease in load ($-\Delta P$) at which two consecutive nanoscale plasticity events, i.e., micro pop-outs, occurred during the unloading cycle. The plasticity events were also greater in number at lower loading rates (e.g., 10^3 μN·s^{-1}) than that at higher loading rates (e.g., 10^5 μN·s^{-1}) (see Figure 17.1c).

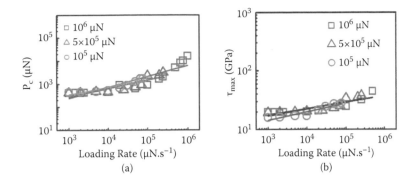

FIGURE 17.2 (See color insert.)
Variations of (a) P_c and (b) τ_{max} with loading rate or $\dot{P}$. (Reprinted with permission of Bhattacharya et al. [1] from Elsevier.)

Now, the critical load, P_c (i.e., the load at which the nanoscale plasticity events had initiated), was found to exhibit a positive power law dependence on the loading rates (Figure 17.2a). Most interestingly, the magnitudes of the corresponding maximum shear stresses (τ_{max}) active just underneath the nanoindenter also exhibited a power law dependence on the loading rates with a positive exponent (Figure 17.2b). These were estimated following Packard and Schuh [3]. The critical load represents the intrinsic contact-deformation resistance against the initiation of plasticity at the nanoscale, which was found to increase with the loading rates, a very first experimental observation made by us for a dense ($\approx$95%), coarse grain ($\approx$20 µm) alumina ceramic. The physical mechanism behind this phenomenon may be that the rate of energy transfer at higher loading rates is so quick that it leads to more localized compressive stress generation. Therefore, a higher critical load is needed to overcome this localized compressive stress to cause the initiation of fresh plasticity events at the nanoscale, e.g., nucleation of further dislocations and/or shear band formation [4, 5].

The maximum shear stress operative underneath the nanoindenter was estimated to be $\approx$20–40 GPa [6, 7], which was much higher than the typical theoretical shear strength, $\tau_{theor} \approx$3 GPa, of alumina ceramics [7]. Therefore, sufficient shear stress was provided by the applied load to cause shear-induced deformation band formation and/or microfracture in the nanoin-dented alumina ceramics leading to the initiation of the nanoscale plasticity events, i.e., nucleation of dislocations. The high resolution FESEM photomicrograph (Figure 17.3) gives direct experimental evidence supporting this view, which shows that a number of shear deformation bands had formed inside the nanoindentation cavities, and even a few microcracks had formed around them, even at such ultralow loads ($\approx$10^6 µN). The initiation of nanoscale plasticity events, i.e., dislocation nucleations, was quite plausible to have occurred in the present coarse-grained alumina ceramic because

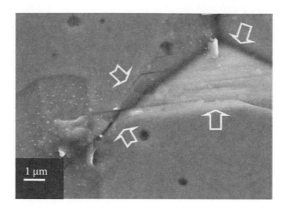

FIGURE 17.3
FESEM photomicrograph showing shear deformation bands and microcracks in and around the nanoindentation cavity (marked by white hollow arrows). (Reprinted with permission of Bhattacharya et al. [1, 6] from Elsevier and from Springer.)

compressive contact stress ($\approx$18–65 GPa) and shear stress ($\approx$20–40 GPa) of very high magnitudes were active in the immediate vicinity of the nanoindents.

Using the relationship proposed by others [8, 9] and our own mathematical modeling [1, 6], it has been predicted that both the instantaneous depth (h) and the reduced depth ($h' = h - h_f$) of penetrations have power law dependencies on the loading rates (dp/dt). Here, h_f stands for the final depth of penetration. The experimental data matched well with the trends obtained from the literature [10] and our own experimental data [1, 6] (see Figures 17.4a–d). It was interesting to note further that the increment in depth ($\Delta h = h_{c2} - h_{c1}$) had itself a power law dependence on the corresponding load increment ($\Delta P = P_{c2} - P_{c1}$) at which two consecutive nanoscale plasticity events (say, 1 and 2) had occurred during the loading cycles [1, 6].

This power law dependence of Δh on ΔP was well in agreement with the trends obtained from literature reports on 4H SiC single crystals [10] at loading rates of 310 and 3100 μN·s^{-1}. In addition, a power law dependence on the critical load, P_c, was observed for (a) the critical depth (h_c) values at which the nanoscale plasticity events had initiated and (b) the corresponding values of depth increment ($\Delta h = h_{c2} - h_{c1}$) at which two consecutive nanoscale plasticity events had occurred during the loading cycles (Figure 17.4a).

This power law dependence was again supported by the data obtained from the literature [4] for polycrystalline alumina. These observations can be explained by the fact that for any given loading rate the experimentally measured instantaneous depth (h) really had a power law dependence on the applied instantaneous nanoindentation load, P [1, 7].

This power law dependence was in accordance with the empirical relationship proposed recently by Ebisu and Horibe [9]. Similarly, the reduced depth of penetration (h') obtained during the unloading cycle of the *P–h* plot exhibited a power law dependence on loading rates (P) (Figure 17.4d),

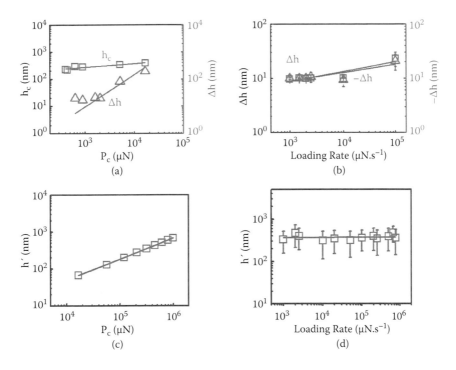

FIGURE 17.4 (See color insert.)
Power law dependencies of (a) h_c and Δh on P_c, (b) Δh and $-\Delta h$ on Loading Rate, (c) h' on P_c, and (d) h' on Loading Rate. (Reprinted with permission of Bhattacharya et al. [1] from Elsevier.)

in accordance with the empirical relationship proposed by Oliver and Pharr [8] and our own mathematical model [1, 6].

17.3 Conclusions

The present work is the very first experimental observation that for a high-density (95% of theoretical), coarse grain (≈20 µm) alumina ceramic, the intrinsic contact deformation resistance, i.e., the critical load (P_c), against the initiation of nanoscale plasticity events increased with the loading rates. The maximum shear stress generated just underneath the nanoindenter had magnitude much higher than the theoretical shear strength of alumina. In Chapter 18 we shall illustrate further how the nanoscale contact resistance ($1/h_f$) of a coarse grain ceramic against the initiation of incipient plasticity events can enhance with the loading rates, following a power law dependence with a positive exponent. It will be also shown that such a process can also cause even a mild enhancement in the nanohardness of coarse-grain ceramics.

References

1. Bhattacharya, M., R. Chakraborty, A. Dey, A. K. Mandal, and A. K. Mukhopadhyay. 2013. New observations in micro-pop-in issues in nanoindentation of coarse grain alumina. *Ceramics International* 39:999–1009.
2. Bradby, J. E., S. O. Kucheyev, J. S. Williams, J. Wong-Leung, M. V. Swain, P. Munroe, G. Li, and M. R. Phillips. 2002. Indentation-induced damage in GaN epilayers. *Applied Physics Letters* 80:383–85.
3. Packard, C. E., and C. A. Schuh. 2007. Initiation of a shear band near a stress concentration in metallic glass. *Acta Materialia* 55:5348–58.
4. Mao, W. G., Y. G. Shen, and C. Lu. 2011. Deformation behaviour and mechanical properties of poly-crystalline and single crystal alumina during nanoindentation. *Scripta Materialia* 65:127–130.
5. Mao, W. G., Y. G. Shen, and C. Lu. 2011. Nanoscale elastic–plastic deformation and stress distributions of the C plane of sapphire single crystal during nanoindentation. *Journal of the European Ceramic Society* 31:1865–71.
6. Bhattacharya, M., R. Chakraborty, A. Dey, A. K. Mandal, and A. K. Mukhopadhyay. 2012. Improvement in nanoscale contact resistance of alumina. *Applied Physics A* 107:783–88.
7. Anstis, G. R., P. Chantikul, B. R. Lawn, and D. B. Marshall. 1981. A critical evaluation of indentation techniques for measuring fracture toughness; I: Direct crack measurements. *Journal of the American Ceramic Society* 64:533–38.
8. Oliver, W. C., and G. M. Pharr. 2004. Measurement of hardness and elastic modulus by instrumented indentation: Advances in understanding and refinements to methodology. *Journal of Materials Research* 19:3–20.
9. Ebisu, T., and S. Horibe. 2010. Analysis of the indentation size effect in brittle materials from nanoindentation load–displacement curve. *Journal of European Ceramic Society* 30:2419–26.
10. Schuh, C. A., and A. C. Lund. 2004. Application of nucleation theory to the rate dependence of incipient plasticity during nanoindentation. *Journal of Materials Research* 19:2152–58.

18

Nanoscale Contact in Ceramics

Manjima Bhattacharya, Riya Chakraborty, Arjun Dey,
and Anoop Kumar Mukhopadhyay

18.1 Introduction

As we have already discussed in Chapter 17, the study of contact-induced
deformations during nanohardness evaluation and the subsequent damage mechanisms in alumina under low loads are significant and important research topics. One can always ask why we are so concerned about
these nanoscale contact events. Let us look at the behind-the-scene features
involved here. What is the typical bond length of alumina? One may also
wonder why bond length is important at all. Well, it is the strength of the
bonds that is distributed along the length of the bond that holds the aluminum and the oxygen atoms in alumina, isn't it? That is why the bond length
is important. The bond orientation is important too. But for the sake of simplicity, we shall confine our discussion for the time being to the issue of bond
length. The bond length of alumina is typically about 0.19 nm. From where,
then, does the basic deformation have to start? It will have to start from this
"bond length" kind of length scale. That is why the deformation at nanoscale
becomes of such crucial importance in the development of materials science
and technology today. What happens, in a nutshell, is this: Very simplistically speaking, the nanoscale deformation process, if it can not be contained
or accommodated at the local microstructural length scale, ultimately can
lead to a permanent residual strain in the system. When at a much localized scale, this strain becomes more than the alumina can bear, very localized microcracks and consequently microdamages can grow. Sustained load
application can eventually lead to the joining of such microdamaged regions,
which leads to coalescence of the microcracks to form a main crack. When
the stress intensity factor at the tip of this main crack reaches a level greater
than or equal to the critical stress intensity factor in mode I, II, or III, then
unstable crack growth occurs, and ultimately that culminates very quickly
into a catastrophic failure. It is undoubtedly true, then, that the genesis of this
macroscopic failure process had its origin at the nano- or submicro-scale deformation followed by the crack initiation. This is why the study of nanoscale

contact deformation and crack initiation is of such paramount importance, more so for a brittle solid like alumina. Therefore, in this chapter we shall ask ourselves whether it is possible to improve the nanoscale contact resistance of alumina. If so, we shall attempt in this chapter to address this question by considering how a coarse grain alumina microstructure responds to sharp contact, i.e., a Berkovich nanoindentation induced deformation.

The same high-density ($\approx$95% of theoretical) coarse grain ($\approx$20 μm) alumina ceramic as discussed in Chapter 17 was utilized for these experiments. To control the extent of damage that such a sharp contact event would create, all of the nanoindentation experiments were conducted at a very low peak load of 10^6 μN. However, the applied loading rates (10^3–10^6 μN·s^{-1}) were deliberately varied in these experiments to see whether this led to a change in the material's response. By varying the loading rates, we are basically affecting the rate of energy transfer from the loading train into the alumina material in question. The effects were explained in terms of shear localization as well as formation of dislocation loops under the Berkovich nanoindenter used on the present alumina samples.

18.2 Evolutions of Pop-ins

The exploded views of the loading and unloading parts of the load-versus-depth (P–h) plots at four typical loading rates are shown in Figures 18.1a and 18.1b, respectively. The plots revealed the presence of a large number of serrations in the P–h plots that varied in number with the loading rates. There were more at the lower loading rates (10^3 μN·s^{-1}) and fewer at higher loading rates (10^6 μN·s^{-1}).

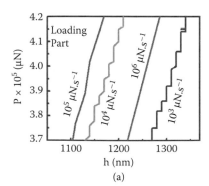
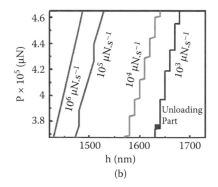

FIGURE 18.1

Exploded views of (a) loading and (b) unloading parts of P–h plots at P_{max} = 10^6 μN. (Reprinted with permission of Bhattacharya and Mukhopadhyay [1] from Hindawi and ISRN.)

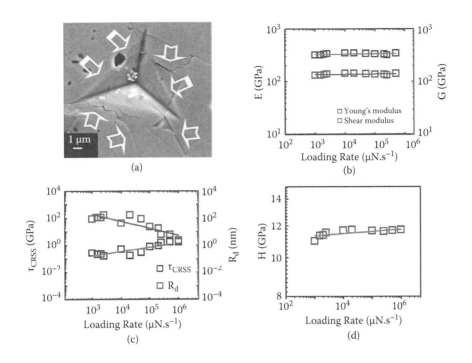

(a)

(b)

(c)

(d)

FIGURE 18.2 (See color insert.)
(a) FESEM photomicrograph showing shear deformations and microcracks in the nanoindentation cavity. Variations of (b) E and G, (c) τ_{CRSS} and R_d, and (d) H with loading rates. (Reprinted with permission of Bhattacharya and Mukhopadhyay [1] from Hindawi and ISRN.)

As discussed in Chapter 17, the serrations implied the occurrences of nanoscale localized plastic deformation signatures expressed in terms of micro pop-in and micro pop-out events in the loading and unloading parts. The high-resolution FESEM photomicrograph (Figure 18.2a) showed shear-induced localized microcracking in the vicinity of the nanoindent and microshear band formation inside the nanoindentation cavity. The genesis of such nanoscale plasticity events has been correlated [2, 3] to shear burst and shear localization that initiates at a critical load (P_c). However, microcrack formation requires a low value of the critical resolved shear stress (τ_{CRSS}). Based on the measured Young's modulus for the present alumina ceramic ($E \approx 344.3$ GPa for 10^6 μN) [4, 5], the values of shear modulus G were estimated from [$E/2(1+v_s)$] assuming v_s as 0.21 [6]. The shear modulus (G) data were not sensitive to variations in loading rate ($\dot{P}$), as expected (Figure 18.2b). Now, following the method suggested by Page, Oliver, and McHargue [7] and using the present experimental data, the corresponding critical resolved shear stress (τ_{CRSS}) values were evaluated (assuming the magnitude of the Burger's vector (ß) as ≈ 0.5 nm [7] for this purpose). These data showed a power law dependence on the loading rates (Figure 18.2c) [1]. The average critical resolved shear stress $\left(\overline{\tau_{CRSS}}\right)$ value was estimated as ≈ 1.23 GPa for the applied load of 10^6 μN. Such a small magnitude of $\left(\overline{\tau_{CRSS}}\right)$ suggests a high possibility of localized microcracking around the nanoindents because of the high stresses generated at the small contact region, particularly at the vertices of the indents due to the steep change in slopes of the supporting contact area. This phenomenon is evident from the FESEM photomicrograph shown in Figure 18.2a.

There is a possibility of dislocation nucleation that might have led to the initiation of nanoscale plasticity events in the present coarse-grained alumina ceramic. Thus, following the method suggested by Gouldstone et al. [8], the possible range of dislocation loop radius (R_d) data was estimated as a function of the loading rates utilized in the present work (assuming the typical magnitude of Burger's vector (ß) ≈ 0.5 nm [7]). The range of dislocation loop radius (R_d) data was estimated as ≈85–2.2 nm, corresponding to the range of loading rates used in the present work (10^6–10^3 μN·s^{-1}). The variation of R_d, however, exhibited a negative power law dependence on P (Figure 18.2c) [1]. The reason behind this may be that at lower loading rates, the indenter spent more time in contact with the alumina microstructure. Hence, the alumina was expected to face more microstructural obstruction during the penetration process. But at higher loading rates, it spent comparatively much less time (1 s) in contact with the alumina microstructure. Therefore, it was highly likely that the amount of microstructural obstruction faced by the penetrating indenter would be much less in this case. Thus, the data presented in Figure 18.2c would suggest that at higher loading rates, the indenter penetrated for relatively shorter amounts of time and, hence, it possibly faced fewer obstructions. As the amount of obstruction faced was less, the radius of the dislocation loop was very small (≈2.2 nm) at higher loading rate, leading to the nucleation of more dislocations per unit volume than those nucleated at lower loading rates, wherein the dislocation loop radii were predicted to be relatively larger (≈85 nm). The greater the number of such nanoscale plasticity events that occur, the more strain there would be in the microstructure, which thereby increased the hardness of the present alumina ceramic by about 6.64% (Figure 18.2d) with $\dot{P}$.

In this context, the estimated average of the maximum shear stress $\left(\overline{\tau_{max}} \approx 30 \pm 14.5\,\text{Gpa}\right)$ as calculated following Packard and Schuh [9] and Bhattacharya et al. [4, 5] developed underneath the indenter, active at depth of about 0.35 times the static contact radius (a), was much greater than the theoretical shear strength (≈3 GPa) of alumina [10], and hence shear-induced deformation and/or fracture were expected (Figure 18.2a).

Further, there was a relationship between the increase in nanohardness and τ_{max}, as shown in Figure 18.3a, which demonstrates that the rate of change of nanohardness of the present alumina ceramic with $\dot{P}$ was linearly dependent on the rate of change of the maximum shear stress with $\dot{P}$. In other words, the higher magnitude of maximum shear stress at higher loading rate could easily have played a pivotal role in nucleating a greater number

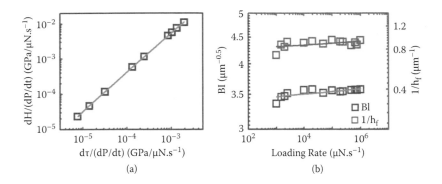

FIGURE 18.3 (See color insert.)
Variations of (a) the rates of change of hardness and maximum shear stress and (b) BI and $(1/h_f)$ with the loading rates. (Reprinted with permission of Bhattacharya and Mukhopadhyay [1] from Hindawi and ISRN.)

of nanoscale plasticity events (because the radii of the dislocation loop were smaller) and, thereby, more localized strain accumulation in the microstructure of the present alumina ceramic, thereby resulting in the apparent small increase of nanohardness $\approx 6.64\%$. The characteristic lower magnitude of the critical resolved shear stress of course had aided some release of stress around the nanoindentation cavity through formation of highly localized microcracks.

Such a picture is further supported by the fact that both the brittleness index (BI), calculated following Lawn and Marshall [11], and the characteristic contact deformation resistance parameter, defined for the first time in the present work as $1/h_f$, increased with the loading rate, following empirical positive power law dependencies (Figure 18.3b). Thus, the data presented in Figure 18.3b implied that the higher the contact deformation resistance of the present alumina, the more prone it became to contact induced by microfracture, as expected for characteristically brittle solids.

18.3 Conclusions

In the case of a high-density (95% of theoretical), coarse grain (≈ 20 µm) alumina ceramic undergoing nanoindentation at an ultralow peak load of 10^6 µN, the nanoscale contact deformation resistance $(1/h_f)$ of the material against the initiation of incipient plasticity events was enhanced by increasing loading rates and followed a power law dependence with a positive exponent. This was reflected in 6% enhancement in nanohardness with loading rate. So far we have studied only the nanoscale contact damage evolution during static contact of ceramics, in as-received or as-sintered conditions.

But we have no idea about how the entire scenario will be affected if the given ceramic is already highly strained, e.g., due to a high strain rate impact at a pressure much higher than its own Hugoniot elastic limit (HEL). This will therefore be our topic of discussion in Chapter 19.

References

1. Bhattacharya, M., and A. K. Mukhopadhyay. 2012. Contact deformation of alumina. *ISRN Ceramics*. http://www.hindawi.com/isrn/ceramics/2012/741042/.
2. Mao, W. G., Y. G. Shen, and C. Lu. 2011. Deformation behaviour and mechanical properties of poly-crystalline and single crystal alumina during nanoindentation. *Scripta Materialia* 65:127–30.
3. Mao, W. G., Y. G. Shen, and C. Lu. 2011. Nanoscale elastic–plastic deformation and stress distributions of the C plane of sapphire single crystal during nanoindentation. *Journal of the European Ceramic Society* 31:1865–71.
4. Bhattacharya, M., R. Chakraborty, A. Dey, A. K. Mandal, and A. K. Mukhopadhyay. 2012. Improvement in nanoscale contact resistance of alumina. *Applied Physics A* 107:783–88.
5. Bhattacharya, M., R. Chakraborty, A. Dey, A. K. Mandal, and A. K. Mukhopadhyay. 2013. New observations in micro-pop-in issues in nanoindentation of coarse grain alumina. *Ceramics International* 39:999–1009.
6. Oliver, W. C., and G. M. Pharr. 2004. Measurement of hardness and elastic modulus by instrumented indentation: Advances in understanding and refinements to methodology. *Journal of Materials Research* 19:3–20.
7. Page, T. F., W. C. Oliver, and C. J. McHargue. 1992. The deformation behavior of ceramic crystals subjected to very low load (nano) indentations. *Journal of Materials Research* 7:450–73.
8. Gouldstone, A., H. J. Koh, K. Y. Zeng, A. E. Giannakopoulos, and S. Suresh. 2000. Discrete and continuous deformation during nanoindentation of thin films. *Acta Materialia* 48:2277–95.
9. Packard, C. E., and C. A. Schuh. 2007. Initiation of a shear band near a stress concentration in metallic glass. *Acta Materialia* 55:5348–58.
10. Anstis, G. R., P. Chantikul, B. R. Lawn, and D. B. Marshall. 1981. A critical evaluation of indentation techniques for measuring fracture toughness; I: Direct crack measurements. *Journal of the American Ceramic Society* 64:533–38.
11. Lawn, B. R., and D. B. Marshall. 1979. Residual stress effects in failure from flaws. *Journal of the American Ceramic Society* 62:106–8.

Section 6

Static Behavior of Shock-Deformed Ceramics

19

Shock Deformation of Ceramics

Riya Chakraborty, Arjun Dey, and Anoop Kumar Mukhopadhyay

19.1 Introduction

Nanoindentation is now proven to be a powerful method for elucidating mechanical properties at the nanoscale level comparable to the mean grain size of ceramic materials (Chapter 6). So, the objective of the present chapter was to evaluate the micromechanical properties, e.g., nanohardness (H) and Young's modulus (E), of the alumina fragments recovered after the shock experiments by the nanoindentation technique and compare these results with those obtained for the as-prepared alumina of about 10 μm grain size. Here, the shock-recovered alumina samples had the identical processing history as that of the as-prepared alumina samples. The details of the experimental methods have already been given in Chapter 9.

The polished and thermally etched microstructure of the as-prepared alumina sample used in the present work is shown in Figure 19.1. The microstructure was dense, with micropores present only at triple grain junctions. However, the grain size distribution was a little heterogeneous. The average grain size was 10.1 ± 0.23 μm. The damaged microstructure of the shock-recovered alumina sample is shown as the inset of Figure 19.1. The damaged microstructure of the shock-recovered alumina prior to nanoindentation experiments exhibited grain boundary microcracking and grain localized slip-band formation in addition to intra- and inter-crystalline cleavage as well as both transgranular and intragranular failure modes.

19.2 Nanoindentation Study

The data on H and E behavior of the as-prepared alumina discs and the alumina fragments recovered after the 6.5 GPa shock experiment are presented as a function of depth in Figures 19.2a and 19.2b. There was an indentation size effect in nanohardness but not in the Young's modulus data.

FIGURE 19.1
FESEM photomicrographs of alumina samples of identical processing history. (a) Polished and thermally etched microstructure in as-prepared condition and (b) fracture surface in shock-recovered condition. (Reprinted/modified with permission of Mukhopadhyay et al. [1] from Elsevier.)

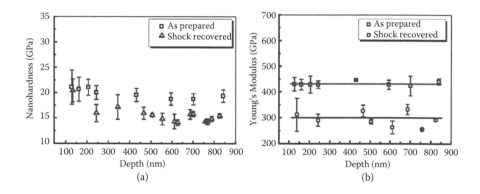

FIGURE 19.2
Nanoindentation data of both as-prepared and shock-recovered alumina of identical process-ing history as a function of depth: (a) nanohardness and (b) Young's modulus. (Reprinted with permission of Mukhopadhyay et al. [1] from Elsevier.)

This aspect shall be dealt with later in this book (Chapter 46). The most impor-tant observation from the nanoindentation experiments was that the average values of H and E of the shock-recovered alumina samples were reduced by about 22% (Figure 19.2a) and 30% (Figure 19.2b), respectively, compared to those of the as-prepared alumina discs. Based on the XRD data analy-sis, it was found that the microstrain in shock-deformed alumina fragments (SDAF) was about 17%, which was more than twice as high as that of the as-sintered alumina [2]. The SEM photomicrograph of a typical nanoindent in as-prepared alumina is shown in Figure 19.3a. There were no microcracks

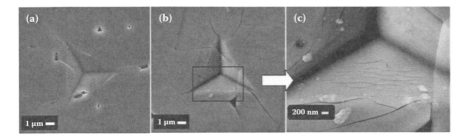

FIGURE 19.3
FESEM photomicrographs of nanoindent in alumina: (a) as-prepared, (b) shock recovered, and (c) details of (b) at higher magnification. (Reprinted with permission of Mukhopadhyay et al. [1] from Elsevier.)

present around the nanoindent impression in the as-prepared alumina sample. In sharp contrast to this observation, the nanoindent impressions at the same load in shocked alumina showed prolific presence of microcracks around the nanoindent impression (Figure 19.3b). Higher magnification view taken from inside this nanoindent (see Figure 19.3c) also showed clear evidence of formation of shear-induced deformation bands inside the nanoindentation cavity. It is well known that microcrack nucleation under sharp contact in brittle solids such as the present alumina ceramic occurs from the interaction of dislocations on two slip planes, both of which are favorably oriented for such interaction to happen. It may also occur when one appropriately oriented slip plane blocks dislocations on a neighboring slip plane [3, 4]. Of course, the interesting part of the present observation is that the microcracks had occurred around the nanoindent that was created at a typical low load of 500 mN, suggesting that there must be enough weakness already existing in the microstructure such that even the small additional loading of 500 mN could cause microcrack formation.

19.3 Occurrence of Pop-ins

The corresponding typical load (P) versus depth (h) plots obtained during the nanoindentation experiments on as-prepared and shock-recovered alumina of identical processing history are shown in Figure 19.4. The data presented in Figure 19.4 showed that for a given constant peak load, e.g., 500 mN, the $P–h$ plot was smooth for the as-prepared alumina, but a pop-in had appeared in the $P–h$ plot of the shock-recovered alumina. The occurrence of the pop-in event strongly suggested the generation of microcracks during nanoindentation in the shock-recovered alumina. It is interesting to note that microcracks around nanoindents in the shock-recovered alumina

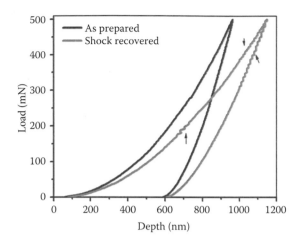

FIGURE 19.4
Load–depth plots obtained during the nanoindentation experiments on both as-prepared and shock-recovered alumina of identical processing history (the black arrows mark the positions of pop-in event that occurred during the nanoindentation experiments in the shock-deformed alumina sample). (Reprinted with permission of Mukhopadhyay et al. [1] from Elsevier.)

were indeed also experimentally observed (Figure 19.3b). Thus, the FESEM observations corroborate the pop-in event observed in the load-depth plot of nanoindentation experiments conducted on the shock-recovered alumina of identical processing history to the as-prepared alumina samples. The details of generation of such pop-ins is discussed in Chapter 47. However, the source of structural weakness in the shock-recovered alumina still needs to be better understood. Therefore, further investigations of the shock-damaged microstructure were conducted using FESEM and TEM as described previously, and the results are presented in the following section.

19.4 Defects in Shock-Recovered Alumina

In the damaged microstructure of the shock-recovered alumina sample, there were many locations that had suffered extensive grain-localized plastic deformation, leading to shear band formation. A typical FESEM photomicrograph of such shear band formation in the fracture surface of the shock-recovered alumina prior to nanoindentation experiments is shown in Figure 19.5a. In addition, TEM-based photomicrographs showed prolific presence of dislocations entanglement (Figure 19.5b). Thus, the FESEM and XRD data corroborate the TEM observation that there was prolific presence of grain-localized plastic deformation that led to generation of dislocations,

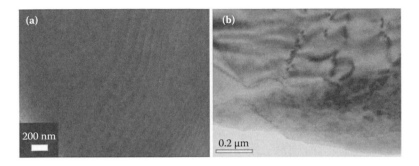

FIGURE 19.5
(a) FESEM photomicrograph of shear band formation in the fracture surface of the shock-recovered alumina, (b) TEM image showing huge entanglement of dislocations. (Reprinted/modified with permission of Mukhopadhyay et al. [1] from Elsevier.)

microcracks, and microfracture zones in the shock-recovered alumina sample prior to the nanoindentation experiment. These were then the causes that made the shock-recovered alumina structurally much weaker than the as-prepared alumina. When the nanoindentation experiments were conducted, the very presence of these structural weakening factors led to formation of further microcracks around nanoindents, even though the load was very small, e.g., 500 mN. Further, as there was extensive grain-localized plastic deformation already present in the microstructure of the shock-recovered alumina, addition of even a very small amount of stress during the contact-induced deformation led to shear-band formation inside the nanoindentation cavity itself. Thus, the shock-induced structural weakness led to lower nanohardness of the shocked alumina as compared to that of the as-prepared alumina of identical processing history. Further, it is suggested that the extensive spatial presence of the weakening factors throughout the microstructure had also offset the expected enhancement in terms of the reduced crystallite size and/or the higher average microstrain in the nanohardness data of the shock-deformed alumina.

19.5 Conclusions

In this chapter, we have been able to look into the nanomechanical properties and related deformation mechanisms of a typical polycrystalline ceramic (alumina) after it has been exposed to high-strain impact in a gas gun. However, we are yet to understand how the load of measurement affects the nanohardness data, especially in comparison to those of an unshocked ceramic. Therefore, in Chapter 20, we shall try to understand the effect of load on nanohardness of shocked ceramics.

References

1. Mukhopadhyay, A. K., K. D. Joshi, A. Dey, R. Chakraborty, A. Rav, A. K. Mandal, J. Ghosh, S. Bysakh, S. K. Biswas, and S. C. Gupta. 2010. Nanoindentation of shock deformed alumina. *Materials Science and Engineering A* 527:6478–83.
2. Mukhopadhyay, A. K., K. D. Joshi, A. Dey, R. Chakraborty, A. Rav, S. K. Biswas, and S. C. Gupta. 2010. Shock deformation of coarse grain alumina above Hugoniot elastic limit. *Journal of Material Science* 45:3635–51.
3. Hagan, J. T. 1979. Micromechanics of crack nucleation during indentations. *Journal of Material Science* 14:2975–80.
4. Cottrell, A. H. 1958. Theory of brittle fracture in steel and similar metals. *Transaction of Materials Society AIME* 212:192–202.

20

Nanohardness of Alumina

Riya Chakraborty, Arjun Dey, and Anoop Kumar Mukhopadhyay

20.1 Introduction

The basic objective of this chapter is to utilize the nanoindentation technique to evaluate nanohardness of shock-recovered fragments of a coarse-grain ($\approx$10 μm), high-density (3.978 g·cc^{-1}) alumina obtained after a carefully conducted flyer-plate shock experiment [1] at a shock pressure of 6.5 GPa and to examine whether the nanohardness was similar to or degraded in comparison to that of the as-sintered alumina ceramics. As a consequence of this effort, we report here for the first time a new explanation for the presence of a strong indentation size effect (ISE) in shock-recovered alumina fragments obtained from an earlier study [1]. The details of the experimental methods have already been discussed in Chapter 9.

Figure 20.1 shows that a much larger number of multiple micro pop-ins appears in the loading part of the load–depth plots of the shock-recovered alumina than those in the as-sintered alumina. These data signify more concrete evidence of the interaction between the penetrating nanoindenter and the preexistent as well as shock-induced microstructural damages that lead to highly localized microcrack formation in the vicinity of the nanoindent. Similar observations were reported by other researchers [2–5] in the cases of quartz, soda-lime glass, fused silica, glassy carbon, etc.

20.2 Indentation Size Effect of Shocked Alumina

At any given load, the shock-recovered alumina ceramics registered a higher depth of penetration (Figures 20.2a and 20.2b) in comparison to those of the as-sintered alumina, suggesting that they had some inherent weakness compared to those of the as-sintered alumina ceramic. In this connection, it may be mentioned that Ebisu and Horibe [6] have recently proposed an empirical equation,

$$P = Ch^n \tag{20.1}$$

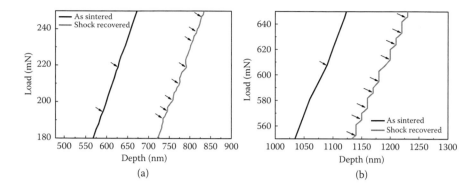

FIGURE 20.1
Typical load-versus-depth plots during loading cycle in nanoindentation experiments on as-sintered and shock-recovered alumina samples at typical loads of (a) 300 mN and (b) 700 mN. (Reprinted with permission of Chakraborty et al. [7] from Springer.)

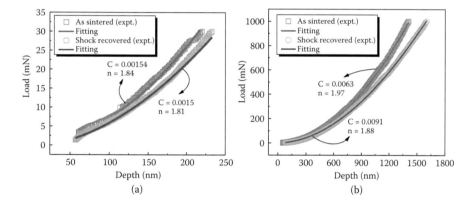

FIGURE 20.2 (See color insert.)
Fitting of the empirical equation $P = Ch^n$ in load-versus-depth data of loading cycle during nanoindentation at peak loads of (a) 30 mN and (b) 1000 mN for shock-recovered and as-sintered alumina. (Reprinted with permission of Chakraborty et al. [7] from Springer.)

where h is the indentation depth at a load (P) in the loading-cycle data, and C and n are constants to explain the presence of ISE in 8Y-FSZ single crystals and 12Ce-TZP polycrystals. It follows very easily that, if $n = 2$ in equation (20.1), the materials should show no ISE, and in such cases the load is proportional to the square of the indentation depth. However, if $n < 2$, then the material should show ISE [6]. Accordingly, following Ebisu and Horibe [6], the second-order differential against indentation depth h in Equation (20.1) yields:

$$\frac{d^2P}{dh^2} = n(n-1)Ch^{n-2} \tag{20.2}$$

It appears obvious from equation (20.2) that if $n = 2$, d^2P/dh^2 will be a constant. On the other hand, if ISE is present, that is, if $n < 2$, d^2P/dh^2 should decrease as h increases. The results of fitting the present experimental load (P) versus depth (h) data in the loading cycle to equation (20.1) are depicted in Figures 20.2a and 20.2b.

In addition, the data on fitting parameters (n and C) and the goodness of fit in terms of the residual sum of squares (R^2) presented elsewhere [7] had showed that, for the cases of both as-sintered alumina and shock-recovered alumina, the goodness of fit was better than 0.99. This information suggested that the empirical equation (20.1) proposed by Ebisu and Horibe [6] provides a reasonably good description of the experimental P–h data in the loading cycle of the present alumina ceramics in both as-sintered and shock-recovered conditions. Further, irrespective of load applied in the nanoindentation experiments, the power law exponent (n) values obtained for the as-sintered alumina were very slightly less than 2, while $n \ll 2$ [7] was observed for the shock-recovered alumina, providing thereby a clue as to why there was the presence of a strong ISE in this case.

Fitting of the experimental P–h data in the loading cycle to equation (20.2) shows that the second-order derivative of load with respect to depth (d^2P/dh^2) decreased continually with depth for both as-sintered and shock-recovered alumina (Figures 20.3a and 20.3b), as $n < 2$. Up to a critical load of 100 mN, the slopes of the regions AB, BC, and CD were nearly equal for both as-sintered and shock-recovered alumina. For example, at 30 mN load, the slopes of the regions AB, BC, and CD are 0.23×10^{-7}, 0.14×10^{-7}, and 0.1×10^{-7}, respectively, for as-sintered alumina, and for shock-recovered alumina the slopes of the regions AB, BC, and CD are 0.22×10^{-7}, 0.13×10^{-7}, and 0.09×10^{-7}, respectively. But at $P > 100$ mN, the slope values of the shock-recovered alumina were much higher than those registered by the as-sintered alumina at comparable intervals of the depths of penetration. For example, at 1000 mN load, the

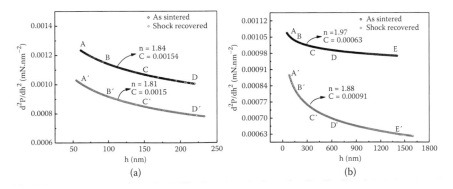

FIGURE 20.3
Variation of d^2P/dh^2 as a function of depth (h) for as-sintered and shock-recovered alumina samples at peak loads of (a) 30 mN and (b) 1000 mN. (Reprinted with permission of Chakraborty et al. [7] from Springer.)

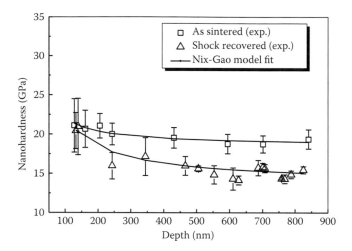

FIGURE 20.4

Nanohardness-versus-depth plots for as-sintered and shock-recovered alumina and the corresponding predicted lines according to the Nix and Gao model. (Reprinted with permission of Chakraborty et al. [7] from Springer.)

slopes of the regions AB, BC, CD, and DE are 0.22×10^{-6}, 0.1×10^{-6}, 0.08×10^{-6}, and 0.03×10^{-6}, respectively, for as-sintered alumina. But, for shock-recovered alumina, the slopes of the regions AB, BC, CD, and DE are 0.94×10^{-6}, 0.36×10^{-6}, 0.16×10^{-6}, and 0.08×10^{-6}, respectively. These data suggested once again the more compliant nature of the shock-recovered alumina. Thus, the present experimental results prove that a given alumina ceramic may exhibit both mild and strong ISE, depending on the extent of microstructural damage that it had or had not withstood already. This also reflects in the nanohardness data, i.e., the nanohardness of the shock-recovered alumina was much lower than that of the as-sintered alumina, and there was a mild ISE present in the as-sintered alumina, while there was a strong ISE present in the shock-recovered alumina (Figure 20.4).

20.3 Deformation of Shocked Alumina

The FESEM- and TEM-based evidence [7] showed the prolific presence of grain-localized microcracks induced by shear-deformation bands, intragrain microcleavages, extensive inelastic deformation at suitably oriented triple-grain junction walls, formation of wing cracks, and intragrain microcleavages at the nanoscale in the damaged microstructure of the shock-recovered alumina ceramic. Further, the TEM-based evidence showed [7] a huge number of dislocations in the shock-deformed alumina fragments. The presence of these microstructural damages led to the existence of highly localized microtensile strain [2], which makes the microstructure more compliant. When these strains cannot be accommodated further through dislocation generations, whose movements are impeded at the grain boundaries, localized microcracking may occur. Consequently, the probability of interaction with a higher number of preexistent as well as shock-induced microstructural damage underneath the indenter also increases. If and when this indeed happens, that would be reflected in a higher depth of penetration by the nanoindenter. Then, at the same comparable load of nanoindentation, the depth of penetration and hence the projected area of contact should be greater in the shock-recovered alumina as compared to those in the as-sintered alumina. As a result of this process, at comparable loads, the nanohardness of shock-recovered alumina should be much lower than that of the as-sintered alumina, as was indeed experimentally observed (Figure 20.4).

Further supportive evidence to such a conjecture stems from the FESEM photomicrographs of the nanoindent, which showed a relatively predominant presence of slip lines, deformation bands, and microcracks in the shock-recovered alumina (Figure 20.5a) and their comparative absence in the as-sintered alumina (Figure 20.5b). The slip bands in shock-recovered alumina are oriented parallel to the nanoindentation cavity's wall surface (Figure 20.5a).

The stress required for the activation of such plastic deformation would have to be provided by the nanoindentation process itself. As a result of this intense local stress distribution, the strain would be significantly high, and its spatial gradient would be quite steep near the base of the nanoindent in the cavity. Therefore, the load at which the first burst is observed in the $P–h$ plot can be utilized to calculate the maximum shear stress, τ_{max}, generated just below the indenter tip given the knowledge of the geometric and elastic properties of the indenter and sample [8]. At ultralow loads causing shallow indentation displacements, the tip radius even for the nominally sharp Berkovich tip (e.g., 150 nm tip radius) will exert a significant contribution to the deformation process. Approximating the tip, for the sake of simplicity, as a spherical

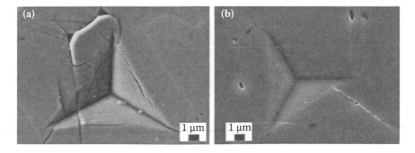

FIGURE 20.5
Typical FESEM photomicrographs of nanoindentation impressions in (a) shock-recovered and (b) as-sintered alumina. (Reprinted with permission of Chakraborty et al. [7] from Springer.)

indenter, the maximum shear stress (τ_{max}) below the tip of the nanoindenter was evaluated [7] for the given peak loads of 100, 500, and 1000 mN as ≈36.7, 40.7, and 36.3 GPa and 20.1, 20.4, and 22.2 GPa, respectively; for the as-sintered and shock-recovered alumina samples. These data were much higher than the theoretical shear strength (τ_{theor}) of the alumina ceramic (≈3 GPa) [9]. Therefore, the applied load provided enough shear stress to cause shear-induced flow and/or deformation in the nanoindented alumina ceramics. Thus, it was likely that the first burst corresponded to the initiation of plastic deformation in the material through the nucleation of shear bands.

20.4 Micro Pop-ins of Shocked Alumina

Thus, based on this experimental evidence and literature data [2–5, 8], it is proposed that the frequent presence of multiple micro pop-in events suggests a generic mechanism responsible for the experimental observation of a relatively strong ISE in the shock-recovered alumina. Such a mechanism is based on the relative frequency of the multiple micro pop-in events that occur as a function of the applied nanoindentation load at the microstructural and nanostructural length scale. The actual physical frequency of such multiple micro pop-in events would be governed by a multitude of factors such as the statistical size and spatial orientation distribution of both preexistent and shock-induced damages with respect to the axis of nanoindenter and the nanoindentation size, the relative orientation of single grain and/or grain assembly containing such damages and/or damage zones, the local spatial distribution of microtensile stresses, etc. Thus, such relative frequencies of the multiple micro pop-in events would govern the relative rate of penetration by the nanoindenter in a given alumina ceramic microstructure and, thereby, the relative extent of the ISE.

20.5 Conclusions

Finally, we suggest that the presence of highly localized microtensile stresses in the suitably oriented grains of the damaged microstructure of shock-recovered alumina was the reason for ISE. In presence of such tensile stress at the same comparable load, the depth of penetration was greater, thereby giving rise to the observed ISE, while the presence of slip lines, deformation bands, and microcracks inside the nanoindentation cavity led to the more frequent occurrence of multiple micro pop-in events in the load–depth plots obtained during the loading cycles in the nanoindentation experiments conducted on the damaged microstructure of the shock-recovered alumina. Our next focus of attention will be the extent to which the existing defects interact with a polycrystalline ceramic that is shocked at a much higher impact pressure (about 12 GPa) at a strain rate of more than 10^4 s^{-1}. These aspects are discussed in Chapter 21.

References

1. Mukhopadhyay, A. K., K. D. Joshi, A. Dey, R. Chakraborty, A. Rav, S. K. Biswas, and S. C. Gupta. 2010. Shock deformation of coarse grain alumina above Hugoniot elastic limit. *Journal of Material Science* 45:3635–51.
2. Subhash, G., S. Maiti, P. H. Geubelle, and D. Ghosh. 2008. Recent advances in dynamic indentation fracture, impact damage and fragmentation of ceramics. *Journal of the American Ceramic Society* 91:2777–91.
3. Morris, D. J., S. B. Myers, and R. F. Cook. 2004. Sharp probes of varying acuity: Instrumented indentation and fracture behavior. *Journal of Materials Research* 19:165–75.
4. Field, J. S., M. V. Swain, and R. D. Dukino. 2003. Determination of fracture toughness from the extra penetration produced by indentation-induced pop-in. *Journal of Materials Research* 18:1412–19.
5. Quinn G. D., P. Green, and K. Xu. 2003. Cracking and the indentation size effect for Knoop hardness of glasses. *Journal of the American Ceramic Society* 86:441–48.
6. Ebisu T., and S. Horibe. 2010. Analysis of the indentation size effect in brittle materials from nanoindentation load-displacement curve. *Journal of the European Ceramic Society* 30:2419–26.
7. Chakraborty, R., A. Dey, A. K. Mukhopadhyay, K. D. Joshi, A. Rav, A. K. Mandal, S. Bysakh, S. K. Biswas, and S. C. Gupta. 2012. Nanohardness of as-sintered and shock-deformed alumina. *Metallurgical and Materials Transaction A* 43:459–70.
8. Bei, H., Z. P. Lu, and E. P. George. 2004. Theoretical strength and the onset of plasticity in bulk metallic glasses investigated by nanoindentation with a spherical indenter. *Physical Review Letters* 93:125504-1–4.
9. Rosenberg, Z., D. Yaziv, Y. Yeshurun, and S. J. Bless. 1987. Shear strength of shock-loaded alumina as determined with longitudinal and transverse Manganin gauges. *Journal of Applied Physics* 62:1120–22.

21

Interaction of Defects with Nanoindents in Shocked Ceramics

Riya Chakraborty, Arjun Dey, and Anoop Kumar Mukhopadhyay

21.1 Introduction

In Chapters 19 and 20, we discussed the nanohardness of shock-recovered fragments of a coarse-grain (≈10 μm), high-density (3.978 g·cc⁻¹) alumina obtained after a carefully conducted flyer-plate shock experiment [1] at a shock pressure of 6.5 GPa, and we also compared the results for as-sintered alumina ceramics. Then, we tried to give an explanation for the presence of a strong indentation size effect (ISE) in shock-recovered alumina fragments obtained in that study [1]. In this chapter, however, we shall discuss the nanohardness and ISE of alumina ceramic shocked at very high gas-gun impact pressure of about 12 GPa at a very high strain rate of more than 10^4 s⁻¹. For the sake of discussion and brevity, we shall call the shocked alumina samples as SA. Further, we shall compare the results with those of the as-received alumina (ARA). The details of the experimental methods have been discussed already in Chapter 9.

The data presented in Figures 21.1a and 21.1b show the presence of a relatively much larger number of serrations signifying multiple micro pop-ins in the load–depth plots during the loading cycle of the SA sample as compared to the presence of those in the ARA sample. This could only have happened due to enhanced interaction between the penetrating nanoindenter and the pre-existent as well as shock-induced microstructural damages that led to highly localized microcrack formation in the vicinity of the nanoindent in the SA sample. Microcrack initiation during nanoindentation has been linked to the observation of pop-ins for a wide variety of materials, e.g., quartz, soda-lime glass, fused silica, and glassy carbon [2–5].

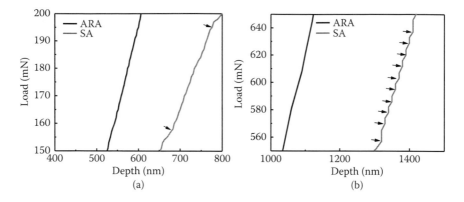

FIGURE 21.1

Typical load (*P*) versus nanoindentation depth (*h*) plots during loading cycle in nanoindentation experiments on as-received and shocked alumina samples showing the details of serrations at (a) low load of 300 mN and (b) high load of 700 mN. (Reprinted with permission of Chakraborty et al. [6] from Elsevier.)

21.2 Indentation Size Effect of Alumina Shocked at High Shock Pressure

Figures 21.2a and 21.2b represent the results of fitting the present experimental load (*P*) versus nanoindentation depth (*h*) data in the loading cycle for the ARA and SA samples to equation (20.1) of Chapter 20. In the case of the ARA sample, the goodness of fit was better than 0.99. A similar statement could be made about the SA sample. This data confirmed that a reasonably good description of the experimental *P–h* data in the loading cycle of the present alumina ceramics in both as-received and shocked conditions was provided by the empirical equation (20.1) of Chapter 20. At any given load, the depth of penetration (Figures 21.2a and 21.2b) in the SA samples was higher than those in the ARA samples. This implied that the SA sampled had more inherent weakness as compared to those of the ARA samples, as mentioned previously (Chapters 19 and 20). The data established further that the power law exponent (*n*) values obtained for the as-received alumina (ARA) samples were very slightly less than 2, irrespective of load applied in the nanoindentation experiments. In contrast, the values of *n* were ≪ 2 for the shocked alumina (SA) samples. This information provided a first-order clue as to why there was a strong presence of ISE in the SA samples used in the present work.

Next, the experimental *P–h* data obtained during the loading cycles of the nanoindentation experiments on ARA and SA samples were fitted to equation (20.2) of Chapter 20. The results of this exercise are shown in Figures 21.3a and 21.3b. The relative percentages of changes in d^2P/dh^2 with respect to *h* at

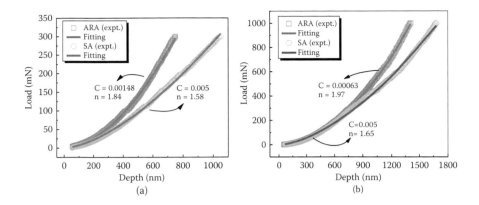

FIGURE 21.2 (See color insert.)
Fitting of the empirical equation $P = Ch^n$ to the load (P) versus nanoindentation depth (h) data of loading cycle for the as-received and shocked alumina samples during nanoindentation at two different peak loads of (a) 300 mN and (b) 1000 mN. (Reprinted with permission of Chakraborty et al. [6] from Elsevier.)

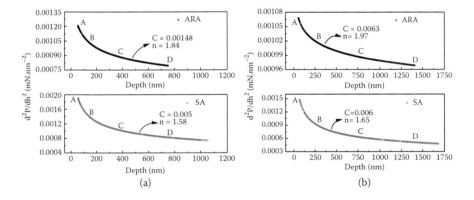

FIGURE 21.3
Variation of d^2P/dh^2 as a function of the nanoindentation depth (h) for as-received and shocked alumina samples at two different peak loads of (a) 300 mN and (b) 1000 mN. (Reprinted with permission of Chakraborty et al. [6] from Elsevier.)

different portions of the plots, i.e., along AB, BC, CD in Figures 21.3a are 15%, 14%, and 10%, respectively, for ARA, and 85%, 46%, and 47%, respectively, for SA. Whereas, for 1000 mN load (Figure 21.3b), the relative percentages of changes in d^2P/dh^2 with respect to h along AB, BC, and CD are 4%, 3%, and 2%, respectively, for ARA and 6%, 3%, and 38%, respectively, for SA. Further, it may be seen from the data presented in Figures 21.3a and 21.3b that for both ARA and SA samples, the second-order derivative of load with respect to the nanoindentation depth (d^2P/dh^2) decreased continuously with depth as n was < 2. Further, it may be noticed from the data of Figures 21.3a and 21.3b that at comparable intervals of the depths of penetration; at regions

AB, BC, and CD the relative percentages of changes in d^2P/dh^2 with respect to h were much higher for the SA samples as compared to those of the ARA samples. These data strongly suggested once again that in comparison to the ARA sample, the SA sample was much more compliant in nature. In search of the shock-induced damages, therefore, extensive SEM and FESEM were conducted on the SA samples and the results are presented in the following section.

21.3 Deformation Due to Shock at High Pressure

The SEM, FESEM, and TEM investigations presented elsewhere [6] showed that numerous defects were present in the SA sample. These include grain-localized microcleavage-induced microcracks and extensive shear-induced complex intragranular microfracture, shear-induced microfracture, intra-grain microcleavages at nanoscale, grain-localized shear band formation, nanoscale slip band formation in a suitably oriented single grain, and the presence of micro wing cracks leading to grain boundary crack formation in the damaged microstructure of the shocked alumina ceramic [6]. The TEM photomicrographs showed the presence of dislocations and dislocation entanglements at a triple-grain junction, at grain edge, and a stacking fault inside a single grain [6]. These photomicrographs provide overall evidence for the extensive presence of microstructural damages in the SA sample. A highly localized microtensile strain can be generated in the microstructure due to the presence of these microstructural damages [2]. The presence of such microtensile strain would make the microstructure more compliant.

One of the means by which the microstructure could accommodate such strains is by the generation of dislocations and their movement. However, when dislocation movements are impeded at the grain boundaries, these strains cannot be accommodated further and, hence, a localized microcracking and/or microfracture may occur. As the spatial density of microstructural damages increases, so does the probability of their interaction with the penetrating nanoindenter during the loading cycle of the nanoindentation experiments conducted in the present work. Such an enhanced interaction probability would be reflected in a higher depth of penetration by the nanoindenter. Therefore, at the same comparable load of nanoindentation, the nanoindentation depth should be more in the SA as compared to those in the ARA samples. When the nanoindentation depth increases, so does the projected area of contact. Hence, at comparable loads, the nanohardness of the SA samples should be much lower than that of the ARA samples, as was indeed experimentally observed (Figure 21.4). Therefore, the nanohardness of ARA had a very slight fall with increase in depth. But the nanohardness of

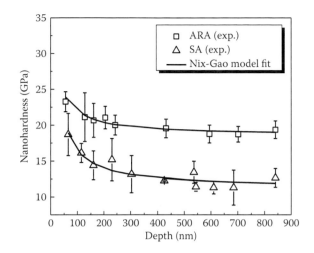

FIGURE 21.4
Nanohardness-versus-depth plots for as-received and shocked alumina. The solid lines depict the predicted variation according to the Nix and Gao model. (Reprinted with permission of Chakraborty et al. [6] from Elsevier.)

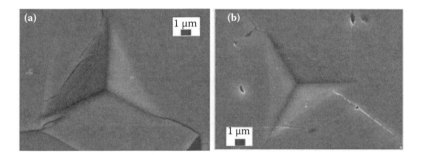

FIGURE 21.5
Typical FESEM of nanoindentation impressions in (a) shock-recovered and (b) as-received alumina. (Reprinted with permission of Chakraborty et al. [6] from Elsevier.)

SA had a very sharp fall with depth (Figure 21.4). The decrease in nanohardness with depth is known as the *indentation size effect* (ISE). Thus the ISE was very mild in ARA but quite strong in SA. As a result, the nanohardness of SA was decreased by as much as 33% with respect to that of the ARA samples.

Further supportive evidence to such a picture also emerged from the FESEM photomicrographs of the nanoindent in the SA sample. These showed a relatively predominant presence of slip lines, deformation bands, and microcracks (Figure 21.5a). However, on a comparative scale, such features were absent in the ARA sample (Figure 21.5b). It is interesting to note that the slip bands in ARA were oriented parallel to the surface of the wall in the nanoindentation cavity (Figure 21.5a). Therefore, based on the result as discussed here, a relatively stronger ISE was expected in the SA sample, which was also experimentally observed (Figure 21.4).

Given the knowledge of the geometric and elastic properties of the indenter and sample [7], the maximum shear stress, τ_{max}, generated just below the nanoindenter tip can be calculated. The load at which the first burst is observed in the $P–h$ plot was utilized for this purpose because it can be directly linked to the initiation of dislocation nucleation [8]. Thus, the appropriate load values at which the first displacement bursts had occurred were obtained from the respective $P–h$ plots of the ARA and the SA samples. It is well appreciated that at ultralow loads causing shallow indentation displacements, the tip radius even for the nominally sharp Berkovich tip (150 nm tip radius) will exert a significant contribution to the deformation process. However, for the sake of simplicity, the tip was approximated as a spherical indenter. Following the procedure given in the literature [7, 8] and taking the appropriate load values, the maximum shear stress (τ_{max}) below the tip of the nanoindenter was evaluated [6] for the given peak loads of 100, 500, and 1000 mN. These values were ≈37, 41, and 36 GPa for the ARA sample at given peak loads of 100, 500, and 1000 mN. In the case of the SA sample, however, the τ_{max} values were reduced by about 50% to ≈19, 20, and 20 GPa, respectively, for the given peak loads of 100, 500, and 1000 mN. Nevertheless, these data were much higher than the theoretical shear strength (τ_{theor}) of the alumina ceramic (≈3) GPa [1]. Therefore, it was evident from this argument that the applied load provided enough shear stress to cause shear-induced flow and/or deformation in the nanoindented alumina ceramics. Thus, there was a strong possibility that the first bursts as observed in the experimental $P–h$ data plots had indeed corresponded, to the initiation of plastic deformation in the present alumina ceramics through the nucleation of shear bands [2–5, 7, 8].

21.4 Conclusions

The reason for ISE was linked to the presence of highly localized microtensile stresses in the suitably oriented grains in the damaged microstructure of the shocked alumina (SA) sample. It is suggested further that, at the same comparable load, the presence of such tensile stress had possibly contributed toward obtaining the higher values of nanoindentation depth. As a result, a strong ISE was observed in the SA sample. It is also proposed that the more frequent occurrence of multiple micro pop-in events in the load depth plots, obtained during the loading cycles in the nanoindentation experiments conducted on the damaged microstructure of the shocked alumina sample, was linked to the presence of slip lines, deformation bands, and microcracks inside the nanoindentation cavity. Up to this chapter, we have no idea about the comparative scenario of indentation size effects in unshocked, 6.5 GPa shocked, and 12 GPa shocked polycrystalline ceramics such as alumina. Better insights into these aspects can help us to design more damage-tolerant polycrystalline ceramics. It will be also interesting to examine how the Nix and Gao model performs when it comes to the comparative ISE scenario mentioned here. This is exactly what is presented in Chapter 22.

References

1. Rosenberg, Z., D. Yaziv, Y. Yeshurun, and S. J. Bless. 1987. Shear strength of shock-loaded alumina as determined with longitudinal and transverse Manganin gauges. *Journal of Applied Physics* 62:1120–22.
2. Subhash, G., S. Maiti, P. H. Geubelle, and D. Ghosh. 2008. Recent advances in dynamic indentation fracture, impact damage and fragmentation of ceramics. *Journal of the American Ceramic Society* 91:2777–91.
3. Morris, D. J., S. B. Myers, and R. F. Cook. 2004. Sharp probes of varying acuity: Instrumented indentation and fracture behavior. *Journal of Materials Research* 19:165–75.
4. Field, J. S., M. V. Swain, and R. D. Dukino. 2003. Determination of fracture toughness from the extra penetration produced by indentation-induced pop-in. *Journal of Materials Research* 18:1412–19.
5. Quinn, G. D., P. Green, and K. Xu. 2003. Cracking and the indentation size effect for Knoop hardness of glasses. *Journal of the American Ceramic Society* 86:441–48.
6. Chakraborty, R., A. Dey, A. K. Mukhopadhyay, K. D. Joshi, A. Rav, A. K. Mandal, S. Bysakh, S. K. Biswas, and S. C. Gupta. 2012. Indentation size effect of alumina ceramic shocked at 12 GPa. *International Journal of Refractory Metals and Hard Materials* 33:22–32.
7. Bei, H., Z. P. Lu, and E. P. George. 2004. Theoretical strength and the onset of plasticity in bulk metallic glasses investigated by nanoindentation with a spherical indenter. *Physical Review Letters* 93:125504-1–4.
8. Packard, C. E., and C. A. Schuh. 2007. Initiation of a shear band near a stress concentration in metallic glass. *Acta Materialia* 55:5348–58.

22

Effect of Shock Pressure on ISE: A Comparative Study

Riya Chakraborty, Arjun Dey, and Anoop Kumar Mukhopadhyay

22.1 Introduction

The basic objective of the present work was to perform a comparative study of indentation size effects in the as-received alumina (ARA) and shocked alumina (SA) by utilizing the nanoindentation technique. Despite the wealth of literature [1–7], such studies are really rare [8]. The SA samples were shock-deformed at 6.5 and 12 GPa and subsequently recovered using a dedicated, specially designed momentum trap. The details of the experimental methods have already been discussed in Chapter 9.

22.2 Comparison of ISE in Alumina Shocked at 6.5 and 12 GPa

At a typical load of 500 mN, the $P–h$ plots for ARA and 6.5 GPa and 12 GPa SA samples (Figure 22.1a) exhibited the characteristic elastoplastic behavior typical of structural ceramics. The frequent presence of multiple micro pop-in events in the $P–h$ plots of both the SA samples (Figure 22.1a) suggest a generic mechanism that must have been responsible for the experimental observation of a relatively stronger ISE in them. Such a mechanism will have its genesis linked to the relative frequency of the multiple micro pop-in events that occur at the microstructural and/or nanostructural length scale as a function of the applied nanoindentation load. In reality, though, a multitude of factors may physically control the frequency of such multiple micro pop-in events, as mentioned previously (Chapters 19–21). The nanohardness of ARA had a very slight fall with increase in depth (Figure 22.1b), whereas that of 6.5 GPa SA had a sharper and that of 12 GPa SA had the sharpest fall with depth (Figure 22.1b). The decrease in nanohardness with depth is termed as the *indentation size effect* (ISE). Thus, the ISE was very mild in ARA but quite strong in the 6.5 GPa SA and strongest in the 12 GPa SA sample.

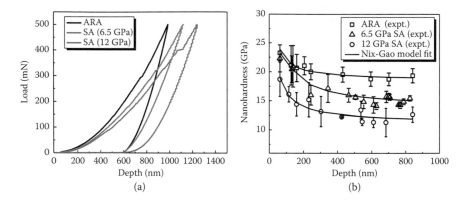

FIGURE 22.1 (See color insert.)
(a) Load–depth plots at a typical illustrative load of 500 mN and (b) nanohardness (H) versus depth (h) plots for ARA, 6.5 GPa SA, and 12 GPa SA samples. The solid lines depict the predicted variation according to the Nix and Gao model. (Reprinted with permission of Chakraborty et al. [8] from Hindwai and ISRN.)

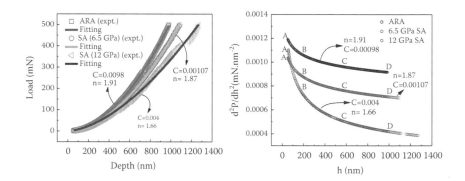

FIGURE 22.2 (See color insert.)
(a) Fitting of the empirical equation $P = Ch^n$ to the load (P) versus depth (h) data of loading cycles during nanoindentations and (b) variation of d^2P/dh^2 as a function of depth (h) for ARA, 6.5 GPa SA, and 12 GPa SA samples at typical load of 500 mN. The solid lines depict the fit. (Reprinted with permission of Chakraborty et al. [8] from Hindwai and ISRN.)

As a result, the nanohardness of 6.5 GPa SA was decreased by as much as 25%, while that of the 12 GPa SA was decreased by as much as 33% with respect to that of the ARA samples. Therefore, depending on the presence or absence of shock-induced microstructural defects, the alumina ceramics may reveal both mild and strong ISE (Figure 22.1b).

Figure 22.2a represents the results of fitting the present experimental load (P) versus nanoindentation depth (h) data in the loading cycle for the ARA, 6.5 GPa SA, and 12 GPa SA samples to equation (20.1) of Chapter 20 at a typical load of 500 mN. The goodness of fit was better than 0.99 for all three samples [8]. This evidence [8] had confirmed that the empirical equation (20.1) of Chapter 20

TABLE 22.1

Relative Percentages of Changes in d^2P/dh^2 versus Nanoindentation Depth (h) Plots of Figure 22.2b along AB, BC, and CD at Typical Loads of 500 mN for ARA, 6.5 GPa SA, and 12 GPa SA Samples

Load (mN)	Sample	AB (%)	BC (%)	CD (%)
500	ARA	10.39	9.42	4.37
	6.5 GPa SA	15.34	13.06	6.80
	12 GPa SA	34.38	30.55	17.24

Source: Reprinted/modified with permission of Chakraborty et al. [8] from Hindwai and ISRN.

gives a good description of the loading cycle of *P–h* plots for ARA, 6.5 GPa SA, and 12 GPa SA samples. It was also observed (Figure 22.2a) that, at a given load, the depths of penetration of both the SA samples were higher than those of the ARA samples. This fact implied that the SA samples had more inherent weakness as compared to those of the ARA sample. Further, the power law exponent (*n*) values were very slightly lesser than 2 in the case of ARA samples at different loads, whereas for SA samples it was observed that the *n* values became << 2 as the shock pressure increases. These data justify the presence of a strong ISE in the SA samples used in the present work.

In Figure 22.2b, equation (20.2) of Chapter 20 is fitted to the experimental *P–h* data obtained during loading cycles of the nanoindentation experiments on ARA, 6.5 GPa SA, and 12 GPa SA samples at a typical load of 500 mN. The relative percentages of changes in the second-order derivative of equation (20.2) of Chapter 20, i.e., d^2P/dh^2, versus depth (*h*) at different portions of the plots, i.e., along AB, BC, and CD in Figure 22.2b, are shown in Table 22.1 for both ARA as well as 6.5 GPa and 12 GPa SA samples. Further, from Figure 22.2b, it is also seen that d^2P/dh^2 decreased continuously with depth for ARA and both the SA samples, as *n* was <2. In addition, it may be observed from Table 22.1 and Figure 22.2b that at comparable intervals of the depths of penetration, i.e., at regions AB, BC, and CD, the relative percentages of changes in d^2P/dh^2 with respect to the depth *h* were much higher for both the SA samples as compared to those of the ARA samples. These data again strongly suggested that the SA samples were more compliant in nature than the ARA sample, and the 12 GPa SA sample showed much more compliance than the 6.5 GPa SA sample.

22.3 Shear Stress and Micro Pop-ins

The theoretical modeling work by Kreuzer and Pippan [1] has suggested that ISE would be more pronounced when slip bands are oriented parallel to the surface as compared to the case when there is nonparallel slip band orientation.

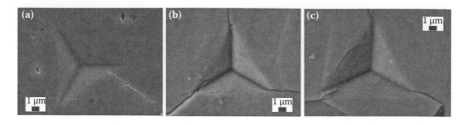

FIGURE 22.3
Typical FESEM photomicrographs of nanoindentation impressions in (a) ARA, (b) 6.5 GPa SA, and (c) 12 GPa SA samples. (Reprinted with permission of Chakraborty et al. [8] from Hindwai and ISRN.)

Further supportive evidence to such a picture emerged from the FESEM photomicrographs of the nanoindents in the SA samples, which showed the relatively predominant presence of slip lines, deformation bands, and microcracks (Figures 22.3b and 22.3c), while on a comparative scale, such features were absent in the as-received alumina (Figure 22.3a). It is interesting to note that the slip bands in the SA samples were again oriented parallel to the surface of the wall in the nanoindentation cavity (Figures 22.3b and 22.3c), as mentioned in Chapters 19–21. Therefore, based on the previously discussed theoretical modeling work [1] and the present FESEM-based evidence (Figure 22.3), a relatively stronger ISE was expected in the SA samples, as was also experimentally observed (Figure 22.1b).

The shear stress required for initiation of the nanoscale displacement jumps will have to be provided by the nanoindentation process itself. Because of the intense local stress distribution inside the nanoindentation cavity, the strain is expected to be very high, and its spatial gradient would be significantly large, especially near the base of the nanoindentation cavity. Given the knowledge of the geometric and elastic properties of the indenter and sample [2–4], the maximum shear stress (τ_{max}) generated just below the nanoindenter tip can be calculated. Thus, following the method adopted by other researchers for nanoindentation of polycrystalline alumina and sapphire [5, 6], the tip was approximated as a spherical indenter. Further, following Packard and Suh [3] and Shang et al. [4], and taking the appropriate critical load (P_c) values from the experimental data (Figure 22.4a), the maximum shear stress (τ_{max}) values active just underneath the tip of the nanoindenter were evaluated (Figure 22.4b) for the nanoindentation load range of 10–1000 mN. The maximum shear stress (τ_{max}) was ≈20–30 GPa for the ARA sample but dropped to ≈20 GPa for both of the SA samples (Figure 22.4b). Nevertheless, these data were much higher than the theoretical shear strength (τ_{theor}) of the alumina ceramic (≈3 GPa) [7]. Therefore, it is evident that the applied load provided enough shear stress to cause shear-induced flow and/or deformation as well as localized microcracking and/or microfracture in the nanoindented alumina ceramics. Thus, there was a strong possibility that the first bursts as observed in the experimental

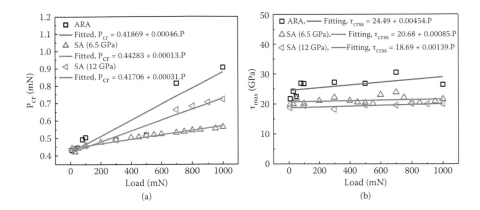

FIGURE 22.4
Variations of (a) critical load (P_c) and (b) maximum shear stress (τ_{max}) active just below the nanoindenter as a function of load (P) for ARA, 6.5 GPa SA, and 12 GPa SA samples. (Reprinted with permission of Chakraborty et al. [8] from Hindwai and ISRN.)

P–h data plots indeed corresponded to the initiation of nanoscale plasticity events in the present alumina ceramics, e.g., through the nucleation of shear bands (Figure 22.3) [2–4].

22.4 Comparison of Deformations in Alumina Shocked at 6.5 and 12 GPa

The generic mechanism of ISE observed as a function of the applied nanoindentation load in the SA samples was linked to the relative frequency of the multiple micro pop-in events that occur at the microstructural and nanostructural length scale. The most important of the factors that may control the frequency of the multiple micro pop-in events can be suggested as:

1. How the microtensile stresses are spatially distributed in the vicinity of the nanoindentation cavity

2. Relative orientation of a single grain and/or an assembly of multiple grains that encompass the microstructural damages and/or damage zones with respect to the nanoindentation loading direction

3. Relative size of the nanoindentation cavity with respect to the size distribution of the pre-existent and shock-induced damages

4. Relative orientation of the nanoindentation loading direction with respect to the spatial distribution of the pre-existent and shock-induced damages

5. Relative size of the characteristic length scale with respect to the grain thickness in the direction of travel of the nanoindenter

6. Presence or absence of residual strain due to shock loading

7. Probability of microcrack generation during loading, etc.

Therefore, the relative frequencies of the multiple micro pop-in events will actually control the relative rate of penetration by the nanoindenter in a given alumina ceramic microstructure. In turn, therefore, the relative extent of the ISE will be controlled by this relative rate of penetration by the nanoindenter.

On the other hand, FESEM photomicrographs showed confirmatory evidence of nanoscale slip band formation (denoted by white hollow arrowhead) in a suitably oriented single grain (Figure 22.5a), shear-induced intragrain microcrack formation (denoted by white solid arrowhead) in a single grain (Figure 22.5b), and extensive shear induced complex intragranular microcleavage (denoted by MC with line arrows) (Figure 22.5c) in SA samples obtained from 6.5-GPa shock experiments [8]. At a higher shock pressure of 12 GPa, the SA samples showed more FESEM-based evidence of extensive grain localized shear band formation (denoted by white hollow arrowhead) and grain boundary microcracking (denoted by white solid arrowhead) (Figures 22.5d and 22.5e) as well as very complex intragranular microcleavage (denoted by MC with line arrows) (Figure 22.5f), where individual cleavage planes are only 10–20 nm thick. All of this evidence (Figure 22.5) points toward the extraordinary importance of nanoscale plasticity, shear induced *localized* deformation, and microfracture events in the SA samples during

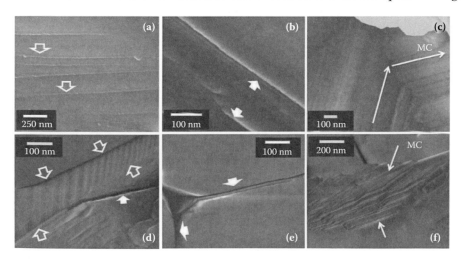

FIGURE 22.5
FESEM photomicrographs of SA samples obtained from 6.5 GPa (a, b, c) and 12 GPa (d, e, f) shock experiments. (Reprinted with permission of Chakraborty et al. [8] from Hindwai and ISRN.)

the present nanoindentation experiments. Additional support for nanoscale plasticity events, e.g., nucleation of dislocations, etc., stems from the TEM-photomicrograph-based evidence presented by Chakraborty et al. [8] for the present SA samples. The TEM-based studies reported by us in Chakraborty et al. [8] for alumina shock-deformed at 6.5 GPa and for alumina shock-deformed at 12 GPa had clearly showed evidence of:

1. Dislocations in a single alumina grain
2. Entanglement of dislocations
3. Dislocation network
4. Dislocations at an impeded grain boundary

all of which proved the extensive presence of microstructural damage in the SA samples.

22.5 Conclusions

There was more frequent presence of multiple micro pop-in events in the $P–h$ plots of the 12 GPa SA samples. This suggests that a generic mechanism was responsible. The experimental observation of ISE was the strongest in 12 GPa SA samples, stronger in the 6.5 GPa SA samples, and mild in the as-sintered alumina samples. The Nix and Gao model could successfully explain the observed ISE in all the alumina samples. Further, the generic mechanism was linked to the relative frequency of the multiple micro pop-in events at the microstructural and nanostructural length scale. Up to this point, we have focused on the nanoscale or microscale static and dynamic contact behavior of only the monolithic ceramics, e.g., alumina. The next question that naturally arises is what the situation might be with different kinds of composite materials. To this end, we present the nanoindentation behavior of both C/C and C/C-SiC composites in Chapter 23.

References

1. Kreuzer H. G. M., and R. Pippan. 2007. Discrete dislocation simulation of nanoindentation: Indentation size effect and the influence of slip band orientation. *Acta Materialia* 55:3229–35.
2. Bei, H., Z. P. Lu, and E. P. George. 2004. Theoretical strength and the onset of plasticity in bulk metallic glasses investigated by nanoindentation with a spherical indenter. *Physical Review Letters* 93:125504-1–4.

3. Packard, C. E., and C. A. Schuh. 2007. Initiation of a shear band near a stress concentration in metallic glass. *Acta Materialia* 55:5348–58.
4. Shang, H., T. Rouxel, M. Buckley, and C. Bernard. 2006. Viscoelastic behavior of a soda-lime-silica glass in the 293–833 K range by micro-indentation. *Journal of Materials Research* 21:632–38.
5. Mao, W. G., Y. G. Shen, and C. Lu. 2011. Deformation behavior and mechanical properties of polycrystalline and single crystal alumina during nanoindentation. *Scripta Materialia* 65:127–30.
6. Mao, W. G., Y. G. Shen, and C. Lu. 2011. Nanoscale elastic-plastic deformation and stress distributions of the C plane of sapphire single crystal during nanoindentation. *Journal of the European Ceramic Society* 31:1865–71.
7. Rosenberg, Z., D. Yaziv, Y. Yeshurun, and S. J. Bless. 1987. Shear strength of shock-loaded alumina as determined with longitudinal and transverse Manganin gauges. *Journal of Applied Physics* 62:1120–22.
8. Chakraborty, R., A. Dey, A. K. Mukhopadhyay, K. D. Joshi, A. Rav, A. K. Mandal, S. Bysakh, S. K. Biswas, and S. C. Gupta. 2012. Comparative study of indentation size effects in as-sintered alumina and alumina shock deformed at 6.5 and 12 GPa. *ISRN Ceramics* 2012:595172-1–11.

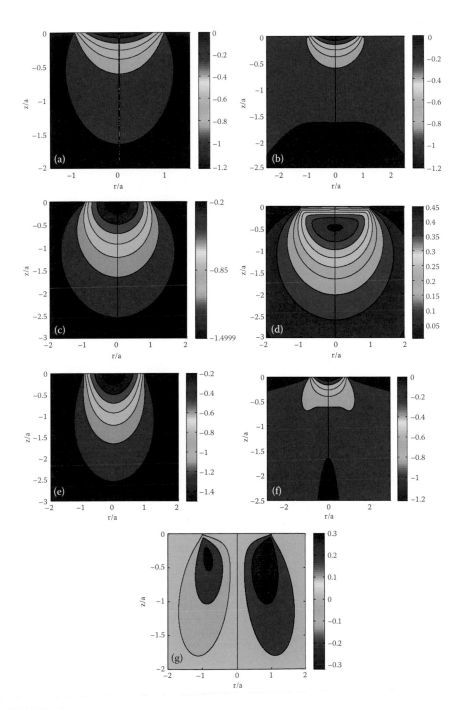

FIGURE 1.2
Normalized stress contours under a Hertzian contact for Poisson's ratio $\nu = 0.34$: (a) principal stress σ_1', (b) principal stress σ_2', (c) principal stress σ_3', (d) shear stress τ, (e) normal stress σ_z', (f) radial stress σ_r', and (g) principal stress acting on the rz plane, σ_{rz}.

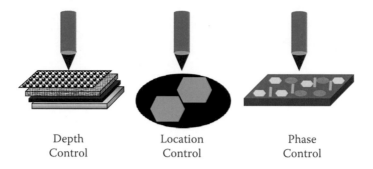

Depth Control	Location Control	Phase Control

FIGURE 5.1
Different modes of nanoindentation technique to characterize materials.

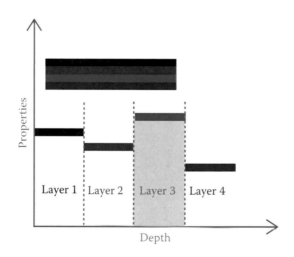

FIGURE 5.2
Schematic of variation of data at each layer in multilayer thin film utilizing CSM mode. (Adapted from Li and Bhushan [1].)

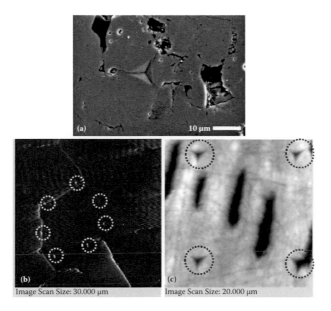

Image Scan Size: 30.000 μm   Image Scan Size: 20.000 μm

FIGURE 5.3

Location-specific nanoindentation: (a) on plasma-sprayed HAp coating showing footprint of indentation avoiding defects, (b) on a grain boundary and very near to a grain boundary region of dense alumina ceramics, and (c) very near, near, and far away from dentin tubules.

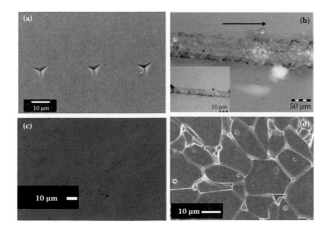

FIGURE 9.1

(a) SEM image of nanoindentation on glass. (b) Optical photomicrograph of the scratch groove made at 10 N load on glass. FESEM photomicrographs of alumina microstructure of (c) 10 μm and (d) 20 μm grain size. ([a]: Reprinted with permission of Chakraborty, R., A. Dey, and A. K. Mukhopadhyay. 2010. Role of the energy of plastic deformation and the effect of loading rate on nanohardness of soda–lime–silica glass. *Physics and Chemistry of Glasses: European Journal of Glass Science and Technology B* 51:293–303, from Society of Glass Technology and Deutsche Glastechnische Gesellschaft.)

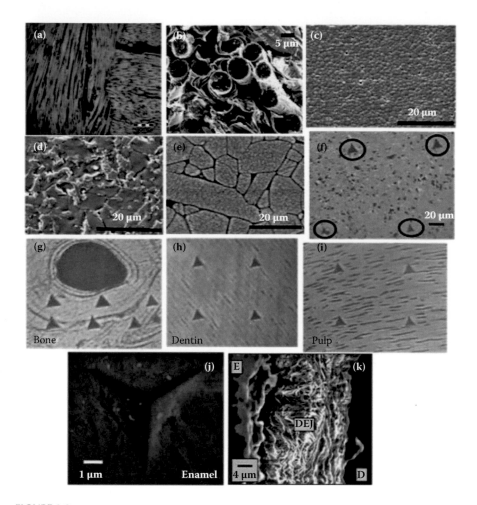

FIGURE 9.2

(a) Optical photomicrograph of plan section of carbon/carbon composite. (b) FESEM photomicrograph of cross section of carbon/carbon composite. (c) Microstructure of the sintered, polished/thermally etched surface of pure HAp (sintered at 1250°C for 2 h). (d) HAp-30wt% mullite (sintered at 1350°C for 2 h). (e) Pure mullite (sintered at 1700°C for 4 h). (f) Optical micrograph of AlN-SiC composite. (g) Optical micrograph of cross section of a human cortical bone. (h) Optical micrograph showing typical indents on human dentin. (i) Optical micrograph showing typical indents on human pulp. (j) FESEM image of a typical nanoindentation cavity on human enamel. (k) FESEM photomicrograph of human dentin–enamel junction (DEJ). ([c–e]: Reprinted with permission of Nath et al. [8] from Elsevier. [k]: Reprinted with permission from Biswas, N., A. Dey, and A. K. Mukhopadhyay. 2013. Mechanical properties of enamel nanocomposite. *ISRN Biomaterials* 2013: Article ID 253761, from Hindawi and ISRN.)

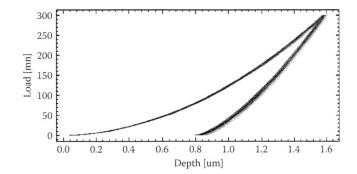

FIGURE 9.4

Load–depth (*P–h*) plots of the Schott BK-7 calibration glass block.

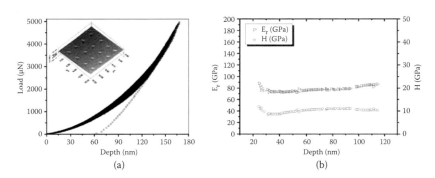

FIGURE 9.6

(a) Partial-unloading–depth plot of the standard fused-quartz sample and (b) hardness (*H*) and reduced modulus (*E*$_r$) as a function of the depth of penetration for the standard fused-quartz sample (inset: SPM image of the indentation array.)

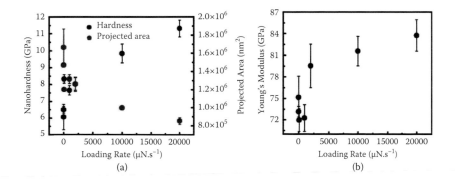

FIGURE 11.1

(a) Nanohardness and projected area of contact. (b) Young's modulus of the SLS glass as a function of loading rate. (Reprinted with permission of Dey, Chakraborty, and Mukhopadhyay [2] from The American Ceramic Society and Wiley).

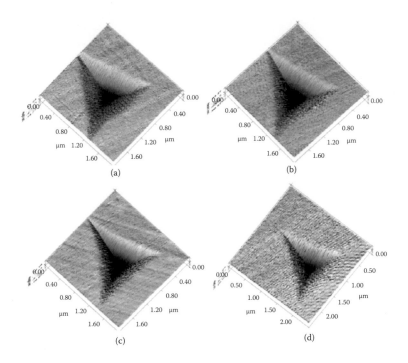

FIGURE 11.3
SPM images of nanoindentations with a constant load of 10,000 μN on SLS glass at loading rates of (a) 10, (b) 100, (c) 2,000, and (d) 20,000 μN·s⁻¹. (Reprinted with permission of Dey, Chakraborty, and Mukhopadhyay [1] from Elsevier.)

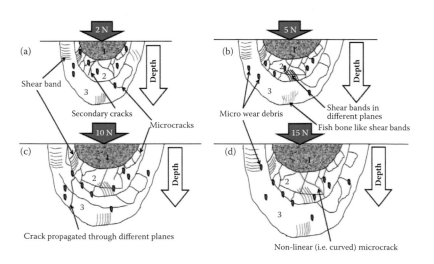

FIGURE 13.10
Schematic diagrams of the surface damage zones for (a) 2 N, (b) 5 N, (c) 10 N, and (d) 15 N. (Reprinted with permission of Bandyopadhyay and Mukhopadhyay [9] from Elsevier.)

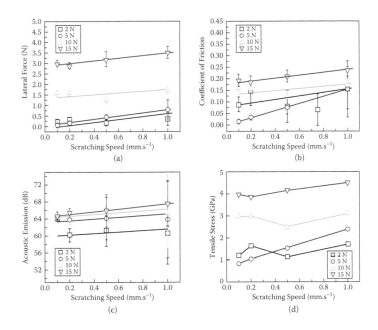

FIGURE 14.1
Variations of (a) lateral force, (b) coefficient of friction, (c) acoustic emission, and (d) tensile stress with the scratching speeds. (Reprinted with permission of Bandyopadhyay, Dey, and Mukhopadhyay [9] from the American Ceramic Society and Wiley.)

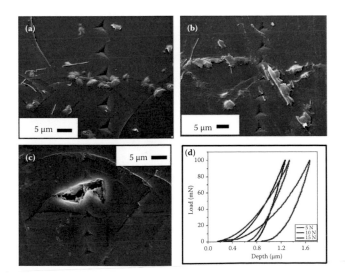

FIGURE 15.2
The nanoindentation conducted through the edge of the scratch grooves for (a) 5N, (b) 10 N, and (c) 15 N; (d) load–depth plots of the indent made at the middle of the damage zone. (Reprinted with permission of Bandyopadhyay et al. [9] from Springer.)

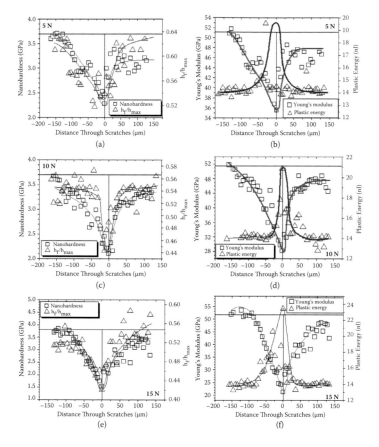

FIGURE 15.3

Plots of nanohardness and elastic recovery (h_f/h_{max}) (a, c, e) and Young's modulus and plastic energy with distance through scratches (b, d, f) for applied loads of 5 N (a, b), 10 N (c, d), and 15 N (e, f). (Reprinted with permission of Bandyopadhyay et al. [9] from Springer.)

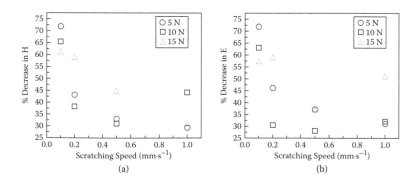

FIGURE 15.4

Percentage decrease in (a) nanohardness and (b) Young's modulus. (Reprinted with permission of Bandyopadhyay, Dey, and Mukhopadhyay [12] from the American Ceramic Society and Wiley.)

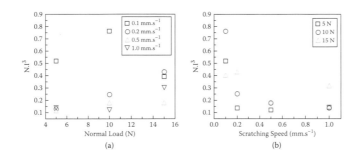

FIGURE 15.5
Variations of microcrack population with (a) applied normal load (b) scratching speed. (Reprinted with permission of Bandyopadhyay, Dey, and Mukhopadhyay [12] from the American Ceramic Society and Wiley.)

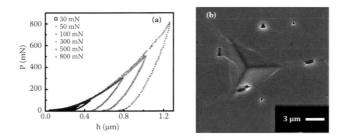

FIGURE 16.2
(a) Load–depth (P–h) plots for various nanoindentation loads and (b) SEM photomicrograph of the residual impression of a typical nanoindent at 10^6 μN load in alumina. (Reprinted with permission of Chakraborty et al. [3] from The Institute of Engineers, India.)

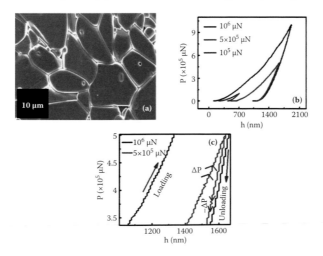

FIGURE 17.1
(a) Surface microstructure of dense, coarse-grain alumina, (b) P–h plots at three different loads of 10^5, 5×10^5, and 10^6 μN, and (c) exploded view of P–h plots. (Reprinted with permission of Bhattacharya et al. [1] from Elsevier.)

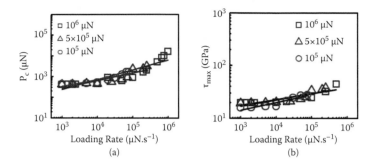

FIGURE 17.2
Variations of (a) P_c and (b) τ_{max} with loading rate or $\dot{p}$. (Reprinted with permission of Bhattacharya et al. [1] from Elsevier.)

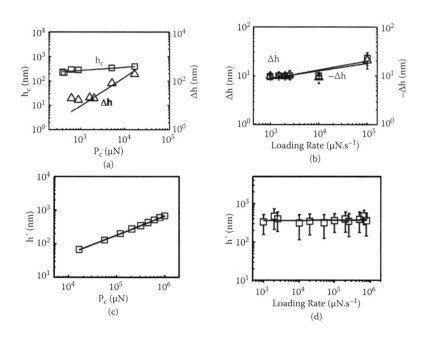

FIGURE 17.4
Power law dependencies of (a) h_c and Δh on P_c, (b) Δh and $-\Delta h$ on Loading Rate, (c) h' on P_c, and (d) h' on Loading Rate. (Reprinted with permission of Bhattacharya et al. [1] from Elsevier.)

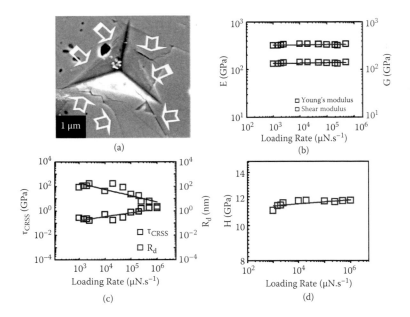

(a)

(b)

(c)

(d)

FIGURE 18.2
(a) FESEM photomicrograph showing shear deformations and microcracks in the nanoindentation cavity. Variations of (b) E and G, (c) τ_{CRSS} and R_d, and (d) H with loading rates. (Reprinted with permission of Bhattacharya and Mukhopadhyay [1] from Hindawi and ISRN.)

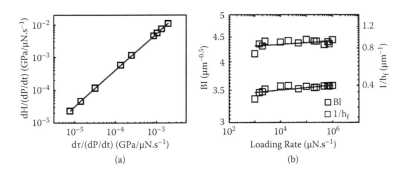

(a)

(b)

FIGURE 18.3
Variations of (a) the rates of change of hardness and maximum shear stress and (b) BI and $(1/h_f)$ with the loading rates. (Reprinted with permission of Bhattacharya and Mukhopadhyay [1] from Hindawi and ISRN.)

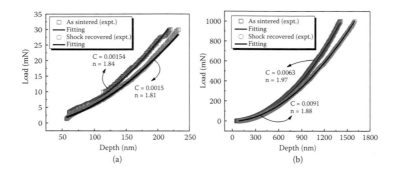

FIGURE 20.2

Fitting of the empirical equation $P = Ch^n$ in load-versus-depth data of loading cycle during nanoindentation at peak loads of (a) 30 mN and (b) 1000 mN for shock-recovered and as-sintered alumina. (Reprinted with permission of Chakraborty et al. [7] from Springer.)

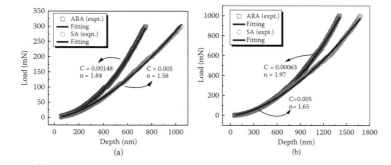

FIGURE 21.2

Fitting of the empirical equation $P = Ch^n$ to the load (P) versus nanoindentation depth (h) data of loading cycle for the as-received and shocked alumina samples during nanoindentation at two different peak loads of (a) 300 mN and (b) 1000 mN. (Reprinted with permission of Chakraborty et al. [6] from Elsevier.)

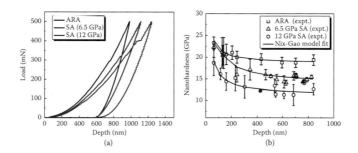

FIGURE 22.1

(a) Load–depth plots at a typical illustrative load of 500 mN and (b) nanohardness (H) versus depth (h) plots for ARA, 6.5 GPa SA, and 12 GPa SA samples. The solid lines depict the predicted variation according to the Nix and Gao model. (Reprinted with permission of Chakraborty et al. [8] from Hindwai and ISRN.)

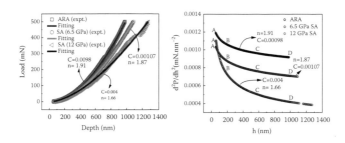

FIGURE 22.2
(a) Fitting of the empirical equation $P = Ch^n$ to the load (P) versus depth (h) data of loading cycles during nanoindentations and (b) variation of d^2P/dh^2 as a function of depth (h) for ARA, 6.5 GPa SA, and 12 GPa SA samples at typical load of 500 mN. The solid lines depict the fit. (Reprinted with permission of Chakraborty et al. [8] from Hindwai and ISRN.)

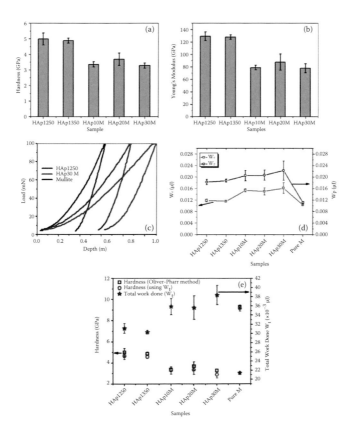

FIGURE 25.2
Phase and composition dependencies of (a) hardness and (b) Young's modulus, (c) $P–h$ plots, (d) elastic and plastic works of nanoindentation, and (e) estimated hardness (Oliver-Pharr method) and the total work done (Wt) during nanoindentation of HAp, mullite, and HAp-mullite composites. (Modified with permission of Nath et al. [11] from Elsevier.)

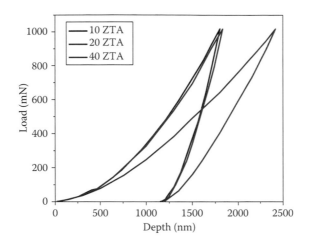

FIGURE 27.3
P–h plots on different ZTA composites.

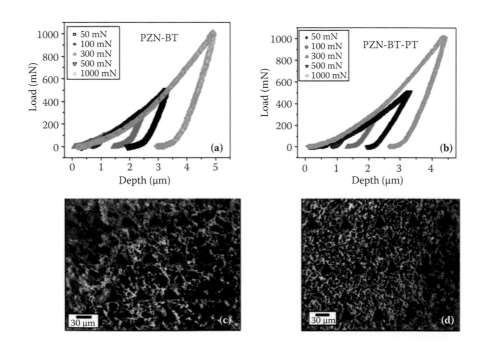

FIGURE 28.1
P–h plots and nanoindentation arrays for (a, c) PZN-BT samples and (b, d) PZN-BT-PT samples. (Reprinted/modified with permission of Himanshu et al. [2] from Council of Scientific & Industrial Research, New Delhi.)

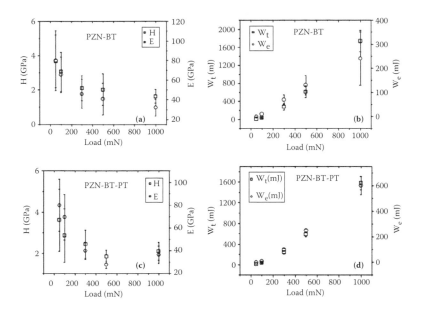

FIGURE 28.2
E, H plots and W_e, W_t plots for (a, b) PZN-BT samples and (c, d) PZN-BT-PT samples. (Reprinted/modified with permission of Himanshu et al. [2] from Council of Scientific & Industrial Research, New Delhi.)

FIGURE 30.1
Anode in (a) pre-reduced, (b) post-reduced, and (c) cross section of dense 8YSZ electrolyte layer; (d) nanoporous LSM cathode layer.

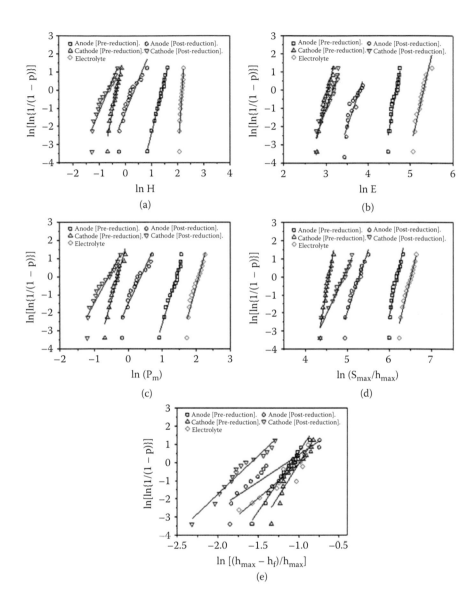

FIGURE 30.3
Weibull distribution plots for: (a) nanohardness, (b) Young's modulus, (c) mean contact pressure, (d) relative stiffness, and (e) relative spring-back data evaluated from nanoindentation experiments performed on different component layers of the single cell in both pre- and post-reduced conditions.

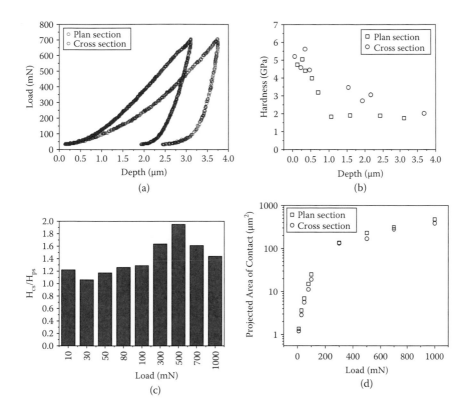

FIGURE 34.1
(a) $P-h$ plots of plan and cross sections. Load dependencies of (b) nanohardness, (c) anisotropy factor, and (d) contacts area. (Reprinted/modified with permission of Dey and Mukhopadhyay [19] from Institute of Materials, Minerals and Mining and Maney.)

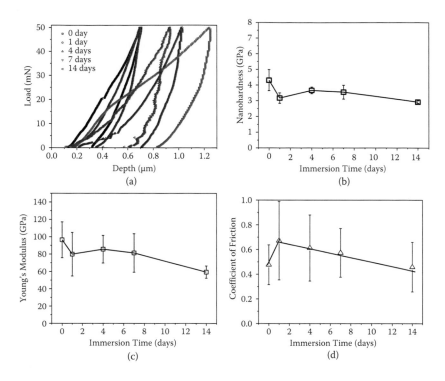

FIGURE 36.1
Effect of immersion time on (a) *P–h* data, (b) nanohardness, (c) Young's modulus, and (d) friction coefficient, µ. (Reprinted with permission of Dey and Mukhopadhyay [7] from the American Ceramic Society and Wiley.)

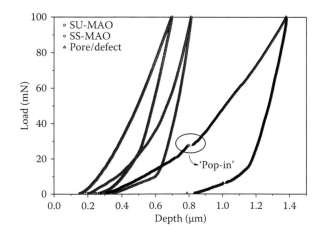

FIGURE 37.1
Typical load–depth (*P–h*) plots of the various MAO coatings. (Reprinted with permission of Dey et al. [5] from Elsevier.)

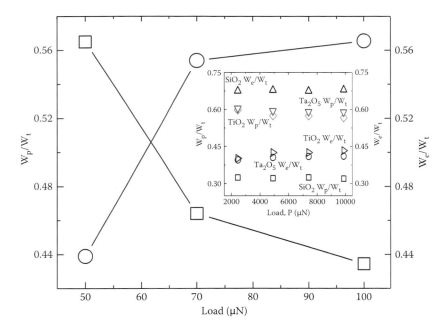

FIGURE 38.2
Energy ratio data as a function of loads. Inset: Literature data [3]. (Reprinted with permission of Das et al. [4] from Institute of Materials, Minerals and Mining and Maney.)

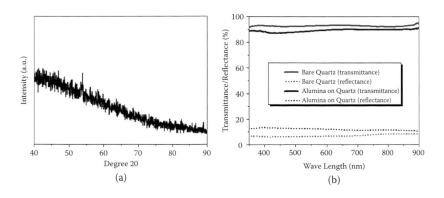

FIGURE 40.1
(a) XRD patterns of alumina films on SS304 and (b) transmittance and reflectance spectra of alumina film deposited on quartz and bare quartz. (Reprinted/modified with permission of Reddy et al. [1] from Elsevier.)

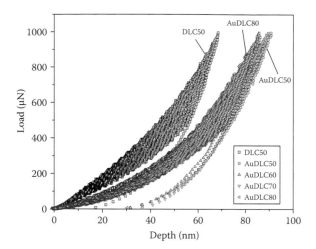

FIGURE 41.1
Typical partial-unloading load-depth plots of AuDLC and pure DLC.

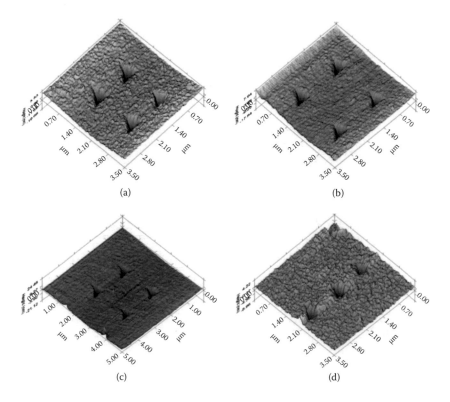

FIGURE 41.2
SPM images of the impression of the indention array: (a) AuDLC50, (b) AuDLC60, (c) AuDLC70, and (d) DLC50.

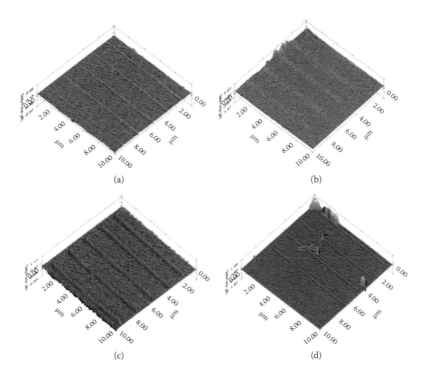

FIGURE 41.4
SPM images of array of nanoscratches: (a) AuDLC50, (b) AuDLC60, (c) AuDLC70, and (d) DLC50.

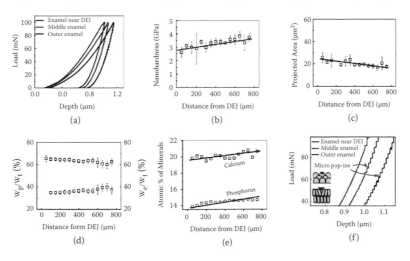

FIGURE 42.1
(a) Load–depth (*P–h*) plots at different regions (outer, middle, and inner) of enamel. Variations in (b) nanohardness; (c) contact area; (d) energy ratios; (e) Ca%, P% as functions of distance from DEJ; and (f) an enlarged view of the *P–h* plots shown in Figure 42.1a. Upper and lower insets show schematically the situation of indenter-microstructure interaction. (Reprinted/modified with permission of Biswas et al. [8] from The Institution of Engineers (India) and Springer.)

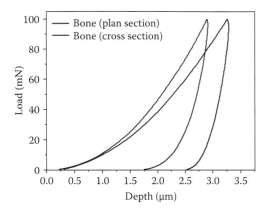

FIGURE 44.5
Typical *P–h* plots on both cross section and plan section of cortical bone sample.

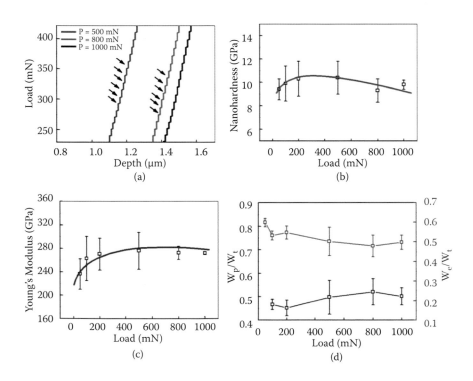

FIGURE 46.2
The presence of both ISE and RISE depending upon load in AlN-SiC composites: variations of
(a) load–depth (*P–h*) plot, (b) nanohardness, (c) Young's modulus, and (d) ratio of energies spent
in elastic and plastic deformations—all as functions of the applied nanoindentation loads.

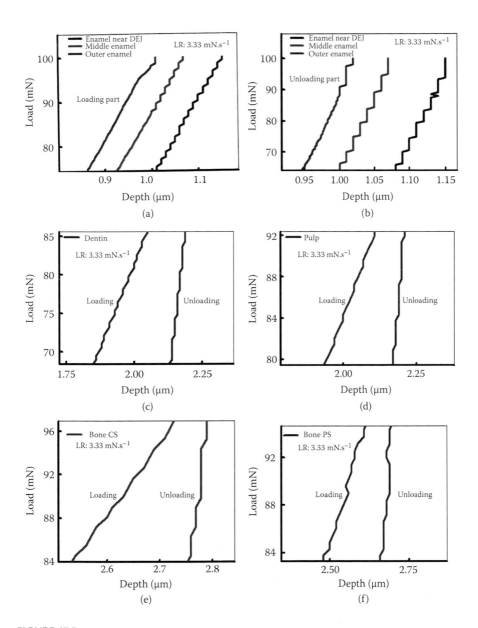

FIGURE 47.5
Examples of micro pop-ins and pop-outs occurring in load–depth (P–h) plots for outer and inner enamel and dentin–enamel junction regions during (a) loading and (b) unloading; for loading and unloading cycles in (c) dentin and (d) pulp regions of human teeth; and also for (e) cross and (f) plan sections of human cortical bones.

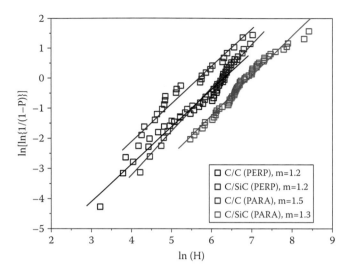

FIGURE 50.2
Typical examples of Weibull statistical fittings for the nanohardness (*H*) data on C/C and
C/SiC composite with fiber orientation of perpendicular and parallel fashions. (Reprinted
(modified) with permission from Sarkar et al. [11].)

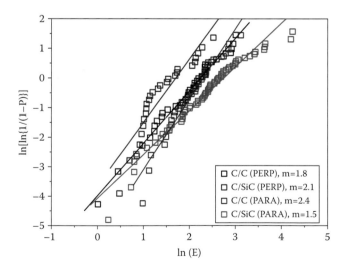

FIGURE 50.3
Typical examples of Weibull statistical fittings for the Young's modulus (*E*) data on C/C and
C/SiC composite with fiber orientation of perpendicular and parallel fashions. (Reprinted
(modified) with permission from Sarkar et al. [11].)

Section 7

Nanoindentation Behavior of Ceramic-Based Composites

23

Nano-/Micromechanical Properties of C/C and C/C-SiC Composites

Soumya Sarkar, Arjun Dey, Probal Kumar Das, Anil Kumar, and Anoop Kumar Mukhopadhyay

23.1 Introduction

There is not much reported work [1–3] on the nano-/micromechanical properties of C/C composites. Usually, the carbon matrix had Young's modulus at least three times lower than that of the carbon fiber, and the two-dimensional (2-D) C/C composite had Young's modulus, nearly two to three times lower than those of the three-dimensional (3-D) C/C composites [2, 3]. It is in this context that this chapter focuses on the nanoindentation behavior of the 2-D C/C as well as the C/C-SiC composites using a Berkovich indenter. The relevant processing details of these composites have been given in Chapter 9. We shall evaluate here the nanohardness, Young's modulus, relative springback, and the relative stiffness of these composites. In addition, we shall try to extract from the experimental data the amount of elastic and plastic energies associated with the nanoindentation processes in these composites. Finally, we shall try to understand how the corresponding microstructural aspects of the composites affect their micromechanical properties. The 2-D C/C composites were prepared with PAN-based woven C-fabric and a graphitic matrix. Some of these were converted to the C/C-SiC composites through the standard LSI method at 1600°C under vacuum [4, 5].

23.2 Nanoindentation Behavior

The density of the base C/C composites ($\approx$1.6–1.8 g·cc^{-1}) was about 40% lower than those of the corresponding C/C-SiC composites (2.2–2.4 g·cc^{-1}) [4, 5]. That is possibly why the flexural strength ($\approx$170 MPa) of the C/C-SiC composites was more than twice as high as those ($\approx$80 MPa) of the C/C composites.

Two loading conditions were made in the nanoindentation experiments [5]. One was that the loading directions were parallel to the direction of C-fabric stacking; the other was that the loading directions were perpendicular to the direction of C-fabric stacking. A typical low load of only 10 mN was used in all the experiments. In the C/C composites, the final depth of penetration in the perpendicular direction of C-fabric stacking was ≈40% higher than that measured in the parallel direction of loading (Figures 23.1 and 23.2). Further, under comparable loading conditions, these composites had final depths of penetration much higher than those of the corresponding C/C-SiC composites. When loaded in a direction parallel to that of the C-fabric stacking, the C/C-SiC composite showed the highest elastic recovery (≈75%–97%) and the least permanent deformation (≈5–10 nm) (Figure 23.1). Therefore, it is not surprising that the C/C-SiC composite had the highest nanohardness (≈1.14 ± 0.84 GPa) under the aforesaid loading condition. But there was a drastic reduction of nanohardness to as low as 0.30 ± 0.22 GPa when loaded in a direction perpendicular to that of the C-fabric stacking [5]. The situation was not that bad in the case of the C/C composites,

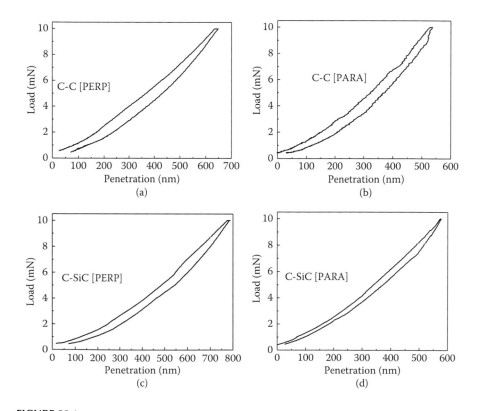

FIGURE 23.1
(a-d): Typical *P–h* plots for C/C and C/C-SiC structures. (Reprinted with permission of Sarkar et al. [5] from the American Ceramic Society and Wiley.)

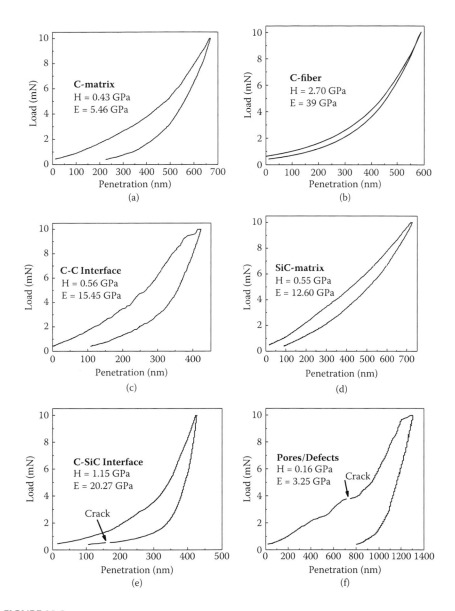

FIGURE 23.2

(a-f): Typical *P–h* plots of various phases of the two composites. (Reprinted with permission of Sarkar et al. [5] from the American Ceramic Society and Wiley.)

which showed nanohardness of 0.50 ± 0.30 GPa and 0.45 ± 0.29 GPa, respectively, when loaded in directions parallel and perpendicular to those of the C-fabric stacking [5]. Evidently, the C/C-SiC composites had nanohardness more than twice as high as those of the C/C composites when loaded in directions parallel to those of the C-fabric stacking.

It is interesting to note that the C/C-SiC composites had Young's modulus of 14.7 ± 13.3 GPa and 5.1 ± 2.9 GPa, respectively, when loaded in directions parallel and perpendicular to the directions of C-fabric stacking. The C/C composites, on the other hand, had Young's modulus of 8.4 ± 5.0 GPa and 8.9 ± 4.6 GPa, respectively, when loaded in directions parallel and perpendicular to the directions of C-fabric stacking [5]. Thus, there was anisotropy in the nanomechanical properties of both the C/C and the C/C-SiC composites. However, the extent of anisotropy was much more pronounced in the case of the C/C-SiC composite. The scatter in data was characteristically very high in the case of the C/C composites because the porosity was very high, about 13%–17% [4, 5]. However, the same logic would not hold for the C/C-SiC composites, which had only about 2%–3% porosity [4, 5].

The large scatter in the measured data only reflects the facts that the nanomechanical properties of both composites have a significant contribution from each of their constituents, like the C-fiber, C-matrix, SiC-matrix, microstructural interfaces, and pores or other defects like surface irregularities, valleys, and hills [2]. The present results compared reasonably with the published literature [1–3, 6]. However, comparison of the results obtained for the C/C-SiC composites was not possible in the absence of reported data in the open literature. The reliability of the nanoindentation data, especially in such heterogeneous microstructures, deserves dedicated attention. This had been dealt with in the application of a weakest link statistical distribution [5] to come up with the more reliable "characteristic values" of a given micro-/nanomechanical property. Further details of the "reliability issues" data are dealt with in Chapter 50. Nevertheless, the characteristic values of both relative spring-back and relative stiffness parameters of the C/C-SiC composites (0.77 and 69.6 GPa) were higher than those of the C/C composites (0.57 and 61.7 GPa) when measured in a direction parallel to the direction of C-fabric stacking [5]. But when measured in a direction perpendicular to the direction of C-fabric stacking, the situation was reversed, i.e., the relative spring-back and stiffness parameters of the C/C-SiC composites (0.66 and 28.3 GPa) were either equal to or lower than those of the C/C composites (0.64 and 49.6 GPa) [5]. These data also further justify why higher amounts of elastic recovery upon unloading happened in the C/C-SiC composites (Figures 23.1 and 23.2) that had initially been loaded in a direction parallel to that of the C-fabric stacking.

23.3 Energy Calculation

The data on the ratios of the elastic energy (W_e) and the plastic deformation work (W_p) spent during the nanoindentation experiments are given in Table 23.1. It is interesting to note that the elastic energy was always higher

TABLE 23.1

Ratios of Elastic-to-Total (W_e/W_t), Plastic-to-Total (W_p/W_t), and Elastic-to-Plastic (W_e/W_p) Deformation Energies of the C/C and C/C-SiC Composites

Sample (loading direction)	W_e/W_t	W_p/W_t	$W_e{:}W_p$
C/C ($\perp$)	0.79	0.21	3.76
C/C ($\parallel$)	0.64	0.36	1.77
C/C-SiC ($\perp$)	0.77	0.23	3.34
C/C-SiC ($\parallel$)	0.94	0.06	15.66

Source: Reprinted with permission of Sarkar et al. [5] from the American Ceramic Society and Wiley.

than that of the plastic deformation work. These data indicated that there were relatively lower amounts of permanent deformation of the composites when the test load was completely removed. As such, W_e was ≈95% of the total work for the C/C-SiC composites when loaded in the direction parallel to that of the C-fabric stacking. This was the highest value of W_e. This most likely happened due to the maximum contribution of highly elastic C-fibers in this direction with lowest involvement of pores or other surface irregularities that could have otherwise contributed to the higher values of the residual depth of penetration. Furthermore, when the C/C-SiC composites were loaded in the direction parallel to that of the C-fabric stacking, the ratio of $W_e{:}W_p$ was found to be ≈5 times higher than that of the C/C-SiC loaded in the direction perpendicular to that of the C-fabric stacking. This also indicated the strong anisotropic nature of the C/C-SiC structures. Thus, the results obtained from the energy partition scenario were in complete conformity with the results obtained for the relative spring-back abilities of the composites.

23.4 Conclusions

The average nanohardness, Young's modulus, and the extent of anisotropy of the aforesaid properties were measured to be the highest for the LSI-derived 2-D C/C-SiC composites along the direction parallel to that of the C-fabric stacking. The critical examination of the relative stiffness and relative spring-back data explained why the extents of anisotropy in nanomechanical properties and elastic recovery were the highest in the C/C-SiC composites compared to those of the C/C composites. The characteristic high scatter in the experimentally measured nanoindentation data was possibly linked to the characteristic heterogeneous microstructure of the composites. In Chapter 24, we learn about the nanoindentation behavior of tape-cast, multilayered ceramic matrix composites. These composites are different types of ceramic matrix composites, where individual layers are tuned up to provide damage tolerance to the overall microstructure.

References

1. Kanari, M., K. Tanaka, S. Baba, M. Eto, and K. Nakamura. 1997. Nanoindentation test on electron beam-irradiated boride layer of carbon-carbon composite for plasma facing component of large Tokamak device. *Journal of Nuclear Materials* 244:168–72.

2. Kanari, M., K. Tanaka, S. Baba, and M. Eto. 1997. Nanoindentation behavior of a two-dimensional carbon-carbon composite for nuclear applications. *Carbon* 35:1429–37.

3. Diss, P., J. Lamon, L. Carpentier, J. L. Loubet, and P. Kapsa. 2002. Sharp indentation behavior of carbon/carbon composites and varieties of carbon. *Carbon* 40:2567–79.

4. Sarkar, S., A. Dey, P. K. Das, A. Kumar, and A. K. Mukhopadhyay. 2007. Depth-sensitive indentation behaviour of C-C and C/C-SiC composites. Paper presented at International Conference on High Temperature Ceramic Matrix Composites (HTCMC-06), New Delhi, India.

5. Sarkar, S., A. Dey, P. K. Das, A. Kumar, and A. K. Mukhopadhyay. 2011. Evaluation of micromechanical properties of carbon/carbon and carbon/carbon-silicon carbide composites at ultra-low load. *International Journal of Applied Ceramic Technology* 8:282–97.

6. Field, J. S., and M. V. Swain. 1996. The indentation characterisation of the mechanical properties of various carbon materials: Glassy carbon, coke and pyrolytic graphite. *Carbon* 34:1357–66.

24

Nanoindentation on Multilayered Ceramic Matrix Composites

Sadanand Sarapure, Arnab Sinha, Arjun Dey,
and Anoop Kumar Mukhopadhyay

24.1 Introduction

The effectiveness of the nanoindentation technique in understanding the mechanical properties of a fiber-reinforced composite at the local microstructural length scale has been established with ample credence in Chapter 23. Now we shall be discussing the nanoindentation study on the multilayer ceramics matrix composites (MLCC), mainly developed by the tape casting method. The MLCC materials are capable of sustaining enhanced strain to failure, i.e., they show damage tolerance [1–4]. In such composites, the sequential failure of each layer is usually accompanied by a load drop, i.e., stress decrement [1–3]. The crack propagation can take place along the interfacial layer, which should be a low-failure-energy material called the *weak interface*. This interface could be weak with a low-strength, low-toughness material such as graphite; porous materials; or a low-strength but more damage-tolerant plastic material such as glass-fiber-reinforced polymer (GFRP) or carbon-fiber-reinforced polymer (CFRP) [1–4]. The interface can also contain a soft, easily deformable phase like $MoSi_2$. Lanthanum phosphate (LP) is reported to form a weak interface with alumina, zirconia, as well as zirconia-toughened alumina (ZTA) [1, 2]. Crack deflection is then expected to take place at the weak interface, leading to an increase in crack path and, hence, an enhancement in the failure energy. Our point of interest here is on the nanoindentation responses of the individual component layers of the tape-cast, sintered alumina/lanthanum phosphate/alumina (A/LP/A) MLCC architecture, which had shown a damage-tolerant characteristic [1–3]. A typical photomicrograph of the MLCC shows an interfacial crack growth during a high-load Vickers microindentation (Figure 24.1). In fact, the particulate A/LP composites have been developed to achieve machinable ceramics [4–7]. However, we have found [1, 2] that nanoindentation studies of the individual elements of tape-cast A/LP/A MLCC are yet to be explored.

FIGURE 24.1
A/LP MLCC showed a crack bifurcation at a weak interface in the LP zone.

In the present case, the lanthanum phosphate and alumina powders were used to fabricate LP and alumina tapes employing the conventional tape casting technique. To make the MLCC, the tapes were placed in the A/LP/A architectural sequence. Then the tapes were laminated and subsequently pressureless-sintered in air at the desired temperatures. Individual lanthanum phosphate (LP) and alumina tapes were also cast and subsequently pressureless-sintered in air [1, 2]. A load range of 10–1000 mN was employed in the present study using a Vickers nanoindentation tip. The nanoindentations were done on the alumina and lanthanum phosphate tapes, which were only about 200–300 microns thick. The individual tapes were of about 10 × 10 × 0.2–0.3 mm in dimension. This is a unique example, where the nanoindentation technique was the *only* technique available to us for the evaluation of the mechanical properties like hardness and Young's modulus.

24.2 Nanomechanical Behavior

24.2.1 Nanoindentation on Lanthanum Phosphate Tape

Figures 24.2a–c show the nanoindentation array in LP at different loads, i.e., 500, 700, and 1000 mN. Typical $P–h$ plots for both low and high loads are shown in Figures 24.3a and 24.3b. At a higher load of 1000 mN, a large pop-in was observed, as shown in Figure 24.3b, plausibly due to microcracking. Further, the data on variation of both the nanohardness (H)

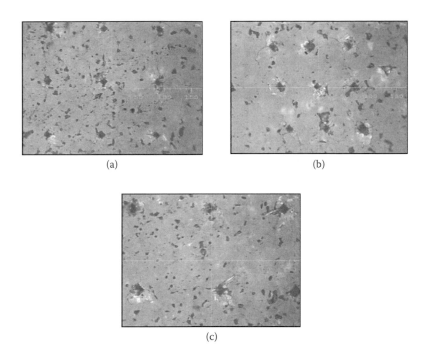

FIGURE 24.2
Nanoindentation array in LP tape at different loads: (a) 500 mN, (b) 700 mN, and (c) 1000 mN.

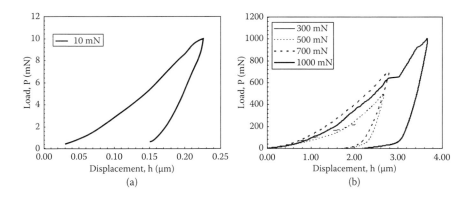

FIGURE 24.3
P–h plots on sintered LP tapes at (a) low load of 10 mN and (b) high loads of 300, 500, 700, and 1000 mN.

and the Young's modulus (*E*) as functions of the nanoindentation load are shown in Figure 24.4. Each of the data points represents an average of at least 15 measurements. The nanohardness was found to be ≈5.3 GPa at 10 mN load, while at 1000 mN load the hardness dropped to ≈3 GPa. Therefore, a strong indentation size effect (ISE) has been observed in the present data.

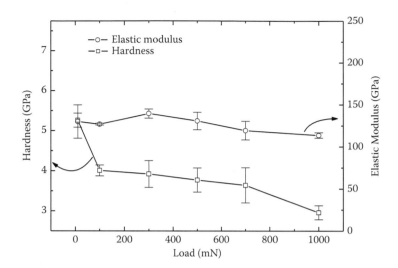

FIGURE 24.4
Nanohardness and Young's modulus for sintered alumina tapes as a function of the nanoindentation load.

The origin of ISE is discussed in detail in Chapter 46. However, the Young's
modulus was insensitive to the nanoindentation load, and the average value
was ≈127 GPa.

In comparison, the A/LP particulate composite is reported to have a Vickers
hardness of ≈1.7 GPa to 10 GPa as the sintering temperature is increased from
1300°C to 1600°C [6]. Further, the hardness value of the same composite was
also sensitive to the volume percentages of the LP contents in the composites [5, 7]. The Vickers hardness of bulk LP at higher sintering temperature
(1600°C) showed a value of ≈5 GPa, while at lower sintering temperature
(1450°C) it showed a value of ≈4.5 GPa [5–7]. The present value of nanohardness measured for the tape-cast LP matched well with the literature data
[5–7] on bulk LP. On the other hand, the Young's modulus of bulk dense LP
was reported to be about 208 GPa [5]. Considering that the residual porosity
was about 16% in the tape-cast material due to the comparatively lower sintering temperature and the pressureless sintering process used in the present
work [1, 2], the predicted Young's modulus would be about 125 GPa [3], which
was a close match with the experimentally measured data of about 127 GPa.

24.2.2 Nanoindentation on Alumina Tape

Further, Figures 24.5a–c show the nanoindentation array in the alumina tape
at different nanoindentation loads of 100, 300, and 1000 mN. The indentation
projected area was larger at higher loads, as expected. The corresponding
typical *P–h* plots are shown in Figures 24.6a and 24.6b for both low- and

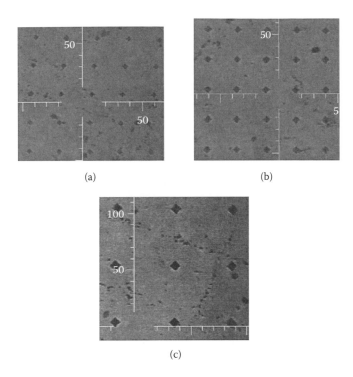

FIGURE 24.5
Nanoindentation array in sintered alumina tapes at different loads: (a) 100 mN, (b) 300 mN, and (c) 1000 mN.

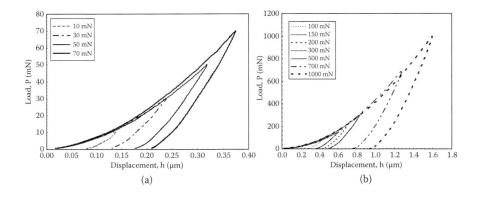

FIGURE 24.6
P–h plots on sintered alumina tapes at (a) low loads of 10–70 mN and (b) high loads of 100–1000 mN.

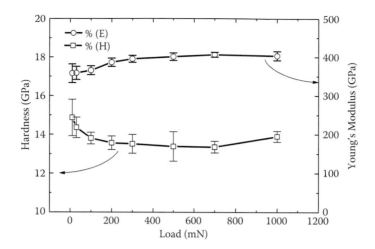

FIGURE 24.7
Nanohardness and Young's modulus for sintered alumina tapes as a function of the nanoindentation load.

high-load regimes, respectively. Further, like the case of the LP tape, the variations of H and E as a function of the nanoindentation load are shown in Figure 24.7. The hardness was found to be ≈15 GPa at 10 mN load, while the hardness at 1000 mN load was obtained as ≈14 GPa. Therefore, a very mild ISE has been observed in the present data on the tape-cast alumina samples. However, in the case of LP tape, the ISE was definitely much stronger. Here also, the Young's modulus was insensitive to the nanoindentation load, and the average value was ≈384 GPa, which is almost as good as those reported by Anstis et al. [7] and Krell and Blank [8] for fine-grain, bulk alumina ceramics.

24.3 Conclusions

Nanoindentation behavior was explained for tape-cast, pressureless-sintered lanthanum phosphate and alumina. These tapes were used for fabricating multilayered ceramic matrix composites. Here the deliberate design was to interface the hard alumina layers with a weak lanthanum phosphate phase. The main idea in designing such an architecture is to deflect the crack through the weak lanthanum phosphate layer, which ultimately will show a high fracture toughness and failure energy. In this fashion, we can develop a damage-tolerant ceramic composite structure that may find further use for structural as well as strategic applications. However, our main interest here was to show how the nanoindentation technique can be used to measure the nanomechanical properties of 200–300 µm thin tapes made of alumina and lanthanum phosphate. To the best of our knowledge, this is the very first report on the nanomechanical properties of pressureless-sintered alumina and lanthanum phosphate tapes. A further extension along this line will be to evaluate the nanoindentation response of a particulate ceramic matrix composite. Therefore, in Chapter 25 we shall try to understand the nanoindentation behavior of hydroxyapatite-based composites.

References

1. Sinha, A. 2005. Preparation and characterization of ceramic nanocomposites. M.Tech. dissertation, Bengal Engineering and Science University.
2. Sarapure, S. 2005. Ceramic nano composites for structural applications. M.Tech. dissertation, Motilal Nehru National Institute of Technology, Allahabad.
3. Dey, A. 2007. Toughening of ceramic matrix composites. M.Tech. dissertation, Bengal Engineering and Science University.
4. Chawla, K. K., H. Liu, J. Janczak-Rusch, and S. Sambasivan. 2000. Microstructure and properties of monazite ($LaPO_4$) coated saphikon fiber/alumina matrix composites. *Journal of the European Ceramic Society* 20:551–59.
5. Wang, R., W. Pan, J. Chen, M. Fang, M. Jiang, and Z. Cao. 2003. Microstructure and mechanical properties of machinable $Al_2O_3/LaPO_4$ composites by hot pressing. *Ceramics International* 29:83–89.
6. Wang, R., W. Pan, J. Chen, M. Jiang, Y. Luo, and M. Fang. 2003. Properties and microstructure of machinable $Al_2O_3/LaPO_4$ ceramic composites. *Ceramics International* 29:19–25.
7. Anstis, G. R., P. Chantikul, B. R. Lawn, and D. B. Marshall. 1981. A critical evaluation of indentation techniques for measuring fracture toughness; I: Direct crack measurements. *Journal of the American Ceramic Society* 64:533–38.
8. Krell, A., and P. Blank. 1995. Grain size dependence of hardness in dense submicrometer alumina. *Journal of the American Ceramic Society* 78:1118–20.

25

Nanoindentation of Hydroxyapatite-Based Biocomposites

Shekhar Nath, Arjun Dey, Prafulla K. Mallik,
Bikramjit Basu, and Anoop Kumar Mukhopadhyay

25.1 Introduction

One of the key issues identified in the failure of orthopedic implants has been the problem of insufficient tissue regeneration (i.e., lack of bioactivity) around the biomaterial immediately after implantation. This has motivated researchers to develop various bioactive composites with bone-mimicking properties for faster bone regeneration as well as to eliminate the problem of elastic modulus mismatch and stress shielding of the implant. The composite approach provides us the flexibility to manipulate such properties as strength, electrical conductivity, and modulus of the composites close to that of natural bone with the addition of second phases [1–3]. Hydroxyapatite (HAp) is the most well-known bioactive ceramic that has evolved in the most significant manner during the last few decades. Despite good bioactivity properties, the inherent brittleness of HAp has triggered widespread research activities to enhance fracture toughness by strengthening HAp with various reinforcements, like alumina and zirconia [4, 5]. Despite many years of research, the toughness of HAp composites could be enhanced to around 5 MPa·m$^{0.5}$ only lately in HAp-Ti systems [6]. However, in many of these cases [1–5], the concern regarding the elastic modulus mismatch resulting in stress shielding was not addressed or solved.

Besides the conventional measurement of flexural strength or Vickers hardness, the nanoindentation technique has been used to probe nanomechanical properties of HAp-based materials. This chapter presents a summary of the results obtained in our research group in HAp-based composites with two different reinforcements: calcium titanate, $CaTiO_3$ (CT) [7], and mullite, $3Al_2O_3 \cdot 2SiO_2$ (M) [8]. The HAp-calcium titanate (40–80 wt% calcium titanate with HAp) composites as well as the pure HAp and calcium titanate were prepared by an optimized, multistage spark plasma sintering (SPS) process [7], while the phase pure HAp, mullite, and HAp-mullite (10–30 wt%

mullite with HAp) composites were prepared by pressureless sintering [8]. The nanoindentations in the HAp-CT composites [9, 10] were done at an ultralow load of 6,000 μN, while those on the pressureless sintered HAp-mullite composites were done at a low load of 100 mN [11]. The further experimental details about the processing of both the HAp-CT and HAp-mullite composites have already been given in Chapter 9.

25.2 HAp-Calcium Titanate Composite

In a HAp-CaTiO$_3$ system, both the phases were predominantly retained. The absence of α- and β-tricalcium phosphate, Ca$_3$(PO$_4$)$_2$ (TCP), and CaO peaks were critically noticed in the XRD patterns of SPS-processed [7] HAp and HAp-CT composites (Figure 25.1a). The stability of phases in HAp-CT composite and the absence of TCP phase during SPS was confirmed [7]. It is known that the presence of TCP is detrimental to the resulting nanomechanical properties. Additionally, a very weak peak of TiO$_2$ was recorded

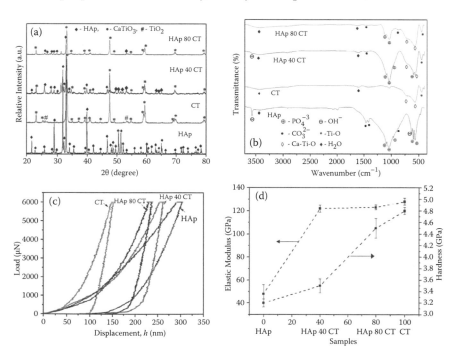

FIGURE 25.1
Characterizations of HAp-CT composites: (a) XRD patterns, (b) FTIR spectra, (c) *P–h* plots, and (d) hardness and elastic modulus of different samples of HAp, calcium titanate, and HAp-CT composites.

in the XRD spectrum of $CaTiO_3$. It is noted that the major peaks of HAp have disappeared, and the characteristic peaks of HAp and the major peaks of $CaTiO_3$ have overlapped in the HAp80CT composite (Figure 25.1a). FTIR analysis (Figure 25.1b) of monolithic HAp, CT, and HAp-CT composites was carried out to complement the results of XRD. Overall, the FTIR results [7] revealed the extent of hydroxylation/dehydroxylation of the present SPS-processed HAp-CT ceramic composites (Figure 25.1b). It has been shown elsewhere [7] that dehydroxylation of HAp-CT ceramics with increasing CT content occurs at a temperature of 1200°C.

The typical load–displacement (*P–h*) curves for monolithic HAp, CT, as well as HAp40CT and HAp80CT composites, as measured using nanoindentation, are shown in Figure 25.1c. It can be observed that the load–displacement (*P–h*) curves get shifted toward the left with an increase in $CaTiO_3$ content, showing stiffer response of the composites than the HAp matrix. In Figure 25.1d, the elastic modulus and hardness are plotted as a function of the wt% of the reinforcing CT phase added to the HAp matrix. A systematic increase in hardness with $CaTiO_3$ addition was recorded in the HAp-CT composites, with little difference between HAp80CT and CT. Compared to HAp, almost a 40% increase in hardness can be realized with 80% $CaTiO_3$ addition. From the data presented in Figure 25.1d, it is also clear that while the 150% increase in elastic modulus with 40% $CaTiO_3$ addition to HAp was estimated from the nanoindentation results, only negligible changes in elastic modulus were measured with 80% $CaTiO_3$ addition or more. In summary, a combination of elastic modulus of 122 GPa and hardness of 4.5 GPa can be obtained with the HAp80CT composite. The most noticeable point here is that we can tailor only the hardness value without compromising modulus by incorporating calcium titanate as a second phase to the HAp. A unique property combination of bonelike electrical conductivity together with proven in vitro and in vivo biocompatibility properties make HAp-$CaTiO_3$ an appealing biomaterial system for further clinical trial [9, 10]. However, to overcome the stress-shielding effect of the bone, we have to maintain the modulus value near to that of bone. On the other hand, in the HAp-mullite system, with incorporation of another second phase, i.e., mullite, we can achieve further lowering of the modulus to the value that is nearer to that of the bone itself.

25.3 HAp-Mullite Composite

The data presented in Figure 25.2a revealed that pure HAp and mullite had much higher hardness values of ≈4.5 GPa and ≈9 GPa, respectively. Similarly, the Young's modulus of HAp was as high as 130 GPa, while that of pure mullite was much higher, at about 220 GPa (Figure 25.2b). The nanomechanical

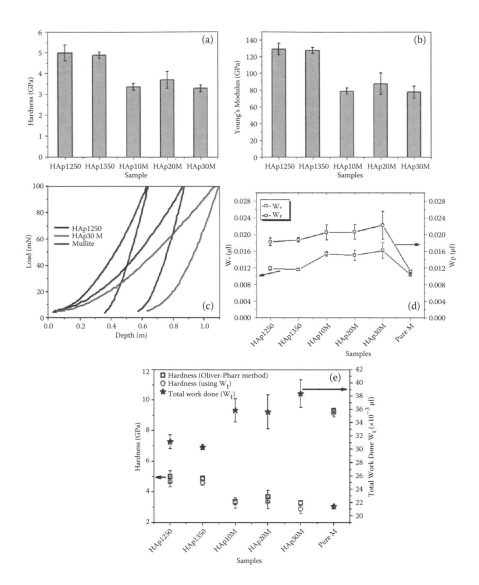

FIGURE 25.2 (See color insert.)
Phase and composition dependencies of (a) hardness and (b) Young's modulus, (c) *P–h* plots, (d) elastic and plastic works of nanoindentation, and (e) estimated hardness (Oliver-Pharr method) and the total work done (W_t) during nanoindentation of HAp, mullite, and HAp-mullite composites. (Modified with permission of Nath et al. [11] from Elsevier.)

properties data of both mullite and HAp compared well with the reported data [12, 13]. It is very interesting to note that when mullite was added, drastic decreases in *H* and *E* data were recorded for these composites (Figure 25.2h). The data on residual depth of penetrations (Figure 25.2c) measured for the pressureless-sintered HAp, mullite, and HAp-(10–30 wt% mullite) composites corroborated well with the data presented in Figure 25.2a-b. As the composites had E of only ≈80 GPa (Figure 25.2b), it shows that the present Young's modulus data did not follow the rule of mixture. It has been proposed elsewhere [11] that this anomalous behavior was possibly linked to the formations of low-hardness phases at grain boundary regions, e.g., CaO and gehlenite, etc. [8].

The typical P–h plots of pure HAp, mullite, and HAp-mullite composite are shown in Figure 25.2c. All the materials depicted a typical elastoplastic nature of deformation, as expected for brittle ceramics. In the earlier HAp-CT system, after incorporating the second phase, i.e., calcium titanate, to HAp, the P–h plot shifted toward the left (Figure 25.1b), which indicated the presence of a stiffer nature in the HAp-CT composites. But here (Figure 25.2c), exactly the opposite trend was noted for the P–h curves, which had shifted to the right, thereby clearly indicating now the more compliant nature of the HAp-mullite composites.

The plastic work (W_p) and elastic work (W_e) data (Figure 25.2d) showed that, in all cases, the plastic work was more than the elastic recovery. All the plastic and elastic works were in the range of 10–20 nJ. These were the lowest for the pure mullite and the highest for the HAp-30 wt% mullite composites. However, for the pure HAp samples, there was not much difference in the W_p and W_e values. W_p and W_e can be correlated with the H and E values of the materials, respectively. Low plastic work and low elastic recovery effectively translate to materials with high hardness and high Young's modulus. Following Tuck et al. [14], the hardness data are predicted (Figure 25.2e), where the work done is considered as the total energy (W_t) rather than as W_p. These values are plotted in Figure 25.2e together with the experimentally measured H data. In the same plot, we have also added the total energy data so that the hardness values obtained from both processes can be correlated. In the present case, the hardness values, calculated considering W as W_p, were found to be almost 3–4 times more than those obtained by the conventional Oliver-Pharr method [15]. Therefore, those values were not included in the plots. Thus, the present results proved beyond doubt that at least for the present HAp-mullite composites, W_t was much more effective than W_p in predicting the hardness values. It is also worthwhile to mention that HAp-mullite composites have the desired combination of in-vitro and in-vivo biocompatibility properties as well as lack of genotoxic properties in-vitro [16–19].

25.4 Conclusions

A combination of elastic modulus of 122 GPa and hardness of 4.5 GPa can be obtained with HAp-80 $CaTiO_3$ composite. In sharp contrast to this behavior, the hardness and Young's modulus values of the HAp-mullite composites were inferior to those of the HAp matrix phase. The reduction of hardness and modulus was linked possibly to the occurrences of relatively softer grain-boundary phases. In predicting the hardness value, the total work done was more effective than only the plastic component of the total work spent during a nanoindentation experiment. Therefore, by interchanging the second phases like calcium titanate and mullite, it may be possible to tailor the mechanical properties such that other important functional properties like electrical conductivity and bioresorption can be retained or may even be improved compared to those of the pure bulk hydroxyapatite. Our focus in Chapter 26 will be the nanoindentation response of Si, a material of immense importance for solar cell, optoelectronics, MEMS, and NEMS device applications.

References

1. Teng, N., S. Nakamura, Y. Takagi, Y. Yamashita, M. Ohgaki, and K. Yamashita. 2001. A new approach to enhancement of bone formation by electrically polarized hydroxyapatite. *Journal of Dental Research* 80:1925–29.

2. Hastings, G., and F. Mahmud. 1988. Electrical effects in bone. *Journal of biomedical engineering* 10:515–21.

3. Coreno, J., and O. Coreno. 2005. Evaluation of calcium titanate as apatite growth promoter. *Journal of Biomedical Materials Research Part A* 75:478–84.

4. Gautier, S., E. Champion, and D. Bernache-Assollant. 1997. Processing, microstructure and toughness of Al_2O_3 platelet-reinforced hydroxyapatite. *Journal of the European Ceramic Society* 17:1361–69.

5. Li, J., B. Fartash, and L. Hermansson. 1995. Hydroxyapatite-alumina composites and bone-bonding. *Biomaterials* 16:417–22.

6. Kumar, A., K. Biswas, and B. Basu. 2013. On the toughness enhancement in hydroxyapatite-based biocomposites. *Acta Materialia* 61:5198–15.

7. Dubey, A. K., G. Tripathi, and B. Basu. 2010. Characterization of hydroxyapatite-perovskite ($CaTiO_3$) composites: Phase evaluation and cellular response. *Journal of Biomedical Materials Research Part B: Applied Biomaterials* 95:320–29.

8. Nath, S., K. Biswas, and B. Basu. 2008. Phase stability and microstructure development in hydroxyapatite-mullite system. *Scripta Materialia* 58:1054–57.

9. Thrivikraman, G., P. K. Mallik, and B. Basu. 2013. Substrate conductivity dependent modulation of cell proliferation and differentiation in vitro. *Biomaterials* 34:7073–85.

10. Mallik, P. K., and B. Basu. Forthcoming. Better early stage osteogenesis of electroconductive HA-$CaTiO_3$ composites in a rabbit animal model. *Journal of Biomedical Materials Research A*. doi: 10.1002/jbm.a.34752

11. Nath, S., A. Dey, A. K. Mukhopadhyay, and B. Basu. 2009. Nanoindentation response of novel hydroxyapatite-mullite composites. *Materials Science and Engineering A* 513-514:197–201.

12. Gong, J., J. Wu, and Z. Guan. 1999. Examination of the indentation size effect in low-load Vickers hardness testing of ceramics. *Journal of the European Ceramic Society* 19:2625–31.

13. Kumar, R. R., and M. Wang. 2002. Modulus and hardness evaluations of sintered bioceramic powders and functionally graded bioactive composites by nano-indentation technique. *Materials Science and Engineering A* 338: 230–36.
14. Tuck, J. R., A. M. Korsunsky, S. J. Bull, and R. I. Davidson. 2001. On the application of the work-of-indentation approach to depth-sensing indentation experiments in coated systems. *Surface and Coatings Technology* 137: 217–24.
15. Oliver, W. C., and G. M. Pharr. 1992. An improved technique for determining hardness and elastic modulus using load and displacement sensing indentation experiments. *Journal of Materials Research* 7:1564–83.
16. Nath, S., S. Kalmodia, and B. Basu. 2013. In vitro biocompatibility of novel biphasic calcium phosphate-mullite composites. *Journal of Biomaterials Applications* 27:497–509.
17. Kalmodia, S., B. Basu, and T. J. Webster. 2013. Gene expression in osteoblasts cells treated with submicron to nanometer hydroxyapatite-mullite eluate particles. *Journal of Biomaterials Applications* 27:891–908.
18. Nath, S., B. Basu, K. Biswas, K. Wang, and R. K. Bordia. 2010. Sintering, phase stability and properties of calcium phosphate-mullite composites. *Journal of the American Ceramic Society* 93:1639–49.
19. Nath, S., B. Basu, M. Mohanty, and P. V. Mohanan. 2009. In vivo response of novel hydroxyapatite-mullite composites: Results up to 12 weeks of implantation. *Journal of Biomedical Materials Research: Part B* 90:547–57.

Section 8

Nanoindentation Behavior of Functional Ceramics

26

Nanoindentation of Silicon

Arjun Dey and Anoop Kumar Mukhopadhyay

26.1 Introduction

In optoelectronics and semiconductor applications, the usage of monocrystalline or polycrystalline silicon (Si), a tetravalent metalloid, is the most dominant controller within the whole industry. Further, thin films of functional materials on Si are often utilized as storage devices for magnetic media as well as the components of micro- and nano-electromechanical systems (MEMS/NEMS). In-depth nanoindentation studies [1–4] can help us to understand the deformation mechanisms of Si surfaces for the sake of design and fabrication purposes. Several researchers [1–4] have reported the occurrence of phase transformations under widely different loading and unloading conditions.

Nanoindentation-induced phase transformations have been studied by Rao et al. [1] for both crystalline and relaxed amorphous silicon samples. Crystalline silicon showed unloading curves with an elbow even for a high load range up to 9,000 μN, while the amorphous silicon offered clear pop-outs at loads beyond 2,000 μN [1]. Nanoindentation experiments were carried out at various loads on single crystal silicon wafers by Yan et al. [2]. A critical transition load (30 mN) has been identified. It was shown by TEM of nanoindent zones that such zones remained mostly amorphous at loads below 30 mN but became nanocrystalline beyond an applied load of 30 mN. Thus, there was a confirmation of the indentation contact pressure-induced phase transformation happening at loads above 30 mN in silicon. Further, very pronounced signatures of elbow formation, i.e., pop-ins, were experimentally found in the unloading curves experimentally obtained for lower loads; in contrast, mainly pop-outs were observed at higher loads [1, 2]. On the other hand, Chang and Zhang [3] reported observation of pop-ins in the loading curve. They proposed that it happened due to transformation of Si-I to Si-II for a Si (100) material even at an ultralow load of 100 μN. A sudden volume change was also observed that looked like an elbow formation, e.g., like what happens in a pop-in event; however, the authors [3] also have reported the same as a pop-out event for the unloading curve. It has been suggested that it can also depend on loading rate and, as such, a lower loading rate may indeed be the conducive situation for an elbow formation [1–3]. Chang and Zhang further continued their research and concluded in subsequent report [4] that the existence of an elbow/pop-out formation does not necessarily always bear a one-to-one correspondence with a straightforward Si-I to Si-II phase transformation phenomenon, as proposed by them in their earlier work [3]. This opinion was also supported through TEM observation conducted earlier by Yan and coworkers [2].

26.2 Nanoindentation Response

In the present chapter, we also want to understand the deformation behavior of Si wafers. The present Si wafer samples were supplied by Fischer, Switzerland. These were subjected to nanoindentation experiments at a wide load range of 10–1000 mN [5]. All nanoindentation experiments were conducted at a temperature of 30°C and a relative humidity of 70%. The nanoindentation arrays are shown in Figures 26.1a and 26.1b. Beyond 300 mN load, a severe cracking was observed, as shown in Figure 26.2. Therefore, we shall be reporting the nanohardness and modulus data up to applied loads of 300 mN only.

The nanohardness and modulus are depicted in Figures 26.3 and 26.4, respectively, as functions of the variations in the nanoindentation loads [5]. The range of variation in nanohardness data of silicon did not exceed 9–11 GPa. The grand average nanohardness data was close to 10 GPa, which matched well with literature data [1–4]. The Young's modulus data also showed a slight increase in trend. The grand average data was about 168 GPa. This value also matched well with data reported for Si wafers [1–4].

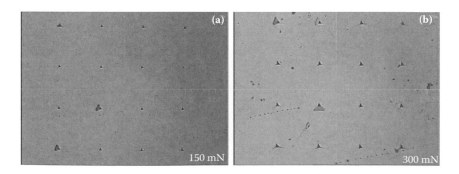

FIGURE 26.1
Micrographs of nanoindentation arrays on Si at lower loads: (a) 150 mN and (b) 300 mN.

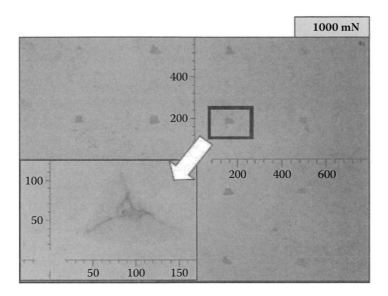

FIGURE 26.2
Micrographs of nanoindentation arrays on Si at higher load showing cracking.

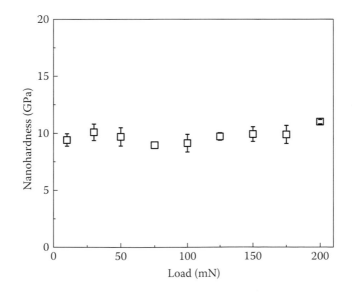

FIGURE 26.3
Variation of nanohardness as a function of nanoindentation loads.

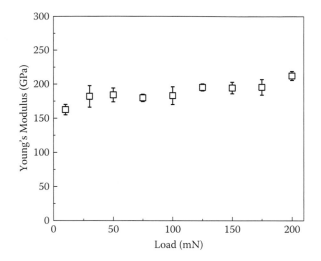

FIGURE 26.4
Variation of Young's modulus as a function of nanoindentation loads.

Next, we want to investigate how the shapes of the *P–h* plots will alter as we increase the indentation loads. The main questions we want to address are of two types: (a) Do the pop-outs or elbows occur only in the unloading parts of the *P–h* plots? (b) Do the pop-ins occur only in the loading parts of the *P–h* plots? Subsequent queries may be: (c) Can the pop-outs or elbows occur both in the loading and in the unloading parts of the *P–h* plots? (d) Can the pop-ins occur both in the loading and in the unloading parts of the *P–h* plots?

In the present experiments, corresponding load versus depth (*P–h*) data plots are shown in Figures 26.5a–d. Actually, these data were also used for calculation of Young's modulus data. Now, if we look to the unloading part of the *P–h* plot at a low load, e.g., 10 mN (Figure 26.5a), we can observe the formation of pop-outs that basically evolved due to increase of volume during phase change that it might have undergone. In contrast, the data of Figure 26.5b show that at a considerably higher load (50 mN), a signature of pop-outs can be present as well. These features were also prominent for even higher loads (150 mN and 300 mN), as shown in Figures 26.5c and 26.5d, respectively. This phenomenon was also seen by other groups of researchers [1–4]. It is obvious that more in-depth studies will be required to understand the actual genesis of the elbows/pop-outs formations. However, in the present study, we have *not* observed any pop-ins in loading curves. So, our observations do not match with those reported by Chang and Zhang [3]. But the experimental conditions were also different. In fact, they have studied it [3] at an ultralow load (100 μN or 0.1 mN), while in the present case, we had started measurements from 10 mN onward.

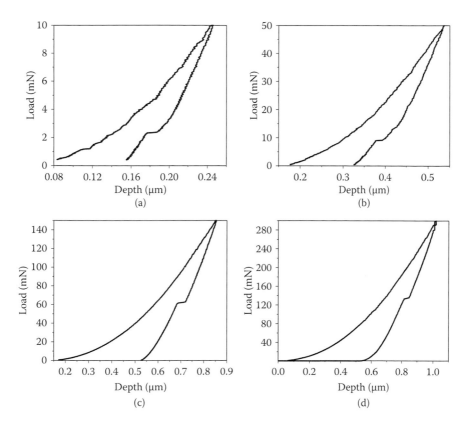

FIGURE 26.5
Typical *P–h* plots of Si showing elbows and pop-outs at different nanoindentation loads:
(a) 10 mN, (b) 50 mN, (c) 150 mN, and (d) 300 mN.

26.3 Conclusions

This chapter focused on the nanoindentation behavior of Si. The deformation behavior of Si in a wide load range from 10 mN onward was investigated in an effort to understand the reason for the formation of elbows and pop-outs. The nanohardness and Young's modulus values were measured for the entire wide load range of 10–300 mN. The experimental data matched well with the reported data. Further, with an increase in the nanoindentation load, the cracks started to generate from all three corners of the nanoindents. In Chapter 27, we shall illustrate the nanoindentation behavior of a tough ceramic, i.e., zirconia-toughened alumina (ZTA).

References

1. Rao, R., J. E. Bradby, S. Ruffell, and J. S. Williams. 2007. Nanoindentation-induced phase transformation in crystalline silicon and relaxed amorphous silicon. *Microelectronics Journal* 38:722–26.
2. Yan, J., H. Takahashi, X. Gai, H. Harada, J. Tamaki, and T. Kuriyagawa. 2006. Load effects on the phase transformation of single-crystal silicon during nanoindentation tests. *Materials Science and Engineering A* 423:19–23.
3. Chang, L., and L. Zhang. 2009. Mechanical behaviour characterisation of silicon and effect of loading rate on pop-in: A nanoindentation study under ultra-low loads. *Materials Science and Engineering A* 506:125–29.
4. Chang, L., and L. Zhang. 2009. Deformation mechanisms at pop-out in mono-crystalline silicon under nanoindentation. *Acta Materialia* 57:2148–53.
5. Dey, A., J. Basu, S. Mukherjee, and A. K. Mukhopadhyay. 2013. Nanoindentation response on silicon.

27

Nanomechanical Behavior of ZTA

Sadanand Sarapure, Arnab Sinha, Arjun Dey,
and Anoop Kumar Mukhopadhyay

27.1 Introduction

In the present chapter, we will be describing the nanoindentation behavior of a tough ceramic, i.e., zirconia-toughened alumina (ZTA). Generally, zirconia is known to be a strong ceramic. It has failure strength ($\approx$650 MPa) nearly 1.5 times that of alumina. Similarly, it has a critical strain energy release rate ($\approx$153 J·m^{-2}) about 7.5 times as high as that of alumina ($\approx$20 J·m^{-2}). To obtain high toughness for structural application purposes, tetragonal zirconia is incorporated in an alumina matrix that gets toughened [1–3]. When the alumina matrix contains the tetragonal phase of appropriate amount, what we get are the ZTA ceramics. These ZTA ceramics often show an R-curve (crack resistance curve) behavior, which means that the material has an intrinsic capacity to exhibit an increase in fracture toughness with indentation crack length [4]. Transformation toughening occurs when the retained metastable tetragonal-ZrO$_2$ (t-ZrO$_2$) transforms to the stable monoclinic-ZrO$_2$ (m-ZrO$_2$) phase in the tensile stress field around a propagating crack. The volume expansion (4%–5%) characteristic of the t $\rightarrow$ m transformation introduces a net compressive stress in the process zone around the crack tip. This reduces the local crack-tip stress intensity and hence the driving force for crack propagation, thereby increasing the effective toughness of the material. The related issues were discussed in detail in an excellent review by Basu [5].

In the present case, 10, 20, and 40 vol% zirconia have been incorporated in an alumina matrix and are named 10ZTA, 20ZTA, and 40ZTA, respectively. The microstructures are shown in Figures 27.1a–c, especially in backscattered mode in SEM, to identify the alumina and zirconia phases. Here, the brighter phase corresponds to the zirconia phase, and the gray phase indicates the alumina phase.

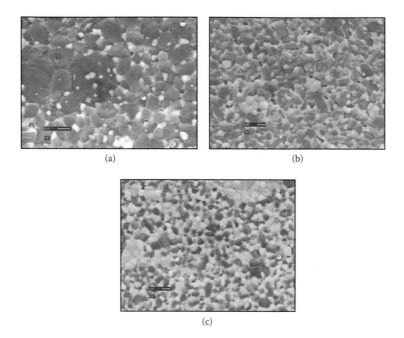

FIGURE 27.1
Microstructures of (a) 10ZTA, (b) 20ZTA, and (c) 40ZTA.

27.2 Nanomechanical Behavior

The nanoindentation experiments have been conducted at 1000 mN in the Fischerscope H100XY$_p$ nanoindentation machine equipped with a Vickers tip. The details of this machine have already been described in Chapter 9. Figure 27.2 shows a typical nanoindentation array in 10ZTA. No signature of cracking has been identified. Further, the typical *P–h* plots of different ZTA composites are shown in Figure 27.3.

The data on nanohardness and Young's modulus are plotted in Figure 27.4 as a function of the vol% of zirconia in the ZTA batch compositions. The nanohardness of the alumina matrix was ≈14 GPa. However, incorporation of 10 vol% zirconia degraded it to ≈11 GPa. In 20ZTA, a subsequent decrement was observed as ≈10 GPa. Further, incorporation of zirconia (40 vol%) in the 40ZTA sample showed almost 40% further reduction in nanohardness value to ≈6 GPa.

In the case of Young's modulus (*E*), parent alumina showed a value of ≈403 GPa, which indicates a dense structure. However, with incorporation of 10 vol% zirconia, the modulus value was drastically decreased to ≈325 GPa. In 20ZTA, further decrement was seen, as *E* was ≈310 GPa. However, with incorporation of 40 vol% of zirconia, almost three times reduction was observed, as *E* was now only about 109 GPa.

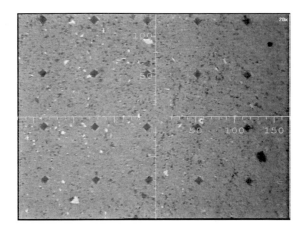

FIGURE 27.2
Nanoindentation array on 10ZTA at 1000 mN.

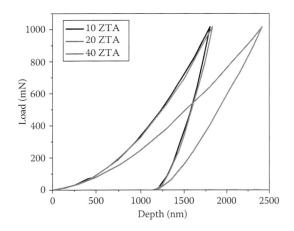

FIGURE 27.3 (See color insert.)
P–h plots on different ZTA composites.

Our work would be incomplete if we didn't measure fracture toughness of such ZTA composites. The main objective is to develop ZTA for the purpose of obtaining damage-tolerant characteristics. However, in the aforesaid nanoindentation studies, our load is not high enough to be able to initiate the cracks so that it would be possible to measure the indentation fracture toughness (IFT). Therefore, we deliberately employed Vickers macroindentation at a relatively higher load of ≈147 N. The indentation fracture toughness data as a function of the volume percent of ZrO_2 in the ZTA batch compositions are shown in Figure 27.5. The matrix alumina showed a very low value of IFT of ≈2.2 MPa·m$^{0.5}$. But even with the small incorporation of 10 vol% ZrO_2 in the ZTA batch composition, IFT increased dramatically by nearly 2.5 times,

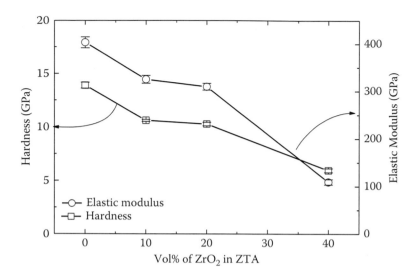

FIGURE 27.4
Hardness and Young's modulus as a function of volume percent of ZrO_2 content in ZTA composite.

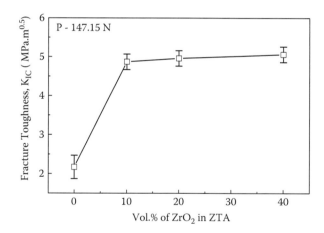

FIGURE 27.5
Indentation fracture toughness as a function of volume percent of ZrO_2 content in ZTA composites.

i.e., ≈4.9 MPa·m$^{0.5}$ in the case of the 10ZTA samples. However, beyond that, if we went on increasing the zirconia content up to 40%, the IFT value was increased a little, to ≈5.1 MPa·m$^{0.5}$. This exciting result depicts that the main contribution to toughening most likely was due to t → m (i.e., tetragonal to monoclinic) transformation of the tetragonal zirconia particles that were retained [1, 2] in the alumina matrix.

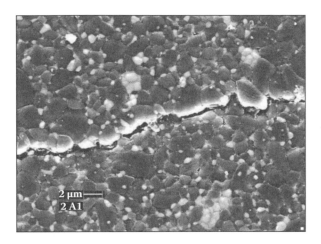

FIGURE 27.6
Grain-localized crack bridging in the 10ZTA sample.

A photomicrograph of a crack generated from the corner of a typical macro Vickers indentation is shown in Figure 27.6. The definite evidence of *grain localized crack bridging* is depicted here. Actually, if you focus on the microstructure of ZTA, it can be seen that the alumina grains are almost uniformly distributed and the zirconia grains are mostly located at the triple points of the alumina grains, almost as if the zirconia grains were trapped between the alumina grains. Therefore, crack deflection by the zirconia particles could also be contributing in the toughness enhancement of the ZTA composites. Faber and Evans [6] also reported such crack deflection in microstructures of a wide variety of phase-transformation-toughened ceramics.

27.3 Conclusions

This chapter investigated enhancement in toughness of ZTA ceramics. The results indicated how the indentation method is useful in understanding deformation mechanisms at both the microstructural and macrostructural length scales. The hardness, Young's modulus, and fracture toughness of the ZTA ceramics were measured here by the nanoindentation and macroindentation techniques. Incorporation of zirconia in alumina matrix showed a decrease in both nanohardness and modulus. However, at the same time, the fracture toughness value of ZTA is 2.5 times as high as that of the brittle alumina matrix. Thorough microstructural investigation was conducted to understand why the toughness had increased. In Chapter 28 we shall learn about the nanoindentation response of actuator ceramics, e.g., PZN-BT-PT and PZN-BT materials, when they are exposed under a Berkovich nanoindenter.

References

1. Sinha, A. 2005. Preparation and characterization of ceramic nanocomposites. M.Tech. dissertation, Bengal Engineering and Science University.
2. Sarapure, S. 2005. Ceramic nano composites for structural applications. M.Tech. dissertation, Motilal Nehru National Institute of Technology, Allahabad.
3. Dey, A. 2007. Toughening of ceramic matrix composites. M.Tech diss., Bengal Engineering and Science University.
4. Awaji, H., S.-M. Choi, and E. Yagi. 2002. Mechanisms of toughening and strengthening in ceramic-based nanocomposites. *Mechanics of Materials* 34:411–22.
5. Basu, B. 2005. Toughening of yttria-stabilised tetragonal zirconia ceramics. *International Materials Reviews* 50:1–18.
6. Faber, K. T., and A. Evans. 1983. Crack deflection processes; I: Theory. *Acta Metallurgica* 31:565–76.

28

Nanoindentation Behavior of Actuator Ceramics

Sujit Kumar Bandyopadhyay, A. K. Himanshu, Pintu Sen,
Tripurari Prasad Sinha, Riya Chakraborty, Arjun Dey,
Payel Bandyopadhyay, and Anoop Kumar Mukhopadhyay

28.1 Introduction

First of all, let us understand what actuators are. Simplistically speaking, the piezotransducers that can generate deformations in the micrometer range or even lower are called the piezoceramic actuators. This behavior has recently opened up very exciting new application possibilities for electromechanical transducers as actuators for hydraulic and pneumatic valves, positioning systems, micromanipulators, and dispensing systems for liquids and gases. To know a little more, we need to know what the piezoceramics are. An example of a simple piezoceramic is barium titanate (BaTiO$_3$)[1]. Another example is lead zirconate titanate (PZT)[2]. Basically, the piezoceramics are used to convert mechanical signals to electrical signals and vice versa. The mechanical signals, e.g., pressure and acceleration, are thus converted into electrical signals. The same piezoceramic can also be used to convert electrical signal into mechanical movement or vibration. So, the piezoceramics are called *transducers* or sometimes *piezotransducers*.

In all kinds of sensors, these transducers convert forces, pressures, and accelerations into electrical signals. Similarly, in sonic and ultrasonic pulse echo machines and ultrasonic flaw detectors, the sonic and ultrasonic transducers are widely used to convert electric voltages into vibrations or deformations. In automotive engineering, these sensors keep passengers safe and ensure intelligent engine management. One piezoceramic application that has been in use for many years is gas ignition. Gas igniters are mass products used in gas heaters and lighters. In ultrasonic applications, piezoceramic components generate great ultrasonic intensities for ultrasonic cleaning and drilling, for ultrasonic welding, or for stimulating chemical processes. Still other piezoceramic components are used as ultrasonic transmitters and receivers in many areas of signal and information processing applications.

One particular type of actuator ceramic is the lead-based relaxor material or its modifications in a predesignated manner, e.g., as in lead zinc niobate $(Pb(Zn_{1/3}Nb_{2/3})O_3\text{-}PZN)$. These and the related issues have already been discussed by us elsewhere [2]. Here we shall bring your attention to the nanoindentation work conducted on two actuator materials, e.g., $(Pb_{0.88}Ba_{0.12})$ $[(Zn_{1/3}Nb_{2/3})_{0.88}Ti_{0.12}]O_3$ (PZN-BT) and $(Pb_{0.8}Ba_{0.2})[(Zn_{1/3}Nb_{2/3})_{0.8}Ti_{0.2}]O_3$ (PZN-BT-PT), as well as discuss how the results correlate with the hysteresis loop measurements in the polar regions of these two actuator materials.

28.2 Nanoindentation Behavior

Figures 28.1a and 28.1b depict the typical *P–h* plots obtained from the nanoindentation experiments in PZN-BT and PZN-BT-PT actuator ceramics, respectively. The corresponding typical nanoindentation arrays generated at an applied load of 1000 mN are shown in Figures 28.1c and 28.1d. The loading parts were elastoplastic, but the initial unloading parts were always elastic in the *P–h* plots (Figures 28.1a and 28.1b). The nanohardness, Young's moduli, elastic energy, and both elastic and plastic energies spent in nanoindentations are shown in Figures 28.2a and 28.2b for the PZN-BT and in Figures 28.2c and 28.2d for the PZN-BT-PT samples. It is evident from the data presented

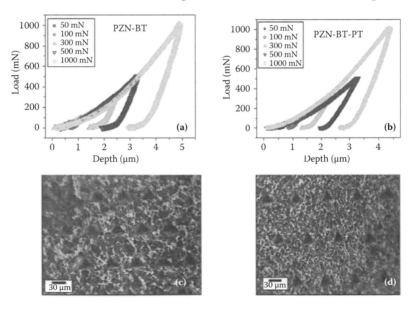

FIGURE 28.1 (See color insert.)
P–h plots and nanoindentation arrays for (a, c) PZN-BT samples and (b, d) PZN-BT-PT samples. (Reprinted/modified with permission of Himanshu et al. [2] from Council of Scientific & Industrial Research, New Delhi.)

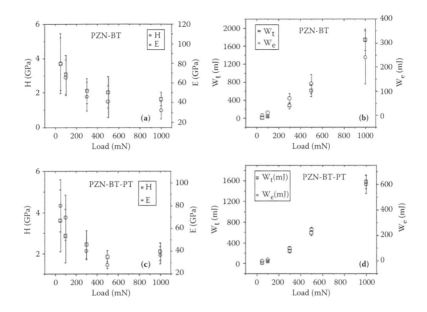

FIGURE 28.2 (See color insert.)
E, H plots and W_e, W_t plots for (a, b) PZN-BT samples and (c, d) PZN-BT-PT samples. (Reprinted/ modified with permission of Himanshu et al. [2] from Council of Scientific & Industrial Research, New Delhi.)

in Figures 28.1 and 28.2 that hardness, Young's modulus, and elastic energy are in general higher for the material PZN-BT-PT as compared to those of the PZN-BT. This had actually happened due to the simple reason that more Ti^{4+} was present in the former. $Ti^{4+}(d^0)$ has an octahedral coordination. It occupies the position at the body center. Here, the Ti-O bond is quite rigid. That is why the permanent distortion of the bond leading to plastic deformation was less than that of the PZN-BT (Figures 28.1a–d). That was why, in general, the PZN-BT-PT materials had Young's modulus and elastic energies higher than those of the PZN-BT samples at all load values (Figures 28.2a–d). But, the plastic deformation in PZN-BT caused its large contribution to plastic energy compared to those of the PZN-BT-PT sample and, hence, the total energies at all load values (Figures 28.2b and 28.2d).

28.3 Polarization Behavior

The previously described nature of the PZN-BT and PZN-BT-PT actuators is also reflected in the polarization behavior (Figures 28.3a and 28.3b). The ferroelectric loops of PZN-BT and PZN-BT-PT were measured at room temperature [2] and are presented in Figures 28.3a and 28.3b. In the case of the PZN-BT sample, a typical ferroelectric polarization saturation had

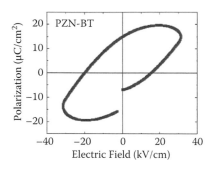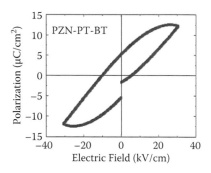

FIGURE 28.3
Polarization plots of (a) PZN-BT samples and (b) PZN-BT-PT samples. (Reprinted with permission of Himanshu et al. [2] from Council of Scientific & Industrial Research, New Delhi.)

occurred at, e.g., $P_s = 19.71\,\mu C/cm^{-2}$, with remnant polarization $P_r \approx 14.281\,\mu C/cm^{-2}$ (Figure 28.3a). For the PZN-BT-PT sample (Figure 28.3b), however, it was found that the magnitude of P_s was as low as $\approx 12.42\,\mu C/cm^{-2}$, which led to the lower value of $P_r \approx 4.98\,\mu C/cm^{-2}$. Therefore, it is concluded from the experimental data that the polarization of PZN-BT-PT was significantly less than that of PZN-BT. In general, the presence of d^0 ion (like Ti^{4+}) causes distortion of the metal–oxygen bond due to semicovalency and, hence, can cause an enhancement of spontaneous ferroelectric polarization [3]. But we have noticed the opposite effect. It has been proposed by us elsewhere [2] that it had happened, possibly, due to the major contribution of the rigid Ti-O bond in PZN-BT-PT actuator, which might have had made it less vulnerable to polarization. This is also supported by the values of elastic energy of PZN-BT-PT being higher than those of PZN-BT, as obtained from the results of the nanoindentation experiments.

28.4 Conclusions

The nanoindentation experiments on PZN-BT-PT and PZN-BT actuator ceramics were done by a Berkovich nanoindenter. The results show that the PZN-BT-PT actuator had more elastic energy compared to that of the PZN-BT actuator. That is why it had higher nanohardness and Young's modulus as compared to those of the PZN-BT sample. In the PZN-BT-PT actuator, Ti^{4+} content was greater than that of PZN-BT actuator, and it is in the B-site of the distorted perovskite structure (ABO_3). As $Ti^{4+}(d^0)$ is at the body center position with octahedral coordination, the Ti-O bond became rigid enough to restrict the plastic deformation. Hence, the presence of more Ti in PZN-BT-PT had caused the relative hike in the elastic energy dissipation with a corresponding decrease in plastic deformation energy. These results also corroborated the results obtained from the polarization studies of these two systems. Due to inherent rigidity of the Ti-O bond, charge separation and hence polarization of the PZN-BT-PT actuator were lower than those of the PZN-BT actuator ceramics. In Chapter 29 we shall shift our focus to the nanoindentation response of a nanostructured magnetoelectric multiferroic material, i.e., the nano bismuth ferrite.

References

1. Zhang, Q., Y. Zhang, T. Yang, S. Jiang, J. Wang, S. Chen, G. Li, and X. Yao. 2013. Effect of compositional variations on phase transition and electric field-induced strain of $(Pb,Ba)(Nb,Zr,Sn,Ti)O_3$ ceramics. *Ceramics International* 39:5403–6.
2. Himanshu, A. K., P. Bandyopadhyay, S. K. Bandyopadhyay, P. Sen, D. C. Gupta, R. Chakraborty, A. K. Mukhopadhyay, B. K. Choudary, and T. P. Sinha. 2010. Dielectric and micromechanical studies of barium titanate substituted (1-y) Pb $(Zn_{1/3}Nb_{2/3})O_3$-yPT ferroelectric ceramics. *Indian Journal of Pure and Applied Physics* 48:349–56.
3. Khomskii, D. I. 2006. Multiferroics: Different ways to combine magnetism and ferroelectricity. *Journal of Magnetism and Magnetic Materials* 306:1–8.

29

Nanoindentation of Magnetoelectric Multiferroic Material

Pintu Sen, Arjun Dey, Anoop Kumar Mukhopadhyay,
Sujit Kumar Bandyopadhyay, and A. K. Himanshu

29.1 Introduction

Bismuth ferrite ($BiFeO_3$) is the only well-known room-temperature magneto-electric multiferroic (MM) material [1, 2]. In MM materials, both ferroelectricity and antiferromagnetism can coexist in a coupled fashion. Such MM materials will be very useful for both magnetic detectors and multistate memory devices. To better exploit the functionalities of the MM materials, the major challenge today has been the synthesis of nano bismuth ferrite (NBFO). Further, ferroelastic ordering has been observed due to plastic deformation in La-based perovskites [3, 4] through studies on mechanical hysteresis. Despite the wealth of literature [5–7], there is no report on the nanomechanical behavior of NBFO samples prepared by sol–gel technique [8], as mentioned in Chapter 9. It is the high porosity of nanomaterials that becomes inhibitive to carry out indentation. Therefore, our main focus in this chapter will be the systematic nanoindentation study at ultralow loads on very low-temperature (300°C), partially annealed green NBFO pellets formed at 600–800 MPa pressure [8]. To the best of our knowledge, this is the first time that nanoindentation has been carried out on a green ceramic nano MM material pellet.

29.2 Nanoindentation Response

The typical load versus depth (P–h) plots for nanoindentations made at three different loads (100, 500, and 1000 μN) are shown in Figure 29.1. The residual depth at higher indentation load was much higher than that measured at lower load. There was no visible damage in the Berkovich nanoindent impressions made at loads of 500 and 1000 μN (Figures 29.2a and 29.2b), except that the area of the impression created at 1000 μN load was a little

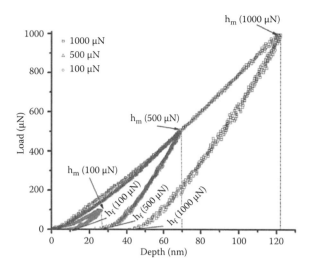

FIGURE 29.1
Load–depth (*P–h*) plots for nanoindentations with constant loads of 100, 500, and 1000 µN on the green NBFO pellet. (Reprinted with permission of Sen et al. [8] from Elsevier.)

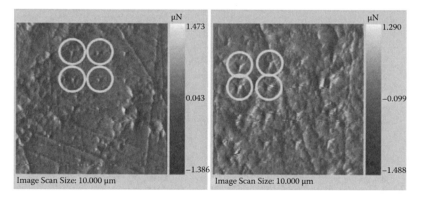

FIGURE 29.2
SPM photomicrographs of nanoindentation array at constant load of (a) 500 µN and (b) 1000 µN. (Reprinted with permission of Sen et al. [8] from Elsevier.)

larger in comparison to that created at 500 µN load. Both nanohardness (Figure 29.3a) and Young's modulus (Figure 29.3b) decreased with increase in load. The decrease in nanohardness with increasing load is well known as indentation size effect (ISE) and can be explained by the Nix and Gao model [9] in terms of statistically stored and geometrically necessary dislocations.

Nanohardness is enhanced as large strain gradients are imposed in ultralow indentation volumes. The residual porosity in the present NBFO sample was really small, at about only 15%. This was found from the analysis of the SPM images (Figure 29.2). The porosity was less, as the packing density was high

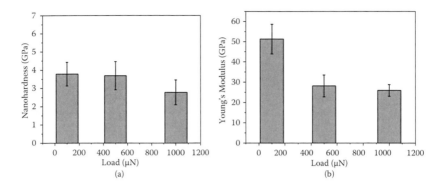

FIGURE 29.3
Load dependencies of (a) nanohardness and (b) Young's modulus. (Reprinted with permission of Sen et al. [8] from Elsevier.)

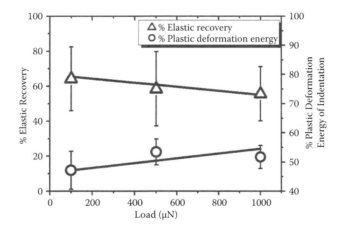

FIGURE 29.4
Load dependencies of ER parameter and the plastic deformation energy. (Reprinted with permission of Sen et al. [8] from Elsevier.)

due to the favorable shape and 15 nm average size of the initial NBFO particles obtained from the sol–gel process [8]. The lower the porosity, the higher is the solid load-bearing area and, hence, the higher is the nanohardness and vice versa [10]. Recent work [11] on brittle porous ($P = 11\%–20\%$) ceramic coatings has identified the exponential porosity–hardness relationship as the best to describe the porosity dependence of nanohardness. Accordingly, it has been shown [8] that the nanohardness (H_0) of a zero-porosity NBFO sample will be about 7.5 GPa, which compares favorably with the nanohardness data reported for Ba-ferrites [12]. The aforesaid trend of experimentally measured data (Figure 29.3) can also be explained in terms of an elastic recovery (ER%) parameter (Figure 29.4) that represents the ratio of elastically recovered depth to maximum depth. A plot of the plastic deformation energy as a function of load (Figure 29.4) indeed corroborates this picture. It also follows, then, that the lower the final depth of penetration, the smaller was the projected contact area, and hence the higher was the nanohardness measured at a lower load (100 μN), thereby giving rise to the ISE, as was observed (Figure 29.3a) in the present experiments.

29.3 Conclusions

The nanohardness value of the NBFO pellet decreased at a rate of only 0.26 GPa·μN^{-1} when the load was increased by an order of magnitude from 100 to 1000 μN. This may be an indication of a highly strain-tolerant microstructure. The decreases in nanohardness and Young's modulus were further explained with elastic recovery and plastic deformation energy concepts. But can we use nanoindentation to evaluate the mechanical integrity of most modern green energy devices, e.g., a solid oxide fuel cell? Well, in this connection, we shall explore in Chapter 30 the mechanical properties of the various component layers of the single solid oxide fuel cell, especially at the local microstructural length scale, because any mechanical disintegration of such a multicomponent system may initiate at its micro-/nanostructural length scales.

References

1. Lebeugle, D., D. Colson, A. Forget, and M. Viret. 2007. Very large spontaneous electric polarization in BiFeO$_3$ single crystals at room temperature and its evolution under cycling fields. *Applied Physics Letters* 91:022907.
2. Lebeugle, D., D. Colson, A. Forget, M. Viret, P. Bonville, J. F. Marucco, and S. Fusil. 2007. Room-temperature coexistence of large electric polarization and magnetic order in BiFeO$_3$ single crystals. *Physical Review B* 76:024116.
3. Orlovskaya, N., N. Browning, and A. Nicholls. 2003. Ferroelasticity in mixed conducting LaCoO$_3$ based perovskites, *Acta Materialia* 51:5063–71.
4. Orlovskaya, N., Y. Gogotsi, M. Reece, B. Cheng, and I. Gibson. 2002. Ferroelasticity and hysteresis in LaCoO$_3$ based perovskites. *Acta Materialia* 50:715–23.
5. Redfern, S. A. T., C. Wang, J. W. Hong, G. Catalan, and J. F. Scott. 2008. Elastic and electrical anomalies at low-temperature phase transitions in BiFeO$_3$. *Journal of Physics: Condensed Matter* 20:452205.
6. Huang, C. W., L. Chen, J. Wang, Q. He, S. Y. Yang, Y. H. Chu, and R. Ramesh. 2009. Phenomenological analysis of domain width in rhombohedral BiFeO$_3$ films. *Physical Review B* 80:140101.

7. Leist, T., K. G. Webber, W. Jo, E. Aulbach, J. Rödel, A. D. Prewitt, J. L. Jones, J. Schmidlin, and C. R. Hubbard. 2010. Stress induced structural changes in La-doped $BiFeO_3$-$PbTiO_3$ high temperature piezoceramics. *Acta Materialia* 58:5962–71.

8. Sen, P., A. Dey, A. K. Mukhopadhyay, S. K. Bandyopadhyay, and A. K. Himanshu. 2012. Nanoindentation behaviour of nano $BiFeO_3$. *Ceramics International* 37:1347–52.

9. Nix, W. D., and H. Gao. 1998. Indentation size effects in crystalline materials: A law for strain gradient plasticity. *Journal of Mechanics and Physics of Solids* 46:411–25.

10. Mukhopadhyay, A. K., and K. K. Phani. 1998. Young's modulus—porosity relations: An analysis based on a minimum contact area model. *Journal of Materials Science* 33:69–72.

11. Dey, A., and A. K. Mukhopadhyay. 2011. Anisotropy in nano-hardness of microplasma sprayed hydroxyapatite coating. *Advances in Applied Ceramics* 109:346–54.

12. Li, Z., and G. Gao. 2011. Chemical bond and hardness of M-,W-type hexagonal barium ferrites. *Can. J. Chem.* 89:573–76.

30

Nanoindentation Behavior of Anode-Supported Solid Oxide Fuel Cell

Rajendra Nath Basu, Tapobrata Dey, Prakash C. Ghosh,
Manaswita Bose, Arjun Dey, and Anoop Kumar Mukhopadhyay

30.1 Introduction

Today, in every stage of life, we are in search of alternative sources of energy because the supply of conventional energies will someday be depleted. In this connection, the research on fuel cell technology is emerging due to its high efficiency and the fact that it is a green process [1, 2]. In this chapter, we concentrate on the planar solid oxide fuel cell (SOFC), which shows more potential than the other types of fuel cells, primarily due to its flexibility of fuels [3]. Actually, in practical application, several single SOFCs are stacked to make the final system. To get high efficiency or reduce the loss, the contact between interconnects and electrodes is subjected to high clamping pressures [4]. Therefore, it is of foremost importance to investigate the mechanical integrity in conditions before and after stack operation. In general, the planar design of the SOFC consists of a PEN (positive-electrolyte-negative) structure. Here, we have chosen the anode, electrolyte, and cathode layers as NiO-8YSZ, 8YSZ, and lanthanum strontium manganite (LSM), respectively [5]. Further, as the single cell is made of different brittle ceramic layers, the elastic mismatch stress may be enough to weaken the structural integrity of the cell. Moreover, thermally induced residual stresses due to the mismatch of thermal expansion between the anode, electrolyte, and cathode layers may take place during the sintering of single-cell assembly. In this connection, investigation of the mechanical properties of the various component layers of the single solid oxide fuel cell, especially at the scale of the local microstructural length scale, becomes an issue of paramount scientific and technological importance, as any mechanical disintegration of a multicomponent system actually initiates at its micro-/nanostructural length scale.

Evaluation of the micromechanical properties like hardness and Young's modulus of individual bulk components of an SOFC cell have been reported by several researchers. Further, a half-cell was investigated by nanoindentation technique [6] at a comparatively higher load of 100 mN. On the other hand, the hardness and Young's modulus were also measured in both pre- and post-reduced conditions, i.e., NiO-8YSZ and Ni-8YSZ [6–8]. Our recently published survey of the literature [9] showed that the measurement of mechanical properties on full single-cell architecture is yet to be explored, and the statistical treatment of such prevalent scatter in the experimentally measured nanoindentation data (as the component's microstructure is heterogeneous and porous) has not yet been addressed by any research group in the world.

Therefore, in this chapter, we shall evaluate the nanohardness and Young's modulus for all three component layers at the microstructural scale utilizing the nanoindentation technique in both pre- and post-reduced conditions. Further, the Weibull model has also been employed to identify the reliability point, which is highly important for further design purposes.

In the present research work, NiO-8YSZ (40–60 v/v), 8YSZ, and LSM were utilized as anode, electrolyte, and cathode layers, respectively. Several layers of green tapes processed by a tape casting method were laminated together with desired thickness under high pressure followed by sintering. The LSM cathode was screen-printed. Finally, the entire single-cell assembly was sintered at ≈1100°C to get the final fabricated cell component [9]. We have reported the processing of the single-cell in detail elsewhere [5].

30.2 Nanomechanical Behavior

The nanohardness (H) and the Young's moduli (E) were measured at an ultralow load of 10 mN. At least 20 to 25 indents were made at five different randomly chosen locations of a given sample, as the layers have characteristically heterogeneous as well as porous microstructures.

The SEM photomicrographs of three different components, e.g., the anode made of NiO-YSZ (i.e., the prereduction structure) and Ni-YSZ (i.e., the postreduction structure), the electrolyte made of dense 8YSZ, and finally the anode prepared by LSM layers of the single-cell, are shown in Figures 30.1a–d. The volume percent of open porosity of pre- and post-reduced anode were found as 27% and 40%, respectively [9]. The cross-sectional image of the whole anode/electrolyte/cathode architecture and nanoporous structure of the LSM cathode are shown in Figures 30.1c and 30.1d, respectively.

Several load-versus-depth (P–h) plots recorded during the nanoindentation experiments are shown in Figures 30.2a–c. Total area under the P–h plot of the anode is much smaller in the case of the pre-reduced condition as compared to that of the post-reduced condition. A similar situation holds true for the cathode layer. However, in both pre- and post-reduced conditions, no significant changes have been seen in the P–h plots of the dense 8YSZ electrolyte; as there is no effect of the reduction heat treatment on either the chemical composition or the microstructure of the layer.

Further, the numerical values of H, E, and other related properties like mean contact pressure (p_m), the relative stiffness (S_{max}/h_{max}), and the relative

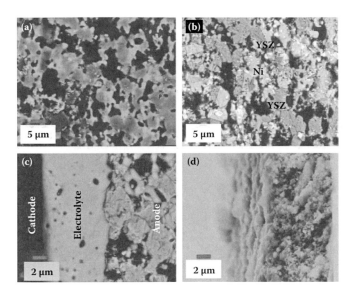

FIGURE 30.1 (See color insert.)
Anode in (a) pre-reduced, (b) post-reduced, and (c) cross section of dense 8YSZ electrolyte layer; (d) nanoporous LSM cathode layer.

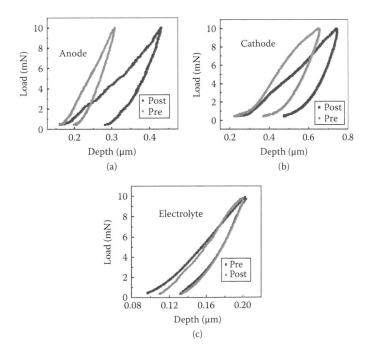

FIGURE 30.2
Typical load–depth (*P–h*) plots of (a) anode, (b) cathode, and (c) electrolyte layers in both pre- and post-reduced conditions.

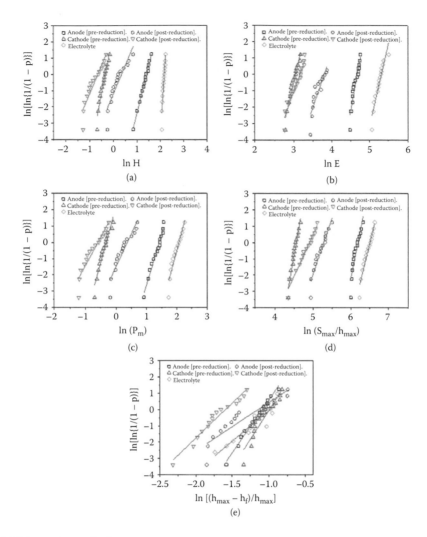

FIGURE 30.3 (See color insert.)
Weibull distribution plots for: (a) nanohardness, (b) Young's modulus, (c) mean contact pressure, (d) relative stiffness, and (e) relative spring-back data evaluated from nanoindentation experiments performed on different component layers of the single cell in both pre- and post-reduced conditions.

spring-back $[(h_{max} - h_f)/h_{max}]$ are evaluated. Moreover, Weibull distribution fittings for the H, E, p_m, (S_{max}/h_{max}), and $[(h_{max} - h_f)/h_{max}]$ data are also evaluated for the different component layers of the single cell in both pre- and post-reduction conditions. These data are shown in Figures 30.3a–e. Based on these Weibull data plots, the corresponding characteristic values of the nanohardness, Young's modulus, mean contact pressure, relative stiffness, and relative spring-back (as well as the subsequent Weibull moduli values) of the different component layers of the single cell are summarized in Table 30.1.

TABLE 30.1

Characteristic Values of Nanohardness, Young's Modulus, Mean Contact Pressure, Relative Stiffness, and Relative Spring-Back as well as the Corresponding Weibull Moduli Data of Anode and Cathode in Both Pre- and Post-reduced Conditions and Electrolyte

Properties and Condition	H (GPa)		E (GPa)		p_m (GPa)		S_{max}/h_{max} (GPa)		$[(h_{max}-h_f)/h_{max}]$	
	Weibull Mod.	Char. Value	Weibull Mod.	Char. Value	Weibull Mod.	Char. Value	Weibull Mod.	Char. Value	Weibull Mod.	Char. Value
Anode										
Pre-reduction	5.74	4.12	12.74	107.00	6.10	4.17	11.38	486.38	6.50	0.34
Post-reduction	3.49	1.41	6.7	44.84	4.04	1.44	6.86	197.50	2.97	0.31
Cathode										
Pre-reduction	7.28	0.71	10.14	21.12	6.84	0.73	13.46	95.85	7.1	0.37
Post-reduction	3.44	0.52	6.62	23.57	3.49	0.55	5.11	134.30	4.25	0.20
Electrolyte	21.04	8.84	10.11	150.58	6.97	7.98	9.36	665.10	3.75	0.36

The 8YSZ electrolyte layer showed the highest characteristic nanohardness and Young's modulus values, while the LSM cathode layer had the lowest values of the same in both pre- and post-reduced conditions. Further, in the case of the cathode layer, the large magnitude of the characteristic scatter, i.e., lower m value, is linked to the presence of a characteristically highly heterogeneous microstructure with a larger average pore size, particularly in the post-reduced condition. A similar observation was also observed for the anode layers (NiO-YSZ and Ni-YSZ), where volume fraction of porosity varied from ≈27% in the pre- to ≈40% in the post-reduced condition.

30.3 Conclusions

In this chapter, the mechanical properties of each individual component layer of a planar single SOFC—the NiO/Ni-YSZ anode, the 8YSZ electrolyte, and the LSM cathode—are evaluated at microstructural length scales by the nanoindentation technique in both pre- and post-reduction conditions. There were very high degrees of characteristic scatter present in the nanomechanical properties data measured for both the cathode and the anode layers, presumably due to their characteristically heterogeneous and nanoporous microstructures. A lot of glass is used in sealing the different components of an effective stack of SOFCs used for power generation. Consequently, we shall investigate the nanoindentation behavior of various glass–ceramic sealants for the anode-supported SOFC in Chapter 31.

References

1. Singhal, S. C., and K. Kendall. 2013. *High temperature solid oxide fuel cells: Fundamentals, design and applications*. London: Elsevier.
2. Stambouli, A. B., and E. Traversa. 2006. Solid oxide fuel cells (SOFCs): A review of an environmentally clean and efficient source of energy. *Renewable and Sustainable Energy Reviews* 6:433–55.
3. Jung, H. Y., S. H. Choi, H. Kim, J. W. Son, J. Kim, H. W. Lee, and J. H. Lee. 2006. Fabrication and performance evaluation of 3-cell SOFC stack based on planar 10 cm×10 cm anode-supported cells. *Journal of Power Source* 159:478–83.
4. Dey, T., D. Singdeo, M. Bose, R. N. Basu, and P. C. Ghosh. 2013. Study of contact resistance at the electrode–interconnect interfaces in planar type solid oxide fuel cells. *Journal of Power Source* 233:290–98.
5. Basu, R. N., A. D. Sharma, A. Dutta, and J. Mukhopadhyay. 2008. Processing of high performance anode-supported planar solid oxide fuel cell. *International Journal of Hydrogen Energy* 33:5748–54.

6. Basu, R., Kumar, S. S., A. K. Mukhopadhyay, and H. S. Maiti. 2007. Improvement in mechanical properties of anode-supported planar SOFC. *ECS Transactions* 7:533–41.
7. Radovic, M., and E. Lara-Curzio. 2004. Mechanical properties of tape-cast nickel-based anode materials for solid oxide fuel cells before and after reduction in hydrogen. *Acta Materialia* 52:5747–56.
8. Nakajo, A., J. Van Herle, and D. Favrat. 2011. Sensitivity of stresses and failure mechanisms in SOFCs to the mechanical properties and geometry of the constitutive layers. *Fuel Cells* 11:537–52.
9. Dey, T., A. Dey, P. C. Ghosh, M. Bose, A. K. Mukhopadhyay, and R. N. Basu. 2014. Influence of microstructure on nano-mechanical properties of SOFC single cell in pre- and post-reduced conditions. *Materials and Design* 53:182–91.

31

Nanoindentation Behavior of High-Temperature Glass–Ceramic Sealants for Anode-Supported Solid Oxide Fuel Cell

Rajendra Nath Basu, Saswati Ghosh, A. Das Sharma,
P. Kundu, Arjun Dey, and Anoop Kumar Mukhopadhyay

31.1 Introduction

In the previous chapter we discussed the importance of solid oxide fuel cells (SOFCs) and why it is important to make SOFC stacks [1]. However, for a planar SOFC stack, gas-tight seals must be applied along the edges of each cell and between the cell stack and gas manifolds to avoid intermixing of fuel gas (on the anode side) and air (on the cathode side). The selection criteria for a good sealant for SOFC are: (a) coefficient of thermal expansion (CTE) match with adjoining components, (b) high ρ (electrical resistivity), and (c) no harmful reaction with joining components [2–4]. Further, the sealant must exhibit (d) high chemical stability and low vapor pressure in both reducing and oxidizing atmospheres, (e) a nonspreading nature to the adjoining fuel cell components at the operating temperature, (f) deformability and the ability to withstand a slight overpressure, and (g) the capacity to survive several thermal cycles during operation at elevated temperature [2–5]. Glass or glass–ceramic sealants can, in principle, meet almost all of these requirements. Most of the works reported in the literature have focused on barium oxide-based borosilicate system, e.g., barium aluminosilicate (BAS) and barium calcium aluminosilicate (BCAS) glasses [6]. Both of these glasses have shown matching CTE values with Crofer 22 APU interconnect and 8YSZ electrolyte, producing a perfect sealing in the SOFC operating environment. But the high ($\approx$35 mole%) BaO content of BAS glass leads to extensive formation of a $BaCrO_4$ phase at 750–800°C. This causes CTE mismatch at the glass–metal interface. As a result, the BAS sealant spalls out. Hence, our aim in this chapter will be to develop an appropriate sealant glass [6–8] and to assess its nanomechanical properties.

TABLE 31.1

Compositions of SOFC Glass–Ceramic Sealants

| Glass Code | Composition In Mole% | | | | Molar Ratio of Network Former (SiO_2:B_2O_3) |
	MgO	BaO	Al_2O_3	La_2O_3	
MA1	22	...	10	15	Only B_2O_3 is present
MA11	22	...	10	15	1
MA12	22	...	10	15	3
BM1	22	25	...	...	Only B_2O_3 is present
BM11	22	25	...	...	0.3
BM12	22	25	...	...	3
BM13	22	25	...	...	12.3

Source: Reprinted/modified with permission of Ghosh et al. [6] from Elsevier.

31.2 Preparation of the Sealant Glass–Ceramic

The details of preparation and characterization techniques of these new SOFC sealant glass–ceramics have been given elsewhere [6], and hence only the mol% compositions are depicted in Table 31.1. It may be noted that in the preparation process [6], B_2O_3 and SiO_2 were chosen as the glass formers. Other ingredients such as BaO and MgO had been used to adjust the CTE with Crofer 22 APU interconnect and 8 YSZ electrolyte. To control the viscosity and improve mechanical properties, La_2O_3 was added. Finally, Al_2O_3 was used to prevent rapid crystallization during heat treatment as well as to control the surface tension of the glass [6].

31.3 Nanomechanical Properties

Figures 31.1a and 31.1b show the data on nanohardness and Young's modulus of the different glass–ceramics developed in the present work [6] based on the base glass compositions mentioned in Table 31.1. All these glasses were converted to suitable glass–ceramics through controlled crystallization at higher temperature [6]. The data of Figures 31.1a and 31.1b show that the magnesium lanthanum alumino borosilicate-based glass–ceramics (MA series: MA1, MA11, MA12, Table 31.1) exhibited better nanomechanical properties as compared to glass–ceramics from the magnesium barium borosilicate–based systems (BM series: BM1, BM11, BM12, BM13, Table 31.1). Young's modulus is a ratio of the applied stress and

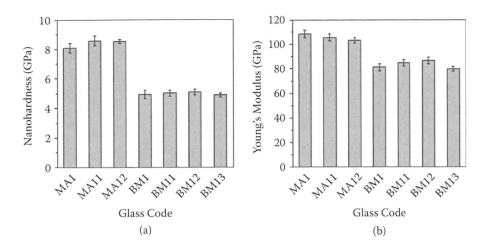

FIGURE 31.1
Composition dependencies of (a) nanohardness and (b) Young's modulus of SOFC glass–ceramic sealants. (Reprinted/modified with permission of Ghosh et al. [6] from Elsevier.)

strain of a material under a constant load. It can therefore be imagined to be inversely related to the bond strength of a material. The bond strength of the respective oxides that were used to make the developed glasses was in the following range: Al-O > La-O > Mg-O > Ba-O. This fact could possibly explain the reason for a relatively higher elastic modulus for the magnesium lanthanum alumino borosilicate samples (Figure 31.1b). Thus, the presence of both La_2O_3 and Al_2O_3 to around 1/4th of the total composition was mainly responsible for such an enhanced nanomechanical property of the developed glass–ceramic sealants. On the other hand, in barium-based compositions where there is no La_2O_3 and Al_2O_3, the values were relatively lower (Figures 31.1a and 31.1b). However, among the glass–ceramics based on BM1, BM11, BM12, and BM13, the BM12 composition showed the highest Young's modulus and nanohardness value, a trend similar to the thermal properties [6]. Thus it can be correlated with the change in B_2O_3 coordination from BO_3^- trigonal to a much more rigid BO_4^- tetrahedral within the glass network. However, all the developed glass–ceramics have relatively higher elastic modulus values as compared to the normal borosilicate glasses, which have a Young's modulus of ≈60–65 GPa [7, 8]. Hence, it seemed that these glass–ceramics could be good candidates for SOFC sealant application. Further, the BM12 composition-based glass–ceramic had the lowest helium gas leak rate of 4.3×10^{-6} (Pa·m²·s⁻¹) [6]. It also satisfied the seamless joining criteria with the interconnect (Figure 31.2a) and the electrolyte (Figure 31.2b).

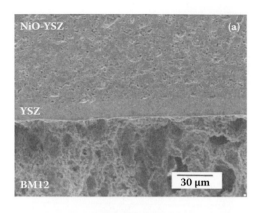

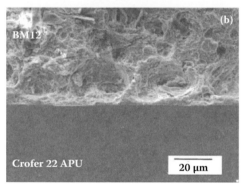

FIGURE 31.2
FESEM photomicrographs of BM12 sealant glass-ceramic in contact with (a) 8YSZ electrolyte
and (b) Crofer 22 APU interconnect. (Reprinted/modified with permission of Ghosh et al. [6]
from Elsevier.)

31.4 Conclusions

The nanomechanical properties of glass–ceramic sealants based on magnesium lanthanum alumino borosilicate and barium magnesium borosilicate
were investigated in order to find their suitability for SOFC sealing applications. Two particular compositions—MA12 and BM12—showed the highest
nanomechanical properties in their respective categories. In general, the
nanomechanical properties of the glass–ceramics from the magnesium lanthanum alumino borosilicate system were much better than those from the
magnesium barium borosilicate system. One particular barium magnesium
borosilicate glass composition (BM12) was found to fulfill all the requirements of an SOFC sealant. So far, we have concentrated on nanoindentation
behavior of bulk glass, ceramics, ceramic matrix composites, and functional
ceramics. In Chapter 32, we shall try to understand the nanoindentation

response of a thick, heterogeneous, porous ceramic coating, in particular, a microplasma-sprayed, bioactive, hydroxyapatite coating, on metallic substrates.

References

1. Basu, R. N. 2006. Materials for solid oxide fuel cells. In *Recent trends in fuel cell science and technology*, ed. S. Basu, 284–31. New York: Springer; New Delhi: Anamaya Publisher.
2. Singhal, S. C., and K. Kendall. 2003. Introduction to SOFCs. In *High temperature solid oxide fuel cells: Fundamentals, design and applications*, Chap. 1. Oxford, UK: Elsevier.
3. Basu, R. N., G. Blass, H. P. Buchkremer, D. Stöver, F. Tietz, E. Wessel, and I. C. Vinke. 2005. Simplified processing of anode-supported thin film planar solid oxide fuel cells. *Journal of the European Ceramic Society* 25:463–71.
4. Basu, R. N., A. D. Sharma, A. Dutta, and J. Mukhopadhyay. 2008. Processing of high-performance anode-supported planar solid oxide fuel cell. *International Journal of Hydrogen Energy* 33:5748–54.
5. Sohn, S. B., and S. Y. Choi. 2004. Suitable glass-ceramic sealant for planar solid-oxide fuel cells. *Journal of the American Ceramic Society* 87:254–60.
6. Ghosh, S., A. D. Sharma, A. K. Mukhopadhyay, P. Kundu, and R. N. Basu. 2010. Effect of BaO addition on magnesium lanthanum alumino borosilicate-based glass-ceramic sealant for anode-supported solid oxide fuel cell. *International Journal of Hydrogen Energy* 35:272–83.
7. Volf, M. B. 1984. *Chemical approach to glass.* Vol. 7 of *series on glass science and technology.* New York: Elsevier.
8. Vogel, W. 1992. *Glass chemistry.* New York: Springer.

Section 9

Static Contact Behavior of Ceramic Coatings

32

Nanoindentation on HAp Coating

Arjun Dey, Payel Bandyopadhyay, Nil Ratan
Bandyopadhyay, and Anoop Kumar Mukhopadhyay

32.1 Introduction

The bioactive hydroxyapatite (HAp) coating on metallic implants provides excellent biocompatibility and biostability to enhance the growth of bone tissues. It is therefore obvious that the in-vivo stability and reliability of the coated implant will depend predominantly upon the local mechanical properties of the coating. Despite the wealth of literature [1–6], there has not been any systematic study of local nanomechanical properties for microplasma-sprayed HAp (MIPS-HAp) coating on SS316L substrates. Therefore, we shall try to understand how the applied load affects the local mechanical properties, specifically the nanohardness (H) and Young's modulus (E), of the MIPS-HAp coatings on SS316L substrates. The details of coating preparation have already been given elsewhere [7] and are summarized in Chapter 9.

32.2 Influence of Load on Nanohardness and Young's Modulus

Figure 32.1 represents the typical load versus depth ($P–h$) plots of the indents made at 100, 500, and 1000 mN, respectively. At higher load, more energy was dissipated and, as a result, a large residual depth of about 3000 nm was observed. The SEM image of a single Berkovich indent at a low load of 80 mN (inset of Figure 32.1) showed smooth indentation area. Any sign of severe contact-induced damage growth or accumulation was absent. The reliability of the nanohardness and Young's modulus data was examined in the light of the weakest-link distribution statistics, i.e., the Weibull distribution fitting (Figures 32.2a and 32.2b). The scatter was sensitive to applied indentation loads. Based on the Weibull statistical treatment, the characteristic values were obtained [7, 8] for both nanohardness (H) and Young's modulus (E),

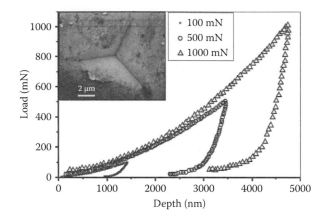

FIGURE 32.1
Load–depth (*P–h*) plots of MIPS-HAp coating at low load (100 mN) and high loads (500 and 1000 mN). Inset: SEM image of a single Berkovich indent at 80 mN load. (Reprinted/modified with permission of Dey et al. [7] from Elsevier.)

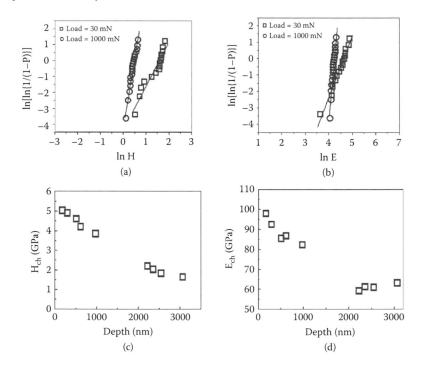

FIGURE 32.2
Plots of (a) Weibull distribution of nanohardness (*H*) data on the plan section of MIPS-HAp coating at loads of 30 and 1000 and (b) Weibull distribution fittings of Young's modulus (*E*) data on the plan section of MIPS-HAp coating at loads of 30 and 1000 mN. Depth dependencies of characteristic values (c) nanohardness (H_{ch}) and (d) Young's modulus (E_{ch}). (Reprinted/modified with permission of Dey et al. [7] from Elsevier.)

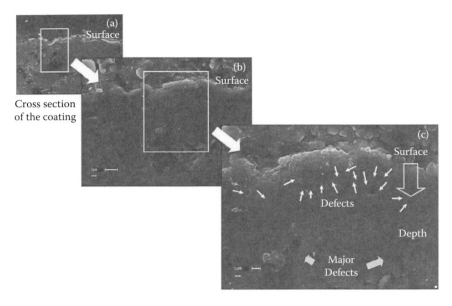

FIGURE 32.3
SEM photomicrographs of the polished cross section of the MIPS-HAp coatings taken at progressively higher magnifications at: (a) 1,000×, (b) 6,000×, (c) 10,000×. (Reprinted with permission of Dey et al. [7] from Elsevier.)

and these are shown respectively in Figures 32.2c and 32.2d. When plotted as a function of depth, both of these data show a strong indentation size effect (ISE). Further, the magnitudes of both characteristic nanohardness (H_{ch}) and characteristic Young's modulus (E_{ch}) were comparable with those reported in the literature [1–3, 5, 6]. The details of the Weibull analysis are discussed in Chapters 33 and 50.

Figure 32.3 represents a collage of the SEM photomicrographs taken from the coating cross section at progressively higher magnifications. It shows a gradual increase of fine micropores and microcracks as well as macroscopic defects, i.e., large macropores and deeper cracks, as one traverses from the surface toward the depth of the coating. When a large load is applied on the sample, the indenter penetrates more deeply into the material and interacts much more with the subsurface defects and pores. These pores and defects apparently increased the depth of penetration, which, in turn, increased the contact area. So, for large applied normal load, lower values of H_{ch} are observed. On the other hand, when the applied nanoindentation load was low (10 mN), the penetration depths were limited to shallow depths of ≈170 nm, so there was less interaction with the pre-existing flaws and the nanoindentation depth was thus very low. This possibly gave rise to a relatively higher value of nanohardness (Figure 32.3c).

32.3 Conclusions

Based on the results presented in this chapter, it is proposed that the variations in the extent of interaction of the indenter with average-size defects across the depth of MIPS-HAp coatings may cause an indentation size effect. When the interaction is at a higher depth of penetration, this results in a larger area of contact and, hence, a lower hardness. At lower load, the depth of penetration is low, and so the nanoindent itself would then have a much lower chance to interact with intrinsic defects, which lie at a much greater depth. As a result, the lower depth of penetration would result in a smaller area of contact and hence a much higher hardness, thus possibly producing the indentation size effect that was observed in the present work. It is evident from this picture that there is an intrinsic presence of stochastic processes that will have a statistical distribution guided by the local microstructural heterogeneity. That is why we have made an attempt to study the Weibull modulus of nanohardness and elastic modulus of HAp coatings in Chapter 33.

References

1. Khor, K. A., H. Li, and P. Cheang. 2003. Characterization of the bone-like apatite precipitated on high velocity oxy-fuel (HVOF) sprayed calcium phosphate deposits. *Biomaterials* 24:769–75.
2. Zhang, C., Y. Leng, and J. Chen. 2001. Elastic and plastic behavior of plasma-sprayed hydroxyapatite coatings on a Ti–6Al–4V substrate. *Biomaterials* 22:1357–63.
3. Wen, J., Y. Leng, J. Chen, and C. Zhang. 2000. Chemical gradient in plasma-sprayed HA coatings. *Biomaterials* 21:1339–43.
4. Nieh, T. G., A. F. Jankowski, and J. Koike. 2001. Processing and characterization of hydroxyapatite coatings on titanium produced by magnetron sputtering. *Journal of Materials Research* 16:3238–45.
5. T. G. Nieh, B. W. Choi, and A. F. Jankowski. 2001. Synthesis and characterization of porous hydroxyapatite and hydroxyapatite coatings. Paper presented at Minerals, Metals, and Materials Society Annual Meeting & Exhibition, New Orleans, LA.
6. Cheng, G. J., D. Pirzada, M. Cai, P. Mohanty, and A. Bandyopadhyay. 2005. Bioceramic coating of hydroxyapatite on titanium substrate with Nd-YAG laser. *Materials Science and Engineering C* 25:541–47.
7. Dey, A., A. K. Mukhopadhyay, S. Gangadharan, M. K. Sinha, D. Basu, and N. R. Bandyopadhyay. 2009. Nanoindentation study of microplasma sprayed hydroxyapatite coating. *Ceramics International* 35:2295–2304.
8. Duan, K., and R. W. Steinbrech. 1998. Influence of sample deformation and porosity on mechanical properties by instrumented microindentation technique. *Journal of the European Ceramic Society* 18:87–93.

33

Weibull Modulus of Ceramic Coating

Arjun Dey and Anoop Kumar Mukhopadhyay

33.1 Introduction

The importance of evaluating the nanomechanical properties for the bioactive HAp ceramic coatings is highlighted by the large amount of data that have been reported [1–16] for the nanohardness [1–4, 7–11, 13–16] and Young's modulus [2–14, 16] of both HAp [1–13, 16] and HAp composite [9, 14] coatings. Such coatings have been deposited by plasma spraying [1–6], laser-assisted process [8–10, 12, 16], sol–gel process [14, 15], sputtering [10, 11], and other thermal spraying techniques [9, 13, 15]. The measurements were made using a wide variety of nanoindenters, e.g., a Berkovich tip [1–7, 9–11, 13–15] or a Vickers tip [8] and a spherical indenter [16]. However, systematic studies of both the nanohardness and elastic modulus as measured by the nanoindentation technique under a variety of applied loads on HAp coating are scarce [2, 3]. It was also found in general that the scatter in data was very high for the plasma-sprayed coatings, presumably due to the highly heterogeneous and porous structure of the coatings [2, 17–19]. However, the quantification of such scatter in the perspective of data reliability has not been attempted for MIPS-HAp coatings. Therefore, in this chapter we shall try to assess how we can quantify the data reliability in terms of a Weibull distribution for the nanohardness and elastic modulus on the plan section of MIPS-HAp coatings as measured by the nanoindentation technique with a Berkovich indenter under a variety of loads in the range of 10–1000 mN [21]. Further details about this aspect are discussed in Chapter 50.

33.2 Data Reliability Issues in MIPS–HAp Coatings

The Weibull distribution fittings for the nanohardness data of the coating are shown for the low (10–100 mN) and high (300–1000 mN) loads in Figures 33.1a and 33.1b, respectively. In a similar fashion, the Weibull distribution fittings for the Young's modulus of the coating, as determined by the nanoindentation experiment, are shown for the low (10–100 mN) and high

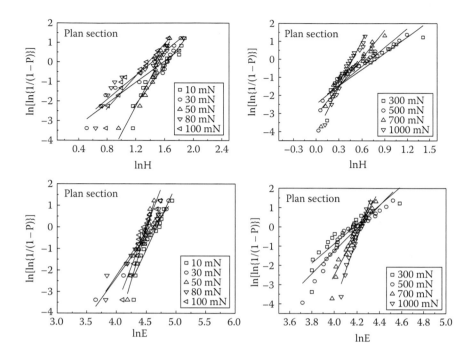

FIGURE 33.1
Weibull statistical plots for MIPS–HAp coatings: (a) low loads, hardness, (b) high loads, hardness, (c) low loads, Young's modulus, and (d) high loads, Young's modulus. (Reprinted/modified with permission of Dey et al. [21] from Springer.)

(300–1000 mN) loads in Figures 33.1c and 33.1d. The scatter in the data was almost characteristically high, especially when measured at relatively lower nanoindentation loads. The FESEM photomicrographs of both plan section and cross section of the MIPS–HAp coating on SS316L substrates are shown respectively in Figures 33.2a and 33.2b. This picture proves beyond doubt that the spatial density of pores, cracks, and defects was very high on the plan section (Figure 33.2a). As a result, these defects contributed to data scatter. The coating also had a thickness of about 230 µm (Figure 33.2b). These features are presented schematically in Figure 33.3 to illustrate that a wide variety of defects may really exist in a MIPS–HAp coating.

The Weibull moduli (m) values of the nanohardness data varied in the range 2–8 (Figures 33.2a and 33.2b) for the plan section. The Weibull moduli (m) values of the Young's modulus data varied in the range 5–15 (Figures 33.2c and 33.2d) for the plan section. Thus, the scatter was more in nanohardness than in Young's modulus. This information implied that microstructural flaws (Figure 33.3) had a more dominant influence on the evolution of the contact area than on the slope of the unloading curves obtained during the nanoindentation experiments conducted on the plan section of the present MIPS–HAp coatings.

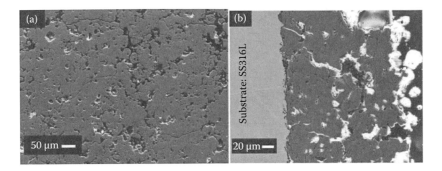

FIGURE 33.2
Microstructure of MIPS-HAp coating for polished (a) plan and (b) cross section. (Reprinted with permission of Dey et al. [21] from Springer.)

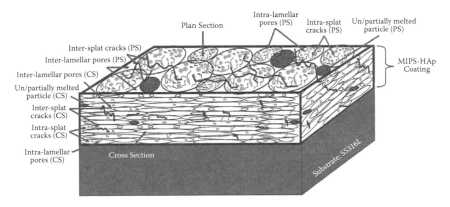

FIGURE 33.3
Schematic illustration of MIPS-HAp coatings on SS316L showing splats and defects. PS = plan section and CS = cross section. (Reprinted with permission of Dey et al. [21] from Springer.)

The use of the Weibull distribution thus provides a means of quantifying the degree of reliability in the present data. It has been shown elsewhere that characteristic values of nanohardness and Young's modulus can provide estimates that are reliable for engineering fail-safe designs [21]. On the plan section, the characteristic nanohardness was about 5 to 1.5 GPa, depending on the nanoindentation load (10–100 mN). Further, the Young's modulus had characteristic values in the range of about 100 to 63 GPa, depending on the load [21].

33.3 Conclusions

Typical Weibull modulus (m) values were about 2–8 for the nanohardness data but increased to about 5–15 for the Young's modulus measured in the plan section of MIPS–HAp coatings. The high spatial density microstructural defects present in the plan section of the coating had contributed in a significant way to the scatter of measured nanomechanical properties. However, this knowledge does not help us to understand whether the coating would perform in the same manner when the contact is in a direction of the plan section of the coating or the cross section of the coating. This issue of anisotropy in nanohardness of microplasma-sprayed hydroxyapatite coating is discussed in Chapter 34.

References

1. Dey, A., A. K. Mukhopadhyay, S. Gangadharan, M. K. Sinha, and D. Basu. 2009. Development of hydroxyapatite coating by microplasma spraying. *Materials and Manufacturing Processes* 24:1249–58.
2. Dey, A., A. K. Mukhopadhyay, S. Gangadharan, M. K. Sinha, D. Basu, and N. R. Bandyopadhyay. 2009. Nanoindentation study of microplasma sprayed hydroxyapatite coating. *Ceramics International* 35:2295–2304.
3. Dey, A., A. K. Mukhopadhyay, S. Gangadharan, M. K. Sinha, and D. Basu. 2008. Mechanical properties of microplasma sprayed HAP coating. In *Proceedings of Interquadrennial Conference of the International Congress on Fracture*, ed. B. K. Raghu Prasad and R. Narasimhan, 311–13. Bangalore: I. K. International Publishing House Pvt. Ltd.
4. Dey, A., A. K. Mukhopadhyay, S. Gangadharan, M. K. Sinha, and D. Basu. 2008. Fracture toughness of microplasma sprayed HAP coating by nanoindentation. In *Proceedings of Interquadrennial Conference of the International Congress on Fracture*, ed. B. K. Raghu Prasad and R. Narasimhan, 217–19. Bangalore: I. K. International Publishing House Pvt. Ltd.
5. Wen, J., Y. Leng, J. Chen, and C. Zhang. 2000. Chemical gradient in plasma-sprayed HA coatings. *Biomaterials* 21:1339–43.
6. Zhang, C., Y. Leng, and J. Chen. 2001. Elastic and plastic behavior of plasma-sprayed hydroxyapatite coatings on a Ti–6Al–4 V substrate. *Biomaterials* 22:1357–63.
7. Khor, K. A., H. Li, and P. Cheang. 2003. Characterization of the bone-like apatite precipitated on high velocity oxy-fuel (HVOF) sprayed calcium phosphate deposits. *Biomaterials* 24:769–75.
8. Cheng, G. J., D. Pirzada, M. Cai, P. Mohanty, and A. Bandyopadhyay. 2005. Bioceramic coating of hydroxyapatite on titanium substrate with Nd-YAG laser. *Materials Science and Engineering: C* 25:541–47.
9. Chen, Y., Y. Q. Zhang, T. H. Zhang, C. H. Gan, C. Y. Zheng, and G. Yu. 2006. Carbon nanotube reinforced hydroxyapatite composite coatings produced through laser surface alloying. *Carbon* 44:37–45.
10. Nieh, T. G., A. F. Jankowski, and J. Koike. 2001. Processing and characterization of hydroxyapatite coatings on titanium produced by magnetron sputtering. *Journal of Materials Research* 16:3238–45.

11. Nieh, T. G., B. W. Choi, and A. F. Jankowski. 2001. Synthesis and characterization of porous hydroxyapatite and hydroxyapatite coatings. Report submitted to Minerals, Metals, and Materials Society Annual Meeting and Exhibition, New Orleans, LA.

12. Arias, J. L., M. B. Mayor, J. Pou, Y. Leng, B. Leon, and M. P. Amora. 2003. Micro- and nano-testing of calcium phosphate coatings produced by pulsed laser deposition. *Biomaterials* 24:3403–8.

13. Gross, K. A., and S. S. Samandari. 2007. Nano-mechanical properties of hydroxyapatite coatings with a focus on the single solidified droplet. *Journal of the Australian Ceramic Society* 43:98–101.

14. Zhang, S., Y. S. Wang, X. T. Zeng, K. A. Khor, W. Weng, and D. E. Sun. 2008. Evaluation of adhesion strength and toughness of fluoridated hydroxyapatite coatings. *Thin Solid Films* 516:5162–67.

15. Gross, K. A., and S. S. Samandari. 2009. Nanoindentation on the surface of thermally sprayed coatings. *Surface and Coatings Technology* 203:3516–20.

16. Pelletier, H., V. Nelea, P. Mille, and D. Muller. 2004. Determination of mechanical properties of pulsed laser-deposited hydroxyapatite thin film implanted at high energy with N^+ and Ar^+ ions using nanoindentation. *Journal of Materials Science* 39:3605–11.

17. Zhou, H., F. Li, B. He, J. Wang, and B. Sun. 2007. Air plasma sprayed thermal barrier coatings on titanium alloy substrates. *Surface and Coatings Technology* 201:7360–67.

18. Basu, D., C. Funke, and R. W. Steinbrech. 1999. Effect of heat treatment on elastic properties of separated thermal barrier coatings. *Journal of Materials Research* 14:4643–50.

19. Guo, S., and Y. Kagawa. 2006. Effect of thermal exposure on hardness and Young's modulus of EB-PVD yttria-partially-stabilized zirconia thermal barrier coatings. *Ceramics International* 32:263–70.

20. Oliver, W. C., and G. M. Pharr. 1992. An improved technique for determining hardness and elastic modulus using load and displacement sensing indentation experiments. *Journal of Materials Research* 7:1564–83.

21. Dey, A., A. K. Mukhopadhyay, S. Gangadharan, M. K. Sinha, and D. Basu. 2009. Weibull modulus of nano-hardness and elastic modulus of hydroxyapatite coating. *Journal of Materials Science* 44:4911–18.

34

Anisotropy in Nanohardness of Ceramic Coating

Arjun Dey and Anoop Kumar Mukhopadhyay

34.1 Introduction

Hydroxyapatite (HAp) is used as a bioactive ceramic coating on metallic implants in (a) total joint prostheses as an alternative to PMMA (polymethyl methacrylate) cement-based fixation, (b) dental implants for bioactive fixation, and (c) fillers for repairing bone defects. Despite the wealth of literature [1–15], however, there has not been much of a systematic study on the nanohardness (*H*) at the local microstructural level of HAp or HAp composite coatings on metallic substrates as measured by the nanoindentation technique. For HAp thin (≈350–650 nm) films [3, 4] on Ti and Si substrates, nanohardness data was a strongly sensitive function of film thickness and the chemical composition of the substrate. For high-velocity oxy fuel (HVOF) sprayed HAp coatings [5], the nanohardness of the coating was greater than that measured at the coating–substrate interface. Just the opposite trend was reported for laser-deposited HAp coatings [10, 11]. Even for laser-deposited HAp coatings, a large variation of nanohardness was observed [6, 10, 11]. HAp–CNT composites were about 30% harder than the HAp matrix material [11]. Flame-sprayed HAp coating also displayed a great variation of nanohardness (≈8–4.5 GPa) measured on both plan and cross sections [13, 14]. The microhardness of macroplasma-sprayed HAp (MAPS–HAp) [1, 2, 7, 8, 12] and MAPS–HAp composite [8, 9] also have been reported. The data were sensitive to choice of indenter shape, e.g., a Vickers [7] or a Knoop indenter [1, 12]. Recently, the author and coworkers reported synthesis characterization along with nanomechanical properties evaluated on the plan section of MIPS–HAp coating on surgical-grade SS316L substrates [15–18].

Most researchers (≈60%) [1–18] used the nanoindentation technique. Barring a few studies [14, 15], the reported nanoindentation data were not systematic. Out of the first 15 references [1–15], the majority (47%) were on cross sections. In the other works, about 33% of the measurements were on plan sections. In some cases [6, 9, 11], measurement locations were not so precisely defined. Therefore, in this chapter, we shall try to characterize the anisotropy, if any, in the nanohardness of microplasma-sprayed HAp (MIPS–HAp) coating on a surgical-grade SS316L substrate. The data were measured by the nanoindentation technique with a Berkovich indenter under loads of 10–1000 mN. The loads were applied onto both plan and cross sections.

34.2 Nanohardness Behavior: Anisotropy

The general nature of the load versus depth (P–h) plots (Figure 34.1a) indicated the presence of an elastic-plastic deformation process, as expected for brittle materials. The nanohardness of cross section (H_{cs}) was greater than that measured in the plan section (H_{ps}) (Figure 34.1b). Thus, there was a strong anisotropy ($H_{cs}/H_{ps} \approx 2$) of nanohardness (Figure 34.1c). To the best of our knowledge, this observation reported by the authors of this chapter [19]

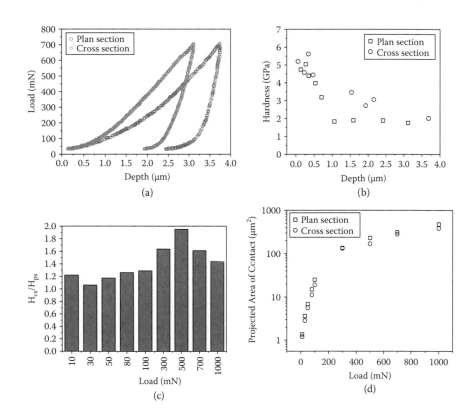

FIGURE 34.1 (See color insert.)
(a) P–h plots of plan and cross sections. Load dependencies of (b) nanohardness, (c) anisotropy factor, and (d) contacts area. (Reprinted/modified with permission of Dey and Mukhopadhyay [19] from Institute of Materials, Minerals and Mining and Maney.)

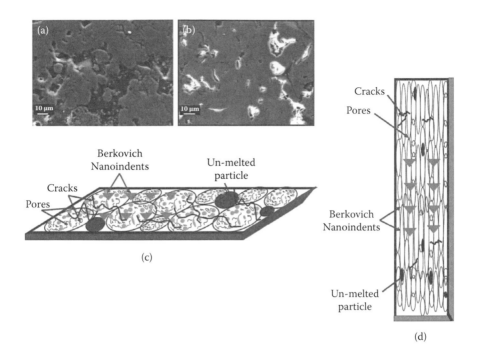

FIGURE 34.2
Microstructures of (a) plan and (b) cross sections. (c, d) Schematics of interactions of the nanoindents with microstructures of (a) plan and (b) cross sections of the MIPS–HAp coatings. (Reprinted/modified with permission of Dey and Mukhopadhyay [19] from Institute of Materials, Minerals and Mining and Maney.)

was the first such experimental observation in a MIPS–HAp coating on SS316L substrate. The data presented in Figure 34.1d showed that at any given load, the projected area of contact was always higher on the plan section. This was what had resulted in higher nanohardness of cross section over plan section. The FESEM photomicrographs of polished plan and cross sections of the coating showed that the densities of pores, cracks, and defects were much more on the plan section (Figure 34.2a) than on the cross section (Figure 34.2b). Similar observations were reported [20] for plasma-sprayed thermal barrier coatings. Based on this evidence, we suggest that the presence of such pores and cracks would certainly reduce the total solid load-bearing contact area [21, 22] (Figures 34.2c and 34.2d). The greater the reduction, the greater is the likelihood of reduction in nanohardness. As a result, nanohardness could be lower on the plan section than on the cross section (Figure 34.1b).

The difference in overall average volume percent porosity in plan (≈ 20 vol%) and cross sections (≈ 11 vol%) of the coating had a direct bearing on the measured nanohardness data. It was shown elsewhere by us [19] that an exponential porosity dependence best explained the nanohardness data from the literature and for the present data. Accordingly, it was found that the predicted values of nanohardness (3.18 and 1.92 GPa for 11 and 20 vol% porosity in plan and cross sections, respectively) matched well with the experimental data for plan (3.08 GPa) and cross sections (1.81 GPa). Hence, there was a good match between the experimental data on average anisotropy index ($H_{cs}/H_{ps} \approx 1.66$) and the predicted data (≈ 1.7).

34.3 Conclusions

At comparable loads, the nanohardness of cross section (H_{cs}) was usually higher than that measured on plan sections (H_{ps}) of the coating, thereby showing a characteristic anisotropy. The anisotropy was linked to the larger volume percent porosity as well as higher spatial density of planar defects, pores, and cracks in the plan section over those in the cross section. In this context, a qualitative model was schematically developed to depict the genesis of anisotropy in nanohardness of the present MIPS–HAp coating. Like the nanohardness, fracture toughness is another important property that defines the intrinsic mechanical resistance of a material against catastrophic crack propagation. This is what we are going to address in Chapter 35, where we shall show how we can use the nanoindentation technique with a Berkovich tip to evaluate the fracture toughness of thick ceramic coatings.

References

1. Kweh, S. W. K., K. A. Khor, and P. Cheang. 2000. Plasma-sprayed hydroxyapatite (HA) coatings with flame-spheroidized feedstock: Microstructure and mechanical properties. *Biomaterials* 21:1223–34.
2. Mancini, C. E., C. C Berndt, L. Sun, and A. Kucuk. 2001. Porosity determinations in thermally sprayed hydroxyapatite coatings. *Journal of Materials Science* 36:3891–96.
3. Nieh, T. G., A. F. Jankowski, and J. Koike. 2001. Processing and characterization of hydroxyapatite coatings on titanium produced by magnetron sputtering. *Journal of Materials Research* 16:3238–45.
4. Nieh, T. G., B. W. Choi, and A. F. Jankowski. 2001. Synthesis and characterization of porous hydroxyapatite and hydroxyapatite coatings. Report submitted to Minerals, Metals, and Materials Society Annual Meeting and Exhibition, New Orleans, LA.
5. Khor, K. A., H. Li, and P. Cheang. 2003. Characterization of the bone-like apatite precipitated on high-velocity oxy-fuel (HVOF) sprayed calcium phosphate deposits. *Biomaterials* 24:769–75.

6. Cheng, G. J., D. Pirzada, M. Cai, P. Mohanty, and A. Bandyopadhyay. 2005. Bioceramic coating of hydroxyapatite on titanium substrate with Nd-YAG laser. *Materials Science and Engineering C* 25:541–47.
7. Chen, Y., Y. Q. Zhang, T. H. Zhang, C. H. Gan, C. Y. Zheng, and G. Yu. 2006. Carbon nanotube reinforced hydroxyapatite composite coatings produced through laser surface alloying. *Carbon* 44:37–45.
8. Arias, J. L., M. B. Mayor, J. Pou, Y. Leng, B. Leon, and M. Perez-Amora. 2003. Micro- and nano-testing of calcium phosphate coatings produced by pulsed laser deposition. *Biomaterials* 24:3403–8.
9. Gross, K. A., and S. S. Samandari. 2007. Nano-mechanical properties of hydroxyapatite coatings with a focus on the single solidified droplet. *Journal of the Australian Ceramic Society* 43:98–101.
10. Gu, Y. W., K. A. Khor, and P. Cheang. 2003. In vitro studies of plasma-sprayed hydroxyapatite/Ti-6Al-4V composite coatings in simulated body fluid (SBF). *Biomaterials* 24:1603–11.
11. Mohammadi, Z., A. A. Z. Moayyed, and A. S. M. Mesgar. 2007. Adhesive and cohesive properties by indentation method of plasma-sprayed hydroxyapatite coatings. *Applied Surface Science* 253:4960–65.
12. Khor, K. A., Y. W. Gu, D. Pan, and P. Cheang. 2004. Microstructure and mechanical properties of plasma-sprayed HA/YSZ/Ti–6Al–4V composite coatings. *Biomaterials* 25:4009–17.
13. Khor, K. A., Y. W. Gu, C. H. Quek, and P. Cheang. 2003. Plasma spraying of functionally graded hydroxyapatite/Ti–6Al–4V coatings. *Surface and Coatings Technology* 168:195–201.
14. Gross, K. A., and S. S. Samandari. 2009. Nanoindentation on the surface of thermally sprayed coatings. *Surface and Coatings Technology* 203:3516–20.
15. Dey, A., A. K. Mukhopadhyay, S. Gangadharan, M. K. Sinha, D. Basu, and N. R. Bandyopadhyay. 2009. Nanoindentation study of microplasma sprayed hydroxyapatite coating. *Ceramics International* 35:2295–2304.
16. Dey, A., A. K. Mukhopadhyay, S. Gangadharan, M. K. Sinha, and D. Basu. 2009. Weibull modulus of nano-hardness and elastic modulus of hydroxyapatite coating. *Journal of Materials Science* 44:4911–18.
17. Dey, A., A. K. Mukhopadhyay, S. Gangadharan, M. K. Sinha, and D. Basu. 2009. Development of hydroxyapatite coating by microplasma spraying. *Materials and Manufacturing Processes* 24:1249–58.
18. Dey, A., A. K. Mukhopadhyay, S. Gangadharan, M. K. Sinha, and D. Basu. 2009. Characterization of microplasma sprayed hydroxyapatite coating. *Journal of Thermal Spray Technology* 18:578–92.
19. Dey, A., and A. K. Mukhopadhyay. 2011. Anisotropy in nanohardness of microplasma sprayed hydroxyapatite coating. *Advances in Applied Ceramics* 109:346–54.
20. Rossi, R. C. 1968. Prediction of the elastic moduli of composites. *Journal of American Ceramic Society* 51:433–40.
21. Mukhopadhyay, A. K., S. K. Datta, and D. Chakraborty. 1991. Hardness of silicon nitride and sialon. *Ceramics International* 17:121–27.
22. Luo, J., and R. Stevens. 1999. Porosity-dependence of elastic moduli and hardness of 3Y-TZP ceramics. *Ceramics International* 25:281–86.

35

Fracture Toughness of Ceramic Coating Measured by Nanoindentation

Arjun Dey and Anoop Kumar Mukhopadhyay

35.1 Introduction

The mode I fracture toughness (K_{Ic}) is the most important property that governs the microplasma-sprayed hydroxyapatite (MIPS–HAp) coating's intrinsic resistance against catastrophic failure. Therefore, it is important to understand the K_{Ic} behavior of these coatings for prospective biomedical implant applications. A critical survey of pertinent literature data [1–11] reveals that most of the K_{Ic} data (average ≈ 0.5 MPa·m$^{0.5}$) reported are evaluated by Vickers microindentation across the cross sections of dense MAPS (macroplasma-sprayed) and high-velocity oxy fuel (HVOF) sprayed HAp coatings on Ti6Al4V substrates and, indeed, very rarely by nanoindentation on MIPS–HAp coatings [7]. Therefore, in this chapter we shall try to show how we can measure fracture toughness and its load dependency by using the nanoindentation technique with a Berkovich indenter such that the measured value of toughness pertains to the microscale, compatible with the local microstructure of the cross section of highly porous MIPS–HAp coatings on SS316L substrates. The relevant details of the MIPS–HAp coatings preparation and the measurement of K_{Ic} by nanoindentation on these MIPS–HAp coatings are given elsewhere [7] and also discussed briefly in Chapter 9.

35.2 Fracture Toughness Behavior

The major experimental finding [7] was that the K_{Ic} data of the present MIPS–HAp coatings were as good as or even slightly higher than those reported (≈ 0.6 MPa·m$^{0.5}$) [1–6] for dense HAp coatings (Figure 35.1a). Further, they had almost negligible (i.e., very slight) degradation with load (Figure 35.1b). The slight degradation can be explained in terms of competitive dependencies of

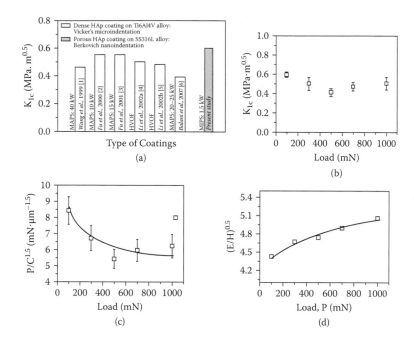

FIGURE 35.1
Dependencies of fracture toughness on (a) coating type and (b) load. Load dependencies of (c) $P/C^{1.5}$ and (d) $(E/H)^{0.5}$ for the MIPS–HAp coatings. (Reprinted with permission of Dey and Mukhopadhyay [7] from the American Ceramic Society and Wiley.)

the terms $(P/C^{1.5})$ (Figure 35.1c) and $(E/H)^{0.5}$ (Figure 35.1d) on the variations in load (P) values.

There was a singular study on load dependence of indentation fracture toughness for dense, post heat-treated HAp coating [10], wherein just a mere 2 N increase in load degraded the K_{Ic} of dense HAp coatings by as much as 16%. In sharp contrast, the load dependency of indentation fracture toughness of the present porous MIPS–HAp coating was minimal (Figure 35.1b) in the sense that even after an increase of load by an order of magnitude (from 100 to 1000 mN), the degradation was not more than 15%, even given the fact that the coating was porous in nature, with about 11 vol% open porosity. The high toughness of the present MIPS-HAp coating could be linked to high crystallinity, phase purity, and a small average size (2.2 μm) of micropores distributed uniformly across the microstructure, as shown elsewhere [7]. The microcracks showed a range of about 1.5 μm to about 5.5 μm and a small average size of only 4.4 μm [7].

It is well known [12–14] that the micropores may effectively blunt the crack. A higher toughness value may thus be imparted as the microstructure resists the catastrophic growth of the crack. What else could be done to toughen a coating? Intrinsic toughening mechanisms are difficult to be operative for such materials. That is why mechanisms like crack bridging

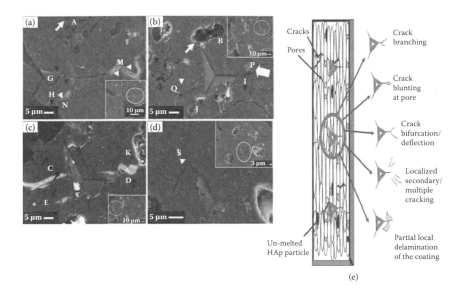

FIGURE 35.2
Crack blunting at (a) micropore, (b) macropore, (c) localized secondary or multiple cracking, and (d) crack branching. (e) Schematic of interaction of the nanoindentation crack with the microstructure. (Reprinted with permission of Dey and Mukhopadhyay [7] from the American Ceramic Society and Wiley.)

and/or microcracking [15, 16] were experimentally observed in the present coating, as shown in Figure 35.2.

When such an extrinsic toughening mechanism operates, two things may happen [15, 16]. There can be formation of a frontal process zone ahead of the growing crack. Along with such a process, a microcracking zone may occur behind the crack, as shown in Figures 35.2a–c. Two physical processes may happen in this zone. The first is that there can be volume dilation. The second is that, due to the microcracking, there is a reduction in local elastic modulus. If the dilation is constrained by surrounding material, there is a localized reduction in crack driving force. This localized reduction in the crack driving force can cause a local drop of stress intensity in the vicinity of the crack tip. Therefore, the crack cannot grow catastrophically any further. In other words, such a mechanism as suggested here may then act to shield the crack tip [17]. This is the most likely explanation for how the extrinsic toughening happens in the MIPS–HAp coating material in the present work.

Finally, there were some additional modes of "crack front" microstructure interaction, i.e., crack branching as well as crack deflection and/or crack bifurcation. In Figure 35.2a, crack branching is denoted by the white arrowheads at regions M and N. Similarly, in Figure 35.2d, the white arrowheads in region S demonstrate crack branching. In a similar note in Figure 35.2b, the large white arrow at regions P and Q depicts the crack deflection and/or bifurcation. The load was 300 mN for the cracks shown in Figure 35.2d.

Similarly, localized secondary microcracking had also happened, as seen in Figure 35.2c. Failure at the surface of the microplasma-sprayed splats, which connect the coating microstructure, had led to the crack branching and crack deflection and/or bifurcation. The stress distribution at the frontal damage zone might no longer correspond to the *K*-field, because now the crack front lies in a narrow zone of connecting splats between pores [18]. Rather, we may plausibly argue that the stress was now possibly distributed over a larger volume of porous, characteristically inhomogeneous material. Therefore, the inhomogeneous spatial distribution of the splats will now govern the inhomogeneities in stress distribution. Under such a situation, localized cracking within a given splat or between the splats may then occur in the regions that lie ahead of the main crack front. If this picture resembles the reality, such a generic process will cause a crack front damage zone. Provided the local stress had crossed the maximum that could be borne by the weakest zone inside a given splat or in between the neighboring splats, a local failure will happen. Thus, Figure 35.2e provides a schematic of the modes of interaction of the indentation cracks with coating microstructure, e.g.,

1. Crack branching
2. Crack blunting at the micropore and macropore levels
3. Crack deflection/bifurcation
4. Localized secondary/multiple cracking
5. Partial local delamination

Such interactions may also facilitate the process of energy dissipation and, hence, toughening of the present MIPS–HAp coating.

35.3 Conclusions

Despite having ≈11 vol% porosity, the MIPS–HAp coating on SS316L substrates showed a high fracture toughness of 0.59 MPa·m$^{0.5}$ at 100 mN load. This value was comparable to that typically reported in the literature (≈0.5 MPa·m$^{0.5}$). Further, the extensive study of the nanoindents by scanning electron microscopy also showed occurrences of crack blunting at micro- and macropores, localized secondary/multiple cracking, partial local delamination, crack branching, and crack deflection/bifurcation in the coating. These phenomena of crack propagation and interaction with the microstructure suggested various means of energy dissipation for achieving higher toughness in the present MIPS–HAp coating. So far so good, but we still do not know if the coating will remain good enough to perform when exposed to the human body environment, i.e., when the implant coated with such a coating is actually used in practice. That is why we shall now try to address the most important effect of a simulated body fluid (SBF) environment on the micro-/nanomechanical properties of a typical bioceramic coating in Chapter 36.

References

1. Wang, M., X. Y. Yang, K. A. Khor, and Y. Wang. 1999. Preparation and characterization of bioactive monolayer and functionally graded coatings. *Journal of Materials Science: Materials in Medicine* 10:269–73.
2. Fu, L., K. A. Khor, and J. P. Lim. 2000. Yttria stabilized zirconia reinforced hydroxyapatite coatings. *Surface and Coatings Technology* 127:66–75.
3. Fu, L., K. A. Khor, and J. P. Lim. 2001. Processing, microstructure and mechanical properties of yttria stabilized zirconia reinforced hydroxyapatite coatings. *Materials Science and Engineering A* 316:46–51.
4. Li, H., K. A. Khor, and P. Cheang. 2002. Young's modulus and fracture toughness determination of high velocity oxy-fuel-sprayed bioceramic coatings. *Surface and Coatings Technology* 155:21–32.
5. Li, H., K. A. Khor, and P. Cheang. 2002. Titanium dioxide reinforced hydroxyapatite coatings deposited by high velocity oxy-fuel (HVOF) spray. *Biomaterials* 23:85–91.
6. Balani, K., R. Anderson, T. Laha, M. Andara, J. Tercero, E. Crumpler, and A. Agarwal. 2007. Plasma-sprayed carbon nanotube reinforced hydroxyapatite coatings and their interaction with human osteoblasts in vitro. *Biomaterials* 28:618–24.
7. Dey, A., and A. K. Mukhopadhyay. 2013. Fracture toughness of microplasma sprayed hydroxyapatite coating by nanoindentation. *International Journal of Applied Ceramic Technology* 8:572–90.
8. Kobayashi, S., W. Kawai, and S. Wakayama. 2006. The effect of pressure during sintering on the strength and the fracture toughness of hydroxyapatite ceramics. *Journal of Materials Science: Materials in Medicine* 17:1089–93.
9. Khor, K. A., Y. W. Gu, C. H. Quek, and P. Cheang. 2003. Plasma spraying of functionally graded hydroxyapatite/Ti–6Al–4V coatings. *Surface and Coatings Technology* 168:195–201.
10. Kweh, S. W. K., K. A. Khor, and P. Cheang. 2000. Plasma-sprayed hydroxyapatite (HA) coatings with flame-spheroidized feedstock: Microstructure and mechanical properties. *Biomaterials* 21:1223–34.
11. Zhang, S., Y. S. Wang, X. T. Zeng, K. A. Khor, W. Weng, and D. E. Sun. 2008. Evaluation of adhesion strength and toughness of fluoridated hydroxyapatite coatings. *Thin Solid Films* 516:5162–67.
12. Callus, P. J., and C. C. Berndt. 1999. Relationships between the mode II fracture toughness and microstructure of thermal spray coatings. *Surface and Coatings Technology* 114:114–28.
13. Tsui, Y. C., C. Doyle, and T. W. Clyne. 1998. Plasma sprayed hydroxyapatite coatings on titanium substrates; Part 1: Mechanical properties and residual stress levels. *Biomaterials* 19:2015–29.

14. Yamada, Y., and R. Watanabe. 1996. Effect of dispersed pores on fracture toughness of HAp/PSZ composites. *Scripta Materialia* 34:387–93.
15. Case, E. D., I. O. Smith, and M. J. Baumann. 2005. Microcracking and porosity in calcium phosphates and the implications for bone tissue engineering. *Materials Science and Engineering A* 390:246–54.
16. Nalla, R. K., J. J. Kruzic, J. H. Kinney, and R. O. Ritchie. 2004. Effect of aging on the toughness of human cortical bone: Evaluation by R-curves. *Bone* 35:1240–46.
17. Nalla, R. K., J. J. Kruzic, and R. O. Ritchie. 2004. On the origin of the toughness of mineralized tissue: Microcracking or crack bridging? *Bone* 34:790–98.
18. Sorensen, B. F., and A. Horsewell. 2001. Crack growth along interfaces in porous ceramic layers. *Journal of the American Ceramic Society* 84:2051–59.

36

Effect of SBF Environment on Nanomechanical and Tribological Properties of Bioceramic Coating

Arjun Dey and Anoop Kumar Mukhopadhyay

36.1 Introduction

The applicability of MIPS–HAp coating in the biomedical field is now accepted [1–9]. However, the comparative study of nanomechanical properties of MIPS–HAp coatings before and after immersion in a simulated body fluid (SBF) solution has rarely been attempted [7]. But such study is really important for the technological advancement in the field of prosthetic replacement because, prior to animal and human trials, the nanomechanical reliability of the coating must be verified in an SBF medium. Therefore, our major focus in this chapter shall be the evaluation of nanomechanical properties before and after immersion in an SBF solution. The details of sample preparation and test techniques have been given elsewhere [7] and discussed briefly in Chapter 9.

36.2 Nano-/Micromechanical Behavior

Typical load–depth plots of the MIPS–HAp coating were obtained from the nanoindentation experiments conducted before and after SBF immersion, and these are shown in Figure 36.1a. The final depth of penetration increased only slightly with increase in immersion time from 1 to 14 days. As a result, there was only a very slight drop in the nanohardness (H) (Figure 36.1b), Young's modulus (E) (Figure 36.1c), and the coefficient of friction (μ) (Figure 36.1d). Similar results were reported by others [10] and explained in terms of enhanced porosity. A similar logic held up in the present study as well [7]. Further, a slight initial decrease followed by increase in the nanomechanical properties had occurred, as the apatite layer deposition process

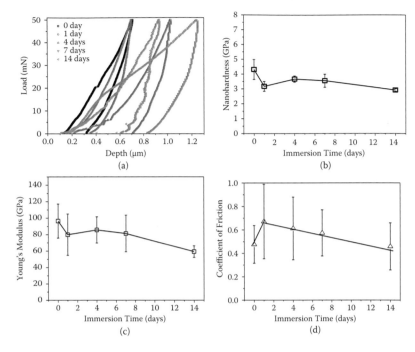

FIGURE 36.1 (See color insert.)
Effect of immersion time on (a) *P–h* data, (b) nanohardness, (c) Young's modulus, and (d) friction coefficient, μ. (Reprinted with permission of Dey and Mukhopadhyay [7] from the American Ceramic Society and Wiley.)

was dominant over the dissolution process during the subsequent 4 to 7 days of immersion in the SBF solution. This has been proved elsewhere [10] by ICP-AES (inductively coupled plasma atomic emission spectroscopy) study and scanning electron micrography. However, after 14 days of immersion, the nanohardness decreased very slightly in comparison to those measured at 4 and 7 days of immersion in the SBF solution (Figure 36.1a). The large number of dissolution sites observed from SEM study [7] ultimately formed a porous microstructure that possibly had slightly degraded the nanohardness of the MIPS–HAp coatings after 7 days of immersion in the SBF solution.

36.3 Tribological Study

The data on variation of average μ of the MIPS–HAp coatings as a function of the immersion time in the SBF solution are shown in Figure 36.1d. The μ increased from 0.47 to 0.67 after just 1 day of immersion in the SBF solution, possibly because much more dissolution took place, instead of apatite deposition, at the early time of immersion in the SBF solution. Afterward, the value of μ settled to 0.46, which was close to the value of μ that was measured for the virgin coating. This happened possibly because the deposition of loose apatite increased as the immersion time progressed from 1 day to 14 days. Further, the loose apatite could produce microwear debris when it contacted the indenter tip. Therefore, the possibility of indenter contact with the surface asperities could be much higher [11]. If such wear debris is engulfed between the indenter sides and the coating surface, the average coefficient of friction should drop, as was indeed experimentally observed. In other words, after 14 days of immersion in the SBF solution, the MIPS–HAp samples showed an excellent surface mechanical property, i.e., better scratch resistance, even when scratched at a ramping load of 10–10.6 N (Figure 36.1d). The data on variations of the μ as a function of sliding distance both before and after immersion in the SBF solution are shown in Figure 36.2. All the μ data increased sharply up to a certain sliding distance. These "critical" values of scratch lengths were 850, 1200, 1100, and 950 μm, respectively, for 0, 4, 7, and 14 days of immersion in the SBF solution. These data would also suggest the presence of a running-in period, in terms of the sliding distance. The corresponding μ values were ≈0.6, 0.8, 0.9, and 0.7. Afterwards, the value of μ remained almost constant within a band. However, within that regime, the strong, local variations of data were noticeable, possibly due to the interaction of the moving indenter tip with deposited loose apatite layer.

The detailed microstructural features of the scratch track are shown in Figures 36.3a–f. After 1 day of immersion, the scratch track showed

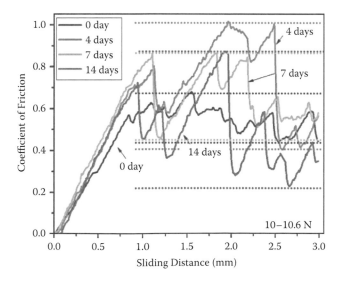

FIGURE 36.2
Variations of the coefficient of friction (μ) as a function of sliding distance. (Reprinted with permission of Dey and Mukhopadhyay [7] from the American Ceramic Society and Wiley.)

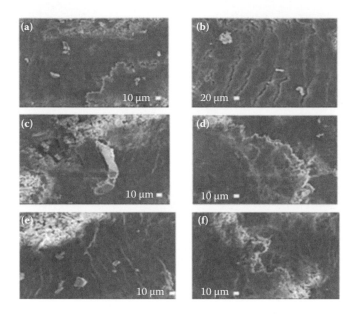

FIGURE 36.3
Scanning electron photomicrographs at higher magnification after different times in SBF solution: (a) 1 day; (b) higher magnification view of (a); (c) 4 days, showing microcracks; (d) 4 days, showing microfracture and microfragments; (e) 7 days, showing microcracks; and (f) 7 days, showing microfracture and microfragments. (Reprinted with permission of Dey and Mukhopadhyay [7] from the American Ceramic Society and Wiley.)

the formation of microcracks generated perpendicular to the scratch path (Figure 36.3a). The microcracks were almost parallel to each other (Figure 36.3b). A small amount of wear debris was also found, as seen in Figures 36.3a and 36.3b. The microfracture occurred at the side of the scratch track (Figure 36.3c), and microfragments were observed in the middle part of the scratch track (Figure 36.3d) after 4 days of immersion. After 7 days of immersion in the SBF solution, the parallel, relatively deeper microcracks were generated perpendicular to the scratch path (Figure 36.3e). These cracks formed as a result of the tensile frictional stress that had acted behind the trailing edge of the indenter [12]. In addition, the combination of microfracture and microfragments was also observed in the middle section of the scratch width (Figure 36.3f). Further, after 14 days of immersion in the SBF solution, the MIPS–HAp coating showed the deeper microcracks perpendicular to the scratch path (Figure 36.4a), peeling failure and microwear debris (Figure 36.4b), local microfracture (Figure 36.4c), and portions microchipped out at the side of the scratch track (Figure 36.4d). Further, scatter in data was observed in Figures 36.1a–d, which is generic for the plasma-sprayed coatings, including HAp [2, 13]. Researchers have reported [14] the delamination of HAp coatings during scratching that were not observed (Figures 36.3 and 36.4) in the present MIPS–HAp coatings.

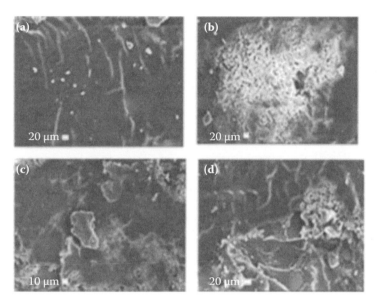

FIGURE 36.4
SEM photomicrographs after 14 days in the SBF solution: (a) microcracks generated perpendicular to the scratch path, (b) peeling failure and wear debris, (c) local microfracture, and (d) microchipped-out portion at the side of the scratch track. (Reprinted with permission of Dey and Mukhopadhyay [7] from the American Ceramic Society and Wiley.)

36.4 Conclusions

After immersion for 14 days in the SBF solution, the nanohardness and Young's modulus values and μ values were only slightly lower than those of the MIPS–HAp coating in the as-sprayed condition, suggesting the possibility of the major dominance of the nascent apatite deposition over the dissolution process. The values μ for the MIPS–HAp coating were immersion-time dependent, with a range of ≈0.5–0.8. However, there was no large-scale delamination or coating peel-off, which proved the stability of the coating after immersion in a synthetically produced body fluid environment.

So far we have learned many things about the nanoindentation responses and nanomechanical properties of bioactive ceramic coatings. But what about the nanomechanical behaviors of other types of coatings, for example, the protective oxide coatings typically used to resist corrosion of metallic alloys? It is in this context that we shall discuss in Chapter 37 the nanohardness and Young's modulus of the micro arc oxidized (MAO) coatings on AZ31B Mg alloys (≈3% Al, 1% Zn, and 96% Mg) evaluated at the microstructural length scale by the nanoindentation technique.

References

1. Dey, A., A. K. Mukhopadhyay, S. Gangadharan, M. K. Sinha, D. Basu, and N. R. Bandyopadhyay. 2009. Nanoindentation study of microplasma sprayed hydroxyapatite coating. *Ceramics International* 35:2295–2304.
2. Dey, A., A. K. Mukhopadhyay, S. Gangadharan, M. K. Sinha, and D. Basu. 2009. Weibull modulus of nano-hardness and elastic modulus of hydroxyapatite coating. *Journal of Materials Science* 44:4911–18.
3. Dey, A., A. K. Mukhopadhyay, S. Gangadharan, M. K. Sinha, and D. Basu. 2009. Development of hydroxyapatite coating by microplasma spraying. *Materials and Manufacturing Processes* 24:1249–58.
4. Dey, A., A. K. Mukhopadhyay, S. Gangadharan, M. K. Sinha, and D. Basu. 2009. Characterization of microplasma sprayed hydroxyapatite coating. *Journal of Thermal Spray Technology* 18:578–92.
5. Dingyong, H. E., Z. Qiuying, Z. Lidong, and S. Xufeng. 2007. Influence of microplasma spray parameters on the microstructure and crystallinity of hydroxyapatite coatings. *Chinese Journal of Materials Research* 21:657–63.
6. Zhao, Q., D. He, L. Zhao, and X. Li. 2011. In vitro study of microplasma sprayed hydroxyapatite coatings in hanks balanced salt solution. *Materials and Manufacturing Processes* 26:175–80.
7. Dey, A., and A. K. Mukhopadhyay. Forthcoming. In vitro dissolution, microstructural and mechanical characterizations of microplasma-sprayed hydroxyapatite coating. *International Journal of Applied Ceramic Technology*.
8. Cheang, P., and K. A. Khor. 1996. Addressing processing problems associated with plasma spraying of hydroxyapatite coatings. *Biomaterials* 17:537–44.
9. Junker, R., P. J. D. Manders, J. G. Wolke, Y. Borisov, and J. A. Jansen. 2010. Bone-supportive behavior of microplasma-sprayed CaP-coated implants: Mechanical and histological outcome in the goat. *Clinical Oral Implants Research* 21:189–200.
10. Yang, Y. C., E. Chang, and S. Y. Lee. 2003. Mechanical properties and Young's modulus of plasma sprayed hydroxyapatite coating on Ti substrate in simulated body fluid. *Journal of Biomedical Materials Research A* 67:886–99.
11. Fernandez, J., M. Gaona, and J. M. Guilemany. 2007. Effect of heat treatments on HVOF hydroxyapatite coatings. *Journal of Thermal Spray Technology* 16:220–28.
12. Sebastiani, M., G. Bolelli, L. Lusvarghi, P. P. Bandyopadhyay, and E. Bemporad. 2012. High resolution residual stress measurement on amorphous and crystalline plasma-sprayed single-splats. *Surface and Coating Technology* 206:4872–80.
13. Wang, L., Y. Wang, X. G. Sun, J. Q. He, Z. Y. Pan, and C. H. Wang. 2012. Microstructure and indentation mechanical properties of plasma sprayed nano-bimodal and conventional ZrO_2-8wt% Y_2O_3 thermal barrier coatings. *Vacuum* 86:1174–85.
14. Gu, Y. W., K. A. Khor, D. Pan, and P. Cheang. 2004. Activity of plasma sprayed yttria stabilized zirconia reinforced hydroxyapatite/Ti–6Al–4V composite coatings in simulated body fluid. *Biomaterials* 25:3177–85.

37

Nanomechanical Behavior of Ceramic Coatings Developed by Micro Arc Oxidation

Arjun Dey, R. Uma Rani, Hari Krishna Thota, A. Rajendra,
Anand Kumar Sharma, Payel Bandyopadhyay,
and Anoop Kumar Mukhopadhyay

37.1 Introduction

Micro Arc Oxidation (MAO) is an electrochemical surface treatment process for generating oxide coatings on metals. Here, the substrate metal is chemically converted into its oxide. The MAO coating grows both inward and outward from the original metal surface. It is a conversion coating. Therefore, unlike a typically deposited coating, e.g., by plasma spraying, it almost obviously has excellent adhesion to the substrate metal. It is similar to anodizing, but it employs much higher potentials of 200 V or so. As a result, discharges occur and the plasma forms. This plasma modifies the structure of the oxide layer. A wide range of substrate alloys can be coated. For example, it can include all wrought aluminum alloys and most cast alloys, although the high levels of silicon can reduce coating quality. On metals such as aluminum, magnesium, and titanium, this MAO process can be used to grow largely crystalline, oxide coatings that are tens or hundreds of micrometers thick. Because the MAO coatings generally possess high hardness and a continuous barrier, they are capable of offering protection against wear, corrosion, or heat as well as electrical insulation. Despite the wealth of literature on Al and Ti alloys [1–4], researchers have very seldom addressed [5] the nanohardness and elastic modulus of MAO coatings on AZ31B Mg alloys ($\approx$3% Al, 1% Zn, and 96% Mg) evaluated at the microstructural length scale by the nanoindentation technique. The as-prepared silicate-based MAO (S-MAO) samples are designated as SU-MAO, while the sealed silicate-based MAO samples are designated as SS-MAO. At least 20 indents were made at five different locations of the sample, as the nature of the surface of the coating was porous and heterogeneous. The measurements were taken well within 10% of the coating thickness, such that there would not be any influence of the substrate's mechanical properties on the measured data.

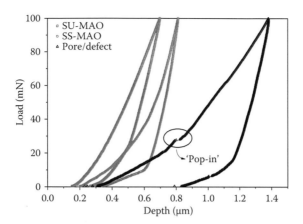

FIGURE 37.1 (See color insert.)
Typical load–depth (*P–h*) plots of the various MAO coatings. (Reprinted with permission of Dey et al. [5] from Elsevier.)

37.2 Nanoindentation Study and Reliability Issue

The typical load-versus-depth (*P–h*) plots of the MAO coatings are shown in Figure 37.1. Generally, the *P–h* plots showed normal behavior in the case of both the SU-MAO and SS-MAO samples. The general nature of these plots indicated the presence of an elastoplastic deformation process, as expected for brittle materials. Further, irregular behavior of the *P–h* plot was also observed due to the influence of defects like pores and microcracks in the MAO coatings, which is characteristic of the MAO process. Moreover, there was also a pronounced presence of large pop-ins due to interaction of pores.

The Weibull distribution fittings for the nanohardness and Young's modulus data of the MAO coatings are displayed in Figures 37.2a and 37.2b, respectively, for the as-grown condition and in Figures 37.2c and 37.2d, respectively, after immersion in the corrosive environment. The characteristic values of nanohardness and Young's modulus were calculated from these data through the application of Weibull statistics following Zhou et al. [6]. The characteristic nanohardness (H_{ch}) value was ≈3 GPa. After immersion in the corrosive environment, the magnitude of (H_{ch}) was marginally reduced for both sealed as well as unsealed MAO coatings (Table 37.1). Further, sealing of MAO samples showed no improvement of nanohardness; however, it showed marginal improvement of the Young's modulus. On the other hand, the characteristic Young's modulus, E_{ch} ≈ 90 GPa, of unsealed MAO coatings was not altered after immersion in the corrosive environment (Table 37.2). In contrast, the sealed MAO coatings showed E_{ch} ≈ 96 GPa, which was marginally decreased to ≈94 GPa when exposed to the corrosive environment. Further, the Weibull modulus, m values of nanohardness for the present MAO

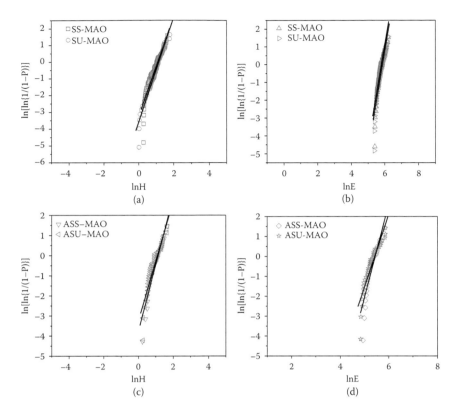

FIGURE 37.2
Weibull plots of (a) nanohardness and (b) Young's modulus data of MAO coatings prior to the corrosion test and (c) nanohardness and (d) Young's modulus data of MAO coatings after corrosion test. (Prefix A stands for after immersion in corrosive environment.) Reprinted with permission of Dey et al. [5] from Elsevier.)

TABLE 37.1

Weibull Analysis Data for the Nanohardness and Young's Modulus of the MAO Coatings before Corrosion Test

Sample Name	H_{ch} (GPa)	m	Correlation Coefficient (R^2)	E_{ch} (GPa)	m	Correlation Coefficient (R^2)
SU-MAO	3.09	3.1	0.97	90.23	5.5	0.85
SS-MAO	3.11	2.9	0.88	96.42	5.6	0.91

Source: Reprinted/modified with permission of Dey et al. [5] from Elsevier.

coatings were ≈3, irrespective of whether a coating was sealed or not and whether the condition was prior to or post immersion in the corrosive environment. However, the m values of the Young's modulus data were in the range of ≈3.5–5.5. These low values signify a high degree of "characteristic" scatter in the data. This was nothing unexpected, because the coatings had

TABLE 37.2

Weibull Analysis Data for the Nanohardness and Young's Modulus of the MAO
Coatings after Corrosion Test

Sample Name	H_{ch} (GPa)	m	Correlation Coefficient (R^2)	E_{ch} (GPa)	m	Correlation Coefficient (R^2)
ASU-MAO	3.04	2.9	0.86	90.02	3.6	0.83
ASS-MAO	3.09	3.3	0.92	93.87	4.6	0.85

Source: Reprinted/modified with permission of Dey et al. [5] from Elsevier. (Prefix A stands for after immersion in corrosive environment.)

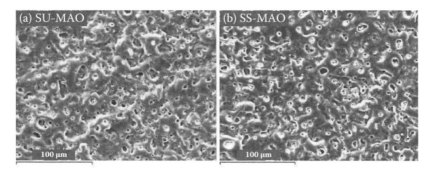

FIGURE 37.3
SEM photomicrographs of the (a) SU-MAO and (b) SS-MAO samples. (Reprinted with permission of Dey et al. [5] from Elsevier.)

"characteristically heterogeneous microstructures" dotted with a "statistical distribution of defects, e.g., pores and cracks/microcracks" [1–3] (Figure 37.3).

In post-immersion condition in the corrosive environment, the m values for both the nanohardness as well as Young's modulus had decreased with respect to those obtained prior to the immersion. This observation means that the defect density increases in post-immersion condition over and above their already existing population in the respective coating microstructures. Due to higher defect density in the post-immersion condition, more scatter is introduced into the data. This process led to the further reduction of the m values in comparison to those obtained prior to immersion in the corrosive environment.

37.3 Conclusions

The characteristic values of nanohardness and Young's modulus of the MAO coatings were evaluated through the Weibull approach to be ≈3 GPa and ≈90 GPa, respectively. Further, sealing of MAO samples showed only a marginal improvement of Young's modulus while the nanohardness remained as good as that of the unsealed coating. However, the sealed coating performed much better than the unsealed one in the corrosive environment. Now that we have gained some knowledge about the nanomechanical behaviors and nanomechanical properties of thick ceramic coatings, our point of attention in Chapter 38 will be the soft ceramic thin films, which are typically used for many electronic, optical, electro-optical, magnetic, energy-conserving, and many other functional applications.

References

1. Curran, J. A., and T. W. Clyne. 2006. Porosity in plasma electrolytic oxide coatings. *Acta Materialia* 54:1985–93.
2. Curran, J. A., and T. W. Clyne. 2005. Thermo-physical properties of plasma electrolytic oxide coatings on aluminium. *Surface and Coatings Technology* 199:168–76.
3. Wheeler, J. M., C. A. Collier, J. M. Paillard, and J. A. Curran. 2010. Evaluation of micromechanical behaviour of plasma electrolytic oxidation (PEO) coatings on Ti–6Al–4V. *Surface and Coatings Technology* 204:3399–3409.
4. Datcheva, M., S. Cherneva, M. Stoycheva, R. Iankov, and D. Stoychev. 2011. Determination of anodized aluminum material characteristics by means of nanoindentation measurements. *Materials Science and Applications* 2:1452–64.
5. Dey, A., R. U. Rani, H. K. Thota, A. K. Sharma, P. Bandyopadhyay, and A. K. Mukhopadhyay. 2013. Microstructural, corrosion and nanomechanical behaviour of ceramic coatings developed on magnesium AZ31 alloy by micro arc oxidation. *Ceramics International* 39:313–20.
6. Zhou, H., F. Li, B. He, J. Wang, and B. Sun. 2007. Air plasma sprayed thermal barrier coatings on titanium alloy substrates. *Surface and Coatings Technology* 201:7360–67.

Section 10

Static Contact Behavior of Ceramic Thin Films

38

Nanoindentation Behavior of Soft Ceramic Thin Films: $Mg(OH)_2$

Pradip Sekhar Das, Arjun Dey, and Anoop Kumar Mukhopadhyay

38.1 Introduction

There is very little experimental data available on the nanomechanical properties of layered double hydroxides such as magnesium hydroxide, $Mg(OH)_2$. As mentioned previously in Chapter 9, such materials have potential future applications in chemo-sensing, catalysis, and energy devices. Therefore, the idea behind this work was to understand the response of the layered structure in an ultralow load application. It is in this context that we report the nanoindentation behavior of a 1.5 μm $Mg(OH)_2$ film chemically deposited on a commercially available soda-lime-silica (SLS) glass substrate at room temperature. To the best of our knowledge, this is the very first time that such an effort has been made.

38.2 Nanoindentation Study

The results presented in this chapter are based on nanoindentation experiments conducted with a Hysitron Tribo Indenter, as mentioned in Chapter 9. During these experiments, the thermal drift was kept at <0.05 nm·s⁻¹. Further details may be found in the literature [1]. The nanohardness and Young's modulus data of the present $Mg(OH)_2$ thin film were calculated using the method proposed by the well-established Oliver-Pharr model [2]. Three different loads (50, 70, and 100 μN) were applied in the present experiments. The unloading part of the experimental load-versus-depth plot was analyzed using the following equation [2]:

$$P = \alpha(h - h_f)^m \tag{38.1}$$

where h is the instantaneous depth of penetration at a nanoindentation load P. The quantity h_f is the final depth of penetration, and both α and m are fitting parameters.

For nanoindentation loads (P) of 50, 70, and 100 µN, the maximum depths of penetration were about 90, 100, and 140 nm. Thus, h_{max} values were well below 10% of film thickness (1.5 µm), as seen in Figure 38.1a. The average elastic recovery was as high as ≈70% (Figure 38.1a). The nanohardness of the Mg(OH)$_2$ thin films was measured as 0.26–0.24 GPa (Figure 38.1b) when measured at loads of 50, 70, and 100 µN. Similarly, the Young's modulus data were 2.83–2.70 GPa (Figure 38.1c) when measured at loads of 50, 70, and 100 µN. The best-fit equations, given as insets of Figures 38.1b and 38.1c, predicted nanohardness of 0.31 GPa and Young's modulus of 2.97 GPa for chemically deposited Mg(OH)$_2$ thin films on the SLS glass substrates. To the best of our knowledge, this is the first such experimental data on chemically deposited Mg(OH)$_2$ thin films on SLS glass substrates.

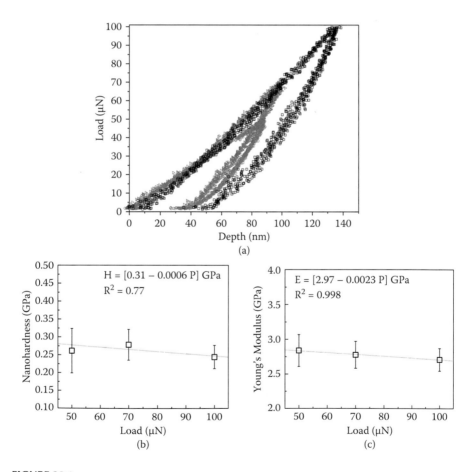

FIGURE 38.1
(a) Typical P–h plots, (b) nanohardness, and (c) Young's moduli as a function of loads. (Reprinted with permission of Das et al. [4] from Institute of Materials, Minerals and Mining and Maney.)

38.3 Energy Calculation

Further, the ratio W_p/W_t—the ratio of plastic deformation work (W_p) to total energy of indentation (W_t)—was varied in the range ≈ 0.43–0.57. In a complementary fashion, the ratio W_e/W_t of elastic (W_e) energy of indentation to W_t was varied in the range ≈ 0.57–0.43 (Figure 38.2). Here, the underlying assumption is that $W_t = W_e + W_p$ [2]. The data in Figure 38.2 matched well with reported experimental data [3] on other materials. The decreasing trend of W_p/W_t (Figure 38.2) with increase in load happened because the ratio (h_f/h_{max}) [2] decreased with increase in load (Figure 38.1a). The increasing trend of W_e/W_t with increase in nanoindentation load (Figure 38.2) possibly reflected the fact that the current Mg(OH)₂ thin films had a microstructure that was inherently capable of more elastic recovery, the more it was deformed. It has been suggested that the layered microstructure of the chemically deposited magnesium hydroxide thin films helped to accommodate the strain due to the applied nanoindentation load and thereby effected the elastic recovery through localized rearrangements upon withdrawal of the nanoindenter during the unloading cycle [4]. Fitting the data from the unloading part of the nanoindentation load-versus-depth plots (Figure 38.1a) to equation (38.1) gave $\alpha \approx 0.017$–0.024 (Figures 38.3a–c), which matched well with the

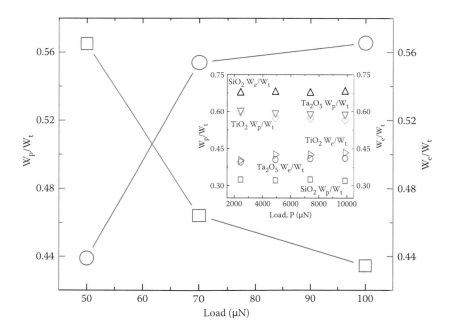

FIGURE 38.2 (See color insert.)
Energy ratio data as a function of loads. Inset: literature data [3]. (Reprinted with permission of Das et al. [4] from Institute of Materials, Minerals and Mining and Maney.)

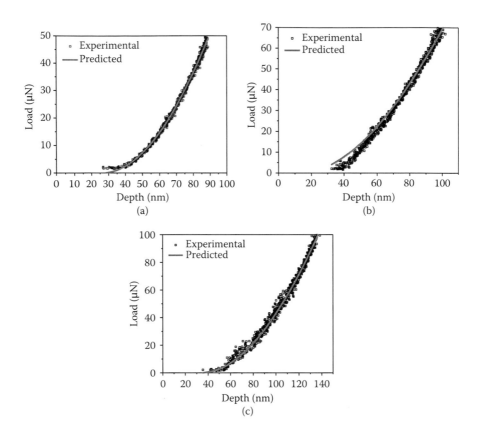

FIGURE 38.3
Fitting of unloading data from Figure 38.1a to equation (38.1) at different peak loads: (a) 50 µN, (b) 70 µN, and (c) 100 µN. Hollow symbols: experimental data; solid lines: predicted trend. (Reprinted with permission of Das et al. [4] from Institute of Materials, Minerals and Mining and Maney.)

experimental data reported for other materials [1], and m was estimated to be ≈1.79–1.92, with an average of 1.86 ± 0.06. These α and m data suggested the conical punch geometry as the best approximation for the present nanoindenter, following Sneddon's analysis [5].

38.4 Conclusions

By virtue of a strain-tolerant layered microstructure, the soft magnesium hydroxide thin films deposited by an inexpensive green chemical deposition technique on SLS glass substrates films showed nanohardness ≈0.3 GPa and Young's modulus ≈3 GPa. Thus, the nanostructured pure magnesium hydroxide film may be useful for protective as well as structural applications. But suppose that we have a hard ceramic thin film instead of a soft thin film. What would be the nanoindentation response of a thin film such as TiN? That will be our topic of discussion in Chapter 39.

References

1. Dey, A., R. Chakraborty, and A. K. Mukhopadhyay. 2011. Nanoindentation of soda lime–silica glass: Effect of loading rate. *International Journal of Applied Glass Science* 2:144–55.
2. Oliver, W. C., and G. M. Pharr. 1992. An improved technique for determining hardness and elastic modulus using load and displacement sensing indentation experiments. *Journal of Materials Research* 7:1564–83.
3. Attaf M. T. 2003. New ceramics related investigation of the indentation energy concept. *Materials Letters* 57:4684–93.
4. Das, P. S., A. Dey, M. R. Chaudhuri, S. Roy, A. K. Mandal, N. Dey, and A. K. Mukhopadhyay. 2012. Chemically deposited magnesium hydroxide thin film. *Surface Engineering* 28:731–36.
5. Sneddon, I. N. 1965. The relation between load and penetration in the axisymmetric Boussinesq problem for a punch of arbitrary profile. *International Journal of Engineering Science* 3:47–57.

39

Nanoindentation Study on Hard Ceramic Thin Films: TiN

Arjun Dey and Anoop Kumar Mukhopadhyay

39.1 Introduction

This chapter discusses the results of a nanoindentation study on a hard ceramic thin film, i.e., Titanium Nitride (TiN). In general, Ti alloys are deployed extensively for biomedical, aerospace as well as spacecraft, automobile, cutting tools, protective surface, etc., applications. The reason is that they offer an attractive combination of higher strength-to-weight ratio, low thermal conductivity, low thermal expansion, and good corrosion resistance. However, the principal limitation of Ti alloys is their poor wear resistance, which for the most part is due to their low surface hardness. In order to solve these problems, different surface modification techniques, such as the plasma-assisted thermochemical treatment, plasma nitriding, ion implantation, and overlay thin-film deposition have been performed to obtain hard TiN thin films and coatings [1–8]. One of the best ways to take advantage of the high hardness of these TiN thin films or coatings is to use them for surface protection of other important materials like SS304 steel. These steels, in their many variants, find applications in machine tools, food processing, chemical production, and most importantly in engineering component industries. Therefore, in developing applications of TiN as a protective, hard, corrosion-resistant coating on SS304, the first thing to be done is to understand the nanomechanical properties of the TiN thin films in sufficient detail.

Kim, Park, and Lee [1] reported that the nanohardness (25–29 GPa) and Young's modulus (330–375) of TiN thin films were sensitive to the flow rates of Ar and N_2. The data may have wide variation, e.g., nanohardness of 28–35 GPa [2]. Karimi et al. [3] showed that depending upon the ratio of [N] and [Ti], a wide range of Young's modulus, e.g., 100–250 GPa, may be obtained for the TiN thin films. In sharp contrast to the higher nanohardness values, the nanoindentation study by Fang, Jian, and Chuu [4] reported 12–14 GPa nanohardness and 170–180 GPa Young's modulus. Still wider ranges have also been reported [5], e.g., about 12–22 GPa nanohardness and 210–350

GPa of Young's modulus. In other work, Chou, Yu, and Huang [6] showed that hardness depends on the thickness of the TiN coating. They reported hardness values in the range of about 15 to 32 GPa while thickness increased from 0.2 μm to 1.7 μm. Kumar et al. [7] showed that the nanomechanical properties of TiN were also strongly dependent on the crystallographic orientation. TiN in (111) orientation showed a comparatively lower hardness and modulus value of about 6 GPa and 120 GPa, respectively. However, TiN in the (200) orientation showed a comparatively higher hardness and modulus value of about 12 GPa and 140 GPa, respectively. The depth dependency of TiN film was also reported by others [8, 9]. For instance, Ma et al. [8] found that the hardness decreases from about 27 GPa to 12 GPa as the depth of indentation increases. Therefore, it is indeed true that the nanoindentation studies of TiN thin films and coatings is a very vibrant research area where many aspects are yet to be well understood, even though there is a wealth of literature on the subject [1–8]. It was therefore our plan to conduct partial-unloading nanoindentation experiments on a TiN film on SS304 substrate.

39.2 Nanoindentation Study

In this study, we performed the partial-unloading nanoindentation experiments at loads up to 10,000 μN with a Berkovich tip on ≈1200 nm thick TiN films deposited on SS304 by magnetron sputtering technique. Typical partial-unloading P–h data plots of the TiN films are shown in Figure 39.1. A total of 53 cycles were employed in the present study. The P–h data plots showed the typical elastoplastic behavior expected for a ceramic material. The penetration depth was kept up to ≈100 nm, which was less than 10% of the total film thickness (120 nm) to avoid the substrate effect. Figure 39.2 depicts the footprints of the nanoindentation array. It is noteworthy that there is no evidence of any cracks or microfracture occurring around the indents. Further, no pileup was found.

The corresponding nanohardness and Young's modulus data are shown in Figures 39.3a and 39.3b, respectively. Three data sets have been included here, which represent the average of data obtained from three nanoindentations conducted in a row-wise fashion. In both nanohardness and Young's modulus data, a strong indentation size effect (ISE) was found up to 35–40 nm. The hardness and Young's modulus were about 45 GPa and 400 GPa, respectively, at a few nanometers of depth, while these dropped to about 20 GPa and 250 GPa, respectively, at about 40 nm depth. Up to this depth, there was a very strong ISE present in the experimental data. The issue of the genesis of ISE, particularly in thin and thick films, is yet to be unequivocally established. This issue is discussed in Chapter 46.

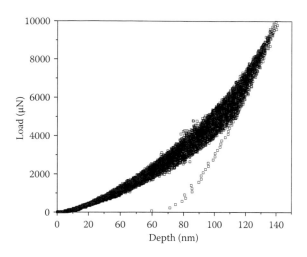

FIGURE 39.1
Typical *P–h* plot on TiN thin film on SS304.

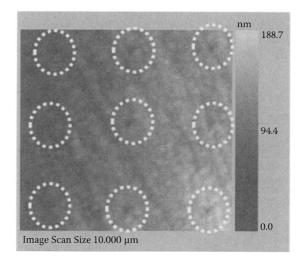

FIGURE 39.2
Typical footprints of nanoindentation on TiN thin film on SS304.

39.3 Depth Dependent Nanomechanical Behavior

There can be two ways to look at the data beyond this critical depth of 40 nm. Since the variations in data are not very high, it may be argued that the data were depth independent at depths beyond 40 nm. Thus, the average data of nanohardness and Young's modulus were about 25 GPa and 270 GPa,

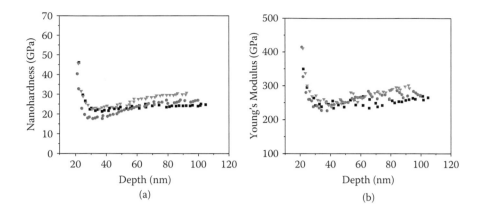

FIGURE 39.3

(a) Nanohardness and (b) Young's modulus as a function of depth for TiN thin film on SS304.

respectively. These data matched well with literature data [1–8]. We believe that there is no influence of substrate in the present nanoindentation data, as we had deliberately kept the nanoindentation depth much less than 10% of the total thickness of the TiN film.

The second or more critical look at the data shown in Figures 39.3.a and 39.3.b might also question if there was a reverse indentation size effect (RISE) present in the data. As of now, both of these outlooks appear interesting. Since the mechanical properties of the substrate did not influence the measurement of nanoindentation data on the TiN thin films, if these observations are really indicative of a RISE, the interesting question that automatically arises is why that would happen, if it happened at all. However, the present authors opined that this is plausibly linked with residual stress which is higher at the region near to the substrate-film interface. Clearly, further research would be necessary to resolve the interesting possibilities posed by the present nanoindentation data on TiN thin films.

39.4 Conclusions

We have presented the data from the partial-unloading nanoindentation experiments conducted with loads up to 10,000 μN with a Berkovich tip on ≈1200 nm thick TiN films deposited on SS304 by magnetron sputtering technique. In all the experiments, the depth of penetration was strictly kept well below 10% of the film thickness to avoid any ambiguity regarding substrate influence. Both nanohardness and Young's modulus data showed a strong ISE up to ≈40 nm. Like TiN, alumina thin films also have many important applications and associated intricacies of the process of nanoindentation-induced damage evolution. Therefore, we shall present the nanoindentation behavior of alumina thin films for space applications in Chapter 40.

References

1. Kim, T.-S., S.-S. Park, and B.-T. Lee. 2005. Characterization of nano-structured TiN thin films prepared by RF magnetron sputtering. *Materials Letters* 59:3929–32.
2. Mayrhofer, P. H., F. Kunc, J. Musil, and C. Mitterer. 2003. A comparative study on reactive and non-reactive unbalanced magnetron sputter deposition of TiN coatings. *Thin Solid Films* 415:151–59.
3. Karimi, A., O. R. Shojaei, T. Kruml, and J. L. Martin. 1997. Characterisation of TiN thin films using the bulge test and the nanoindentation technique. *Thin Solid Films* 308–309:334–39.
4. Fang, T.-H., S.-R. Jian, and D.-S. Chuu. 2004. Nanomechanical properties of TiC, TiN, and TiCN thin films using scanning probe microscopy and nanoindentation. *Applied Surface Science* 228:365–72.
5. Shojaei, O. R., and A. Karimi 1998. Comparison of mechanical properties of TiN thin films using nanoindentation and bulge test. *Thin Solid Films* 332:202–8.
6. Chou, W. J., G. P. Yu, and J. H. Huang. 2002. Mechanical properties of TiN thin film coatings on 304 stainless steel substrates. *Surface and Coatings Technology* 149:7–13.
7. Kumar, A., D. Singh, R. Kumar, and D. Kaur. 2009. Effect of crystallographic orientation of nanocrystalline TiN on structural, electrical and mechanical properties of TiN/NiTi thin films. *Journal of Alloys and Compounds* 479:166–72.
8. Ma, L. W., J. M. Cairney, M. J. Hoffman, and P. R. Munroe. 2006. Deformation and fracture of TiN and TiAlN coatings on a steel substrate during nanoindentation. *Surface and Coatings Technology* 200:3518–26.
9. An, J., and Q. Y. Zhang. 2005. Structure, morphology and nanoindentation behavior of multilayered TiN/TaN coatings. *Surface and Coatings Technology* 200:2451–58.

40

Nanoindentation Study on Sputtered Alumina Films for Spacecraft Application

I. Neelakanta Reddy, N. Sridhara, V. Sasidhara Rao,
Anju M. Pillai, Anand Kumar Sharma, V. R. Reddy,
Anoop Kumar Mukhopadhyay, and Arjun Dey

40.1 Introduction

This chapter describes the nanomechanical properties of alumina thin film, particularly developed by the sputtering technique, because of its superiority among other deposition methods in terms of uniformity with high deposition rates. Alumina thin film is a very popular material for biomedical, automobile, aerospace, etc., applications. For spacecraft application, metallic foils, e.g., SS304, Ti, Ta, etc., are often employed as radiation heat shields [1]. The basic idea of the present research is that the introduction of thin and high transmittance alumina films may act as a protective shield on these metallic foils. Thus, it is proposed that thin, transparent alumina ceramic films could provide surface protection for a substrate while also retaining the thermo-optical properties of the metallic substrate.

In the present study, alumina thin films were deposited at room temperature on SS304 thin foils as well as quartz substrates by the direct rf magnetron sputtering method [1]. To measure the optical properties, the quartz substrate was used because of its transparent nature. For evacuating the sputtering chamber and to obtain a high vacuum of better than 5.0×10^{-6} mbar, a combination of rotary and turbo molecular pumps was utilized, while during deposition, a pressure of only 1.2×10^{-2} mbar was maintained.

40.2 Optical Behavior

The deposited alumina films gave an excellent adhesion with both SS and quartz substrates when measured by the standard, conventional tape peel-off test according to the ASTM D903 protocol. Figure 40.1a shows a typical

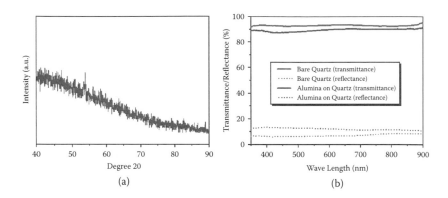

FIGURE 40.1 (See color insert.)
(a) XRD patterns of alumina films on SS304 and (b) transmittance and reflectance spectra of alumina film deposited on quartz and bare quartz. (Reprinted/modified with permission of Reddy et al. [1] from Elsevier.)

XRD spectrum of the as-deposited alumina thin films, which are amorphous in nature [1]. Figure 40.1b shows the transmittance and reflectance spectra of alumina on quartz and bare quartz. A high transmittance value of about 93% was found for bare quartz substrate, while after alumina coating, the transmittance value had dropped only slightly to ≈89% [1]. The reflectance behaviors for both cases were found to be in correspondence with each other. Thus, it is proven that the present alumina thin films can retain almost all of the optical properties of the quartz substrate. For the purpose of measuring nanoindentation, we grew alumina for several hours and got a thicker film. The idea behind using the higher thickness film was to avoid the influence of the substrate's mechanical properties on the measured nanomechanical properties of the alumina thin films. The implication of substrate effect in nanoindentation-based measurement is explained in greater depth in Chapter 51. Of course, one can ask about the transmittance value for such a thick (≈1.2 µm) coating. Actually, we have also measured the transmittance value of codeposited alumina on quartz. The data showed around ≈85% transmittance [2]. So, the main objective to retain the optical properties of the substrate was indeed achieved to a reasonable level of satisfaction.

40.3 Nanomechanical Behavior

Next, for the nanoindentation study, a continuous stiffness measurement (CSM) mode was adapted to produce variation of stiffness as a function of depth in a single indentation experiment. Here, the depth of experiment was kept up to ≈160 nm to avoid the influence of the substrate's mechanical properties on the measured nanomechanical properties of the alumina

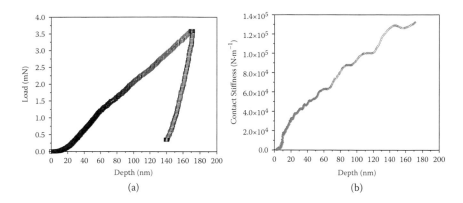

FIGURE 40.2
(a) Typical *P–h* plot on alumina film and (b) variation of contact stiffness as a function of depth.

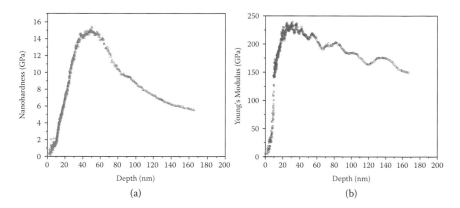

FIGURE 40.3
Variation of (a) nanohardness and (b) Young's modulus as a function of depth.

thin films. The drift value was maintained to as low as <0.1 nm·s⁻¹. Figure 40.2a shows the typical *P–h* plot data obtained for the CSM mode, and Figure 40.2b shows the variations of the contact stiffness as a function of indentation depth. Stiffness values increased with increasing depth of penetration. The variation of nanohardness and the Young's modulus as a function of the indentation depth are shown in Figures 40.3a and 40.3b, respectively. Initially, up to about 20–30 nm, both nanohardness and modulus were continuously increased, possibly due to inappropriate contact between the tip and sample surface at such ultralow depth. Beyond that point, however, both nanohardness and Young's modulus showed a strong indentation size effect (ISE). The nanohardness was ≈15 GPa at lower depth and dropped to ≈6 GPa at a relatively greater depth of ≈160 nm. In a similar manner, the Young's modulus value of ≈240 GPa was measured at shallower depth and subsequently dropped to ≈150 GPa at a relatively greater depth of ≈160 nm.

Thus, there was strong ISE in the data. The detailed reasoning about the origin of the ISE is discussed in Chapter 46.

If we consult the literature about the nanoindentation of alumina thin films, we can understand that it is really rich. In fact, it was already documented by us [1, 2] that nanoindentation behavior of the alumina thin films is strongly dependent on the deposition method, substrate temperature, post-deposition annealing temperatures and times, thickness, stoichiometry, phases present, the presence of the residual stress, etc. We found that hardness data of the alumina thin films measured by the nanoindentation technique were reported by other researchers to be in the range of ≈6 to 27 GPa, whereas the Young's modulus data of the alumina thin films measured by the nanoindentation technique were reported by other researchers to be in the range of ≈40 to 219 GPa [3–16]. Thus, the present experimental value matched well with the data range reported in the literature. Comparatively lower range of the preset data was plausibly linked with lack of crystallinity of the alumina film as was also depicted from XRD data (Figure 40.1a).

40.4 Conclusions

In the present study, alumina thin film was deposited on SS304 thin foils as well as quartz substrates by the direct rf magnetron sputtering method at room temperature. The basic idea of the present research was to study the feasibility of using thin and high transmittance alumina films as a protective shield on metallic foils without distorting the optical properties of the metallic thin foil substrates that are used as radiation heat shields in spacecraft propulsion systems. Using a continuous stiffness measurement (CSM) mode, the nanoindentation study gave indentation size effect (ISE) in both nanohardness and Young's modulus. Further, the measured data were within the range reported in the literature. Apart from conventional hard ceramic films like alumina and TiN, the metal nanoparticle encapsulated diamond-like carbon (M-DLC) thin films, prepared by a wide variety of techniques, have also evolved in recent times as a new class of materials for advanced tribological applications. Therefore, in Chapter 41, our focus will be the nanoscale static and dynamic contact responses of such M-DLC films.

References

1. Reddy, I. N., V. R. Reddy, N. Sridhara, S. Basavaraja, A. K. Sharma, and A. Dey. 2013. Optical and microstructural characterizations of pulsed RF magnetron sputtered alumina thin film. *Journal of Materials Science and Technology*, 29: 929–936.

2. Reddy, I. N., V. R. Reddy, N. Sridhara, V. S. Rao, M. Bhattacharya, A. K. Mukhopadhyay, A. K. Sharma, and A. Dey. 2013. Development of alumina thin film by magnetron sputtering: Study on optical, microstructural and mechanical properties. Unpublished work.
3. Musil, J., J. Blazek, P. Zeman, S. Proksova, M. Sasek, and R. Cerstvy. 2010. Thermal stability of alumina thin films containing γ-Al$_2$O$_3$ phase prepared by reactive magnetron sputtering. *Applied Surface Science* 257:1058–62.
4. Schneider, J. M., W. D. Sproul, R. W. J. Chia, M. S. Wong, and A. Matthews. 1997. Very-high-rate reactive sputtering of alumina hard coatings. *Surface and Coatings Technology* 96:262–66.
5. Bobzin, K., E. Lugscheider, M. Maes, and C. Pinero. 2006. Relation of hardness and oxygen flow of Al$_2$O$_3$ coatings deposited by reactive bipolar pulsed magnetron sputtering. *Thin Solid Films* 494:255–62.
6. Khanna, A., D. G. Bhat, A. Harris, and B. D. Beake. 2006. Structure-property correlations in aluminum oxide thin films grown by reactive AC magnetron sputtering. *Surface and Coating Technology* 201:1109–16.
7. Zywitzki, O., and G. Hoetzsch. 1997. Correlation between structure and properties of reactively deposited Al$_2$O$_3$ coatings by pulsed magnetron sputtering. *Surface and Coating Technology* 94–95:303–8.
8. Chiba, Y., Y. Abe, M. Kawamura, and K. Sasaki. 2008. Formation process of Al$_2$O$_3$ thin film by reactive sputtering. *Vacuum* 83:483–85.
9. Schneider, J. M., W. D. Sproul, and A. Matthews. 1997. Phase formation and mechanical properties of alumina coatings prepared at substrate temperatures less than 500°C by ionized and conventional sputtering. *Surface and Coatings Technology* 94–95:179–83.
10. Koski, K., J. Holsa, and P. Juliet. 1999. Properties of aluminium oxide thin films deposited by reactive magnetron sputtering. *Thin Solid Films* 339:240–48.
11. Sridharan, M., M. Sillassen, J. Bøttiger, J. Chevallier, and H. Birkedal. 2007. Pulsed DC magnetron sputtered Al$_2$O$_3$ films and their hardness. *Surface and Coating Technology* 202:920–24.
12. Fietzke, F., K. Goedicke, and W. Hempel. 1996. The deposition of hard crystalline Al$_2$O$_3$ layers by means of bipolar pulsed magnetron sputtering. *Surface and Coatings Technology* 86–87:657–63.
13. Edlmayr, V., M. Moser, C. Walter, and C. Mitterer. 2010. Thermal stability of sputtered Al$_2$O$_3$ coatings. *Surface and Coatings Technology* 204:1576–81.
14. Soppa, E., S. Schmauder, G. Fischer, J. Brollo, and U. Weber. 2003. Deformation and damage in Al/Al$_2$O$_3$. *Computational Materials Science* 28:574–86.
15. Cremer, R., M. Witthaut, D. Neuschutz, G. Erkens, T. Leyendecker, and M. Feldhege. 1999. Comparative characterization of alumina coatings deposited by RF, DC and pulsed reactive magnetron sputtering. *Surface and Coatings Technology* 120–121:213–18.
16. Zywitzki, O., and G. Hoetzsch. 1996. Influence of coating parameters on the structure and properties of Al$_2$O$_3$ layers reactively deposited by means of pulsed magnetron sputtering. *Surface and Coatings Technology* 86–87:640–47.

41

Nanomechanical Behavior of Metal-Doped DLC Thin Films

Arjun Dey, Rajib Paul, A. K. Pal, and Anoop Kumar Mukhopadhyay

41.1 Introduction

In this chapter, the tribological and surface nanomechanical properties of metal doping in diamond-like carbon (M-DLC) films will be discussed. Actually, M-DLCs are a new class of materials, and they are typically prepared by chemical vapor deposition techniques. They show great promise for advanced tribological applications. The synthesis of nanoparticulate gold films encapsulated in different matrices has been reported by a number of groups using numerous techniques, including precipitation, vapor deposition, and organic encapsulation. The main hurdle to the further widespread application of metal nanoparticles is their agglomeration due to high surface energy [1]. This makes it difficult, but not impossible, to achieve the proper dispersion of the metal in nanocrystalline form homogeneously in the DLC matrix. In recent times, incorporation of gold and silver nanoparticles in a DLC matrix and the related surface plasmon resonance effects were extensively reported by our research collaborators Paul et al. [2–4]. Recently, we have also showed that silver-doped DLC thin film deposited by a CVD (chemical vapor deposition) technique offers a reasonable combination of good surface nanomechanical and nanotribological properties [3] along with the concomitant corroborative enhancement in optical characteristics as well as the ratio of sp^2/sp^3 hybridization. Beyond our experiments involving silver nanoparticles, we have recently extended the approach to the incorporation of nanocrystalline gold particles in similar DLC matrices.

Nanocrystalline gold incorporated DLC (AuDLC) composite films of 70–120 nm were deposited by a capacitatively coupled rf (13.56 MHz) plasma CVD technique. AuDLC composite films as well as pure DLC thin films were developed on quartz glass. The films had thickness of about 70–120 nm. The AuDLC nanocomposite films were deposited with 50%, 60%, 70%, and 80% of argon in the argon–methane mixture. The corresponding thin films shall be denoted here as AuDLC50, AuDLC60, AuDLC70, and AuDLC80. The DLC thin films were also similarly deposited with 50%, 60%, 70%, and 80% of argon in the argon–methane mixture. The corresponding thin films shall be denoted here as DLC50, DLC60, DLC70, and DLC80.

41.2 Nanoindentation Study

In the present investigation, the partial-unloading experiments of 53 cycles were conducted at a peak load of 1000 µN at a loading rate of 0.09 µN·s⁻¹ by using a Berkovich indenter with a tip radius of ≈150 nm. Further, single-pass nanoscratch experiments were conducted at a constant load of 100 µN with the same Berkovich indenter mentioned previously. The scratch lengths were 8 µm. The typical partial-unloading load–depth curves are shown in Figure 41.1 for AuDLC thin films. The partial-unloading load–depth plot of DLC50 is also incorporated only for comparison (Figure 41.1). The slope of the loading curve in the plots of partial unloading versus depth for DLC50 was much stiffer than those in the partial unloading versus depth plots of AuDLC. Typical in situ SPM images of the impression of Berkovich indentation on the array of AuDLC nanocomposites and a typical DLC are shown in Figures 41.2a–d. There was no visible damage in the nanoindent impressions of the base DLC50 and AuDLC nanocomposite films. A comparison of the nanoindent impressions in DLC50 and AuDLC50 nanocomposite films

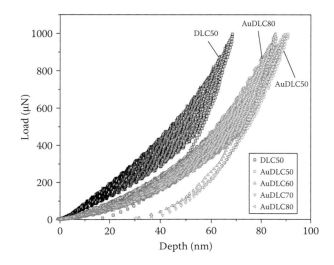

FIGURE 41.1 (See color insert.)
Typical partial-unloading load-depth plots of AuDLC and pure DLC.

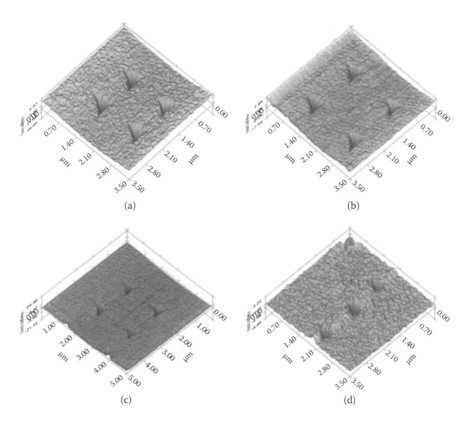

FIGURE 41.2 (See color insert.)
SPM images of the impression of the indention array: (a) AuDLC50, (b) AuDLC60, (c) AuDLC70, and (d) DLC50.

shows that at the same comparable load, the area of the impression was much larger in AuDLC50 in comparison to that in DLC50.

To avoid the effect of the mechanical properties of the substrate on the measured nanomechanical properties data of the AuDLC nanocomposite and the DLC thin films, a low depth regime of ≈20 nm was maintained. The average data for nanohardness and Young's modulus are shown in Figures 41.3a–d for AuDLC and pure DLC thin films as a function of the argon (Ar) percentage in the premixed methane–argon gas mixture as mentioned here and also in Chapter 9. The general message conveyed by all of these data was that, for both the AuDLC and DLC films, the average values of both nanohardness and Young's moduli had decreased because the extent of sp^3 hybridization had decreased as the percentage of argon was increased from 50% to 80%. This data also explains why both nanohardness and Young's moduli of AuDLC50 were much higher than those of the AuDLC80 films. This was a direct consequence of the increase in sp^2 content, which means the loss of the favorable sp^3 hybridization. The same explanation justifies why both nanohardness

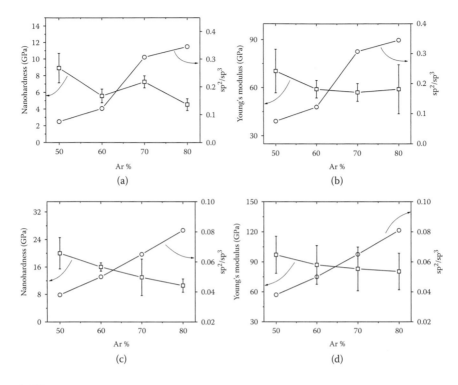

FIGURE 41.3
(a) Nanohardness* and (b) Young's modulus* as a function of sp²/sp³ content for AuDLC.
(c) Nanohardness and (d) Young's modulus as a function of sp²/sp³ content for DLC.
(*Nanohardness and Young's modulus measured at ~20 nm depth.)

and Young's moduli of DLC50 were much higher than those of the DLC80 films. It has been reported that for both DLCs and metal-doped DLCs, higher hardness and Young's moduli values were obtained for higher sp³ content in the films [5]. Further, the present AuDLC nanocomposite thin films showed nanohardness and Young's modulus data that were comparatively lower than those of the corresponding pure DLC films. This had happened simply because the nanometallic cluster incorporation reduced the hardness and Young's modulus values of the corresponding DLC thin films [6].

41.3 Nanotribological Study

Figures 41.4a and 41.4b show typical SPMs of the scratch trails of AuDLC nanocomposite thin films and the similar one for a typical DLC thin film. These data prove that there was no evidence of permanent deformation, suggesting that the scratching was predominantly elastic. Figure 41.5 presents

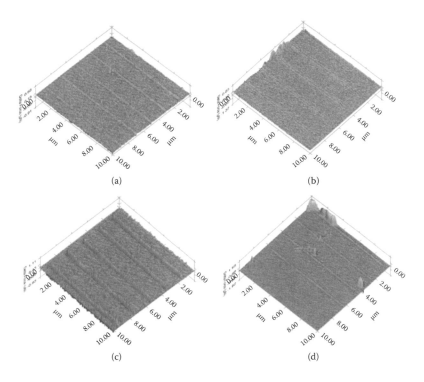

(a)   (b)

(c)   (d)

FIGURE 41.4 (See color insert.)
SPM images of array of nanoscratches: (a) AuDLC50, (b) AuDLC60, (c) AuDLC70, and (d) DLC50.

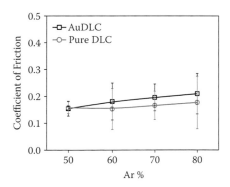

FIGURE 41.5
Coefficient of friction as a function of percentage of argon for both AuDLC and pure DLC.

the data on the coefficient of friction for both AuDLC nanocomposite and DLC thin films as a function of the argon (Ar) percentage in the premixed methane–argon gas mixture, as mentioned previously.

The coefficient of friction (μ) values for both the AuDLC nanocomposite and the DLC thin films had evidently increased as the percentage of argon was increased from 50% to 80%. The μ value of AuDLC50 was much lower (≈0.15), while AuDLC80 showed the highest value (≈0.2). Further, similar trends were also observed in the data for the DLC thin films. Akin to the cases of AuDLC50 and AuDLC80, here also the μ value of DLC50 was much lower (≈0.16), while that of DLC80 was the highest value (≈0.18). This trend was expected because the extent of sp^3 hybridization had decreased as the percentage of argon was increased from 50% to 80%. The sp^3-bonded carbon exhibits the properties of diamond, such as a lower friction coefficient and higher hardness [7]. Except for AuDLC50, the coefficient of friction values of all the AuDLC nanocomposite thin films were only just marginally higher than those of the corresponding DLC thin films. Therefore, the increase in μ of DLC films had happened due to nano-gold incorporation in the DLC matrices. This would mean that, generally, the frictional force was relatively slightly higher for the AuDLC films.

41.4 Adhesion Mechanisms

There are two mechanisms from which the frictional force can originate. These mechanisms are (a) the adhesion associated with the elastic deformation of the surfaces and (b) the abrasion of the surface by simple plowing, which involves only plastic deformation. The in situ SPM images confirmed that plastic plowing occurred in both the AuDLC nanocomposite and DLC thin films. This is where the nanomechanical properties of AuDLCs came into the picture. The AuDLC nanocomposite films already had lower strength because they had lower nanohardness and Young's modulus values compared to those of the DLC films. Therefore, the AuDLC films involved a larger plowing effect and, thus, had generated a relatively higher magnitude of the frictional force. It is this relatively slightly larger magnitude of the frictional force that had caused very slightly higher friction coefficient in the AuDLC nanocomposite thin films compared to those of the DLC films [8]. In the present experiments, a comparatively higher normal load of 100 μN was used. Despite that, the average friction coefficient data (μ̄) were low, less than 0.2. Thus, the present experimental data were in tune with the prediction of the classical adhesion theory of friction [9].

It has been reported that slight degradation in nanohardness and elastic modulus of metal (Ag, Cu, Ti, etc.)-doped DLCs can be compensated by the reduction of the internal compressive stress, which in turn enhances the tribological performance of the coatings [10]. Among all the AuDLC nanocomposite films investigated in the present work, the AuDLC50 nanocomposite thin films showed the best performance in terms of nanohardness, Young's modulus, and coefficient of friction. Further, the μ data (0.15) of the AuDLC50 nanocomposite thin films were very close to those (0.16) of the DLC50 thin films.

These data had clearly indicated that the incorporation of gold in DLC had definitely reduced significantly the internal stresses in the DLC thin film matrix. Therefore, we can conclude from the present experimental nanomechanical and nanotribological data that although the AuDLC nanocomposite thin films showed lower nanohardness and elastic modulus than the DLC thin films, in terms of nanotribological properties, these nanocomposite thin films behaved in a fashion almost similar to those of the DLC thin films.

41.5 Conclusions

The major message that we have derived from the present experimental nanomechanical and nanotribological data is that the AuDLC nanocomposite thin films showed lower nanohardness and elastic modulus but, in terms of nanotribological properties, these nanocomposite thin films behaved in a fashion almost similar to those of the DLC thin films. The AuDLC nanocomposite thin films showed lower nanohardness and elastic modulus compared to those of the DLC thin films because metal incorporation increased the ratio of sp^2/sp^3 hybridization due to the increase in the percentage of argon from 50% to 80% in the premixed methane–argon gas mixture. The most interesting observation was that in terms of tribological property, the AuDLC nanocomposite thin films behaved almost similar to those of the DLC thin films.

So far, we have tried to understand the intricacies in nanoindentation behavior of glass, ceramics, ceramic matrix composites, and thin films as well as thick ceramic coatings. But there is one thing we have missed so far. Mother Nature has provided us with natural nanocomposites like bone and teeth, which have a hierarchical architectural pattern of microstructure where the elemental blocks can start from dimensions as small as a few nanometers and can be as large as a few tens of millimeters. Therefore, our next aim in Chapter 42 will be to study the nanoindentation behavior of dental enamel nanocomposites, especially the orientational effect of enamel prisms on the overall nanohardness.

References

1. Lerme, J., B. Palpant, B. Prevel, M. Pellarin, M. Trielleux, J. L. Vialle, A. Perez, and M. Broyer. 1998. Quenching of the size effects in free and matrix-embedded silver clusters. *Physical Review Letter* 80:5105–8.
2. Paul, R., S. Hussain, S. Majumder, S. Varma, and A. K. Pal. 2009. Surface plasmon characteristics of nanocrystalline gold/DLC composite films prepared by plasma CVD technique. *Materials Science and Engineering B* 164:156–64.

3. Paul, R., A. Dey, A. K. Mukherjee, S. N. Sarangi, and A. K. Pal. 2012. Effect of nanocrystalline silver impregnation on mechanical properties of diamond-like-carbon films by nano-indentation. *Indian Journal of Pure and Applied Physics* 50:252–59.

4. Dey, A., R. Paul, A. K. Pal, and A. K. Mukhopadhyay. 2012. Nanoindentation and nanotribological studies on Au-DLC nanocomposite film.

5. Charitidis, C. A. 2010. Nanomechanical and nanotribological properties of carbon-based thin films: A review. *International Journal of Refractory Metals and Hard Materials* 28:51–70.

6. Wei, Q., R. J. Narayan, A. K. Sharma, J. Sankar, and J. Narayan. 1999. Preparation and mechanical properties of composite diamond-like carbon thin films. *Journal of Vacuum Science and Technology* A17:3406–14.

7. Li, X., and B. Bhushan. 1998. Micromechanical and tribological characterization of hard amorphous carbon coatings as thin as 5 nm for magnetic recording heads. *Wear* 220:51–58.

8. Hong, S., and M. Y. Chou. 1998. Effect of hydrogen on surface energy anisotropy of diamond and silicon. *Physical Review B* 57:6262–65.

9. Bowden, F. P., and D. Tabor. 1950. *Friction and lubrication of solids*, Part I. Oxford, UK: Clarendon Press.

10. Wei, Q., J. Sankar, and J. Narayan. 2001. Structure and properties of novel functional diamond-like carbon coatings produced by laser ablation. *Surface and Coatings Technology* 146–147:250–57.

Section 11

Nanoindentation Behavior on Ceramic-Based Natural Hybrid Nanocomposites

42

Orientational Effect in Nanohardness of Tooth Enamel

Nilormi Biswas, Arjun Dey, and Anoop Kumar Mukhopadhyay

42.1 Introduction

Natural biological materials have developed hierarchical and heterogeneous composite structures over millions of years of evolution in order to sustain the mechanical loads experienced in their specific environments. To perform diverse mechanical, biological, and chemical functions, these nanobiohybrid composite materials, e.g., tooth, bone, scale, shell, horn [1], etc., provide a varied arrangement of material structures at various length scales that work in concert. The science behind these structures and correlation with the property are yet to be well understood [1–10]. In this connection, it is now well known that the tooth is, in practicality, a ceramic-organic nanohybrid composite comprised of a hard enamel, the more ductile dentin, and a soft connective tissue, the dental pulp. Further, enamel is the hardest and the most highly mineralized structure in the human body, with approximately 96 wt% HAp being the primary mineral. The enamel and dentin regions of the tooth are interlocked by an irregular interface called the dentin-enamel junction (DEJ). Furthermore, the tooth has a hierarchical architecture spanning from macrostructure to microstructure to nanostructure [1]. The microstructure of the enamel reveals closely packed enamel rods or prisms encapsulated by an organic protein called *enamel sheath*. The prisms or rods consist of nanostructured inorganic HAp crystals, the nanocrystals being oriented at various angles inside the rod. The macrostructural properties are often determined by the respective micro- and/or nanostructures. Further, HAp is a brittle ceramic and, hence, is prone to suffer from contact damage. Therefore, in this chapter we aim to study the nanohardness of enamel at each region, starting from the DEJ to the end of the outer enamel region, by the nanoindentation technique [2]. We shall try to correlate the nanomechanical property with its microstructural and compositional properties.

Polished longitudinal sections of a freshly extracted premolar tooth sample from an Indian male (65 years old) were used. The nanohardness (H)

measurements were done on the enamel region of the premolar tooth using a commercial nanoindenter, as mentioned in Chapter 9.

42.2 Nanomechanical Behavior and Energy Issues

Figure 42.1a shows the typical load-versus-depth (P–h) plots obtained from nanoindentation experiments conducted at a load of 100 mN on different regions of enamel. The final depth of penetration (h_f) of outer enamel (E_0), i.e., far away from the DEJ at ≈0.64 μm, was slightly less than that of the inner enamel (E_i) very near to the DEJ at ≈0.77 μm. Further, the slope of the loading curve of the P–h plot of E_i was less stiff than that of the P–h plot of E_0. SEM photomicrographs taken at lower (Figures 42.2a–c) and higher magnifications (Figures 42.2d–f) shows micro-rupturing or a relatively smoother residual impression (Figure 42.2e), depending on location, i.e., outer or middle enamel zone. Small distortions of the shape of the nanoindent at the inner enamel (E_i) zone near to the DEJ were also observed (Figure 42.2f).

The nanohardness data increased with distance from the DEJ zone to the outer enamel zone (Figure 42.1b) as the projected contact area decreased

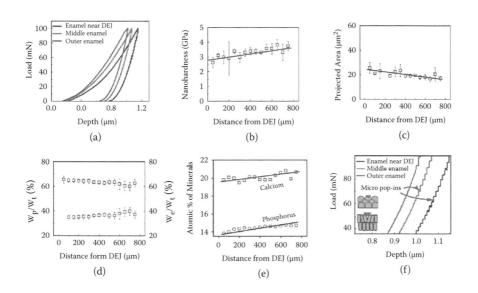

FIGURE 42.1 (See color insert.)
(a) Load–depth (P–h) plots at different regions (outer, middle, and inner) of enamel. Variations in (b) nanohardness; (c) contact area; (d) energy ratios; (e) Ca%, P% as functions of distance from DEJ; and (f) an enlarged view of the P–h plots shown in Figure 42.1a. Upper and lower insets show schematically the situation of indenter-microstructure interaction. (Reprinted/modified with permission of Biswas et al. [8] from The Institution of Engineers (India) and Springer.)

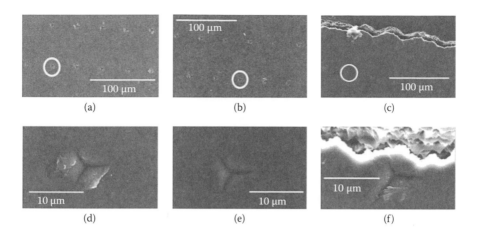

FIGURE 42.2
Microstructures of different regions of enamel nanocomposite at lower magnifications [(a) outer, (b) middle, and (c) inner enamel] and at higher magnifications [(d) outer, (e) middle, and (f) inner enamel]. (Reprinted/modified with permission of Biswas et al. [8] from the Institution of Engineers (India) and Springer.)

(Figure 42.1c). The less energy spent in plastic deformation (Figure 42.1d), the smaller was the final depth of penetration and, consequently, the projected contact area. In other words, the amount of elastic recovery decreased from the outer enamel zone very gradually to the DEJ, as was indeed reflected from the trend of W_e/W_t data obtained in the present work. Further, the EDX data plotted in Figure 42.1e show that the atomic percentages of both calcium (Ca) and phosphorus (P) increased as a function of distance starting from the DEJ zone to the enamel end [3]. This data also explains why the nanohardness (Figure 42.1b) increased in a similar fashion.

But the exact situation may not be so simple and straightforward [4, 5]. Frank, Capitant, and Goni [5] have argued that the increasing trend of hardness of the outer enamel layer might not simply be related to an increase in calcium and/or phosphorus content but, perhaps, could be linked to the fine differences in the ultrastructural orientations of the HAp crystals. On the other hand, the nanoindentation studies by other researchers [5] have established that the microstructural orientation (e.g., perpendicular or parallel to the DEJ) of enamel rods can really change the nanohardness values. In this connection, it may be mentioned that in the present work, SEM (Figures 42.3.a–d) and SPM (Figures 42.3.e–h) photomicrographs also showed the existence of different orientations of enamel prisms at different zones of the enamel. The average length of the sausage-shaped enamel prisms was ≈16.18 ± 4.32 μm, with a maximum length of ≈22.57 μm and a minimum of ≈11.84 μm. Similarly, the average diameter of circular-shaped enamel prisms was ≈6.97 ± 0.59 μm, with a minimum of ≈6 μm and a maximum of ≈7.41 μm. The aforesaid dimensions measured for the enamel rods were comparable to

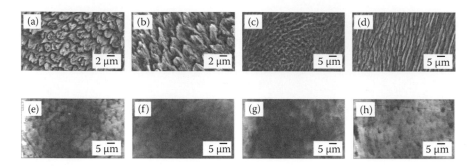

FIGURE 42.3
SEM (a–d) and corresponding SPM images (e–h) of different regions of enamel: (a, e) outer enamel, (b, f) 45° alignment of enamel prisms, (c, g) middle enamel, and (d, h) inner enamel. (Reprinted/modified with permission of Biswas et al. [8] from The Institution of Engineers (India) and Springer.)

other reported data [1]. The dissimilar orientations of the enamel prisms or rods were well documented in several other studies as well [3, 6].

Based on the results of the present study, we therefore propose the following simplified yet unified picture to explain the observed variations in the nanomechanical properties of the enamel nanocomposite. The orientational differences of the enamel rods happened naturally as the distance from a region close to the DEJ to the outer enamel zone is traversed. As a result, there happened an orientational dependence in the extent of biomineralization [7] and, as a consequence, in the chemical compositions in the zones close to and away from the DEJ. This corroborated a functionally graded microstructure as we traversed from the region close to the DEJ to the outer enamel zone [7].

It tacitly implies that the region close to the DEJ becomes lean in calcium (Ca) and phosphorus (P) contents, while the region away from it becomes relatively richer in calcium (Ca) and phosphorus (P) contents (Figure 42.1d). Hence, the elastic recovery is lesser in the region close to the DEJ zone, while elastic recovery is relatively greater in the region away from the DEJ (Figure 42.1d). Therefore, nanohardness registers a relatively lower magnitude in the region close to the DEJ and increases with a small positive gradient [8] to register a relatively higher magnitude in the region away from the DEJ (Figure 42.1b).

42.3 Micro Pop-in Events

Figure 42.1f shows an enlarged view of the *P–h* plots of Figure 42.1a. Almost no signature of micro pop-in events in the *P–h* plot (Figure 42.1a) was seen for the data corresponding to the outer enamel zone. However, the number of the micro pop-in events increased as the indenter entered more toward the DEJ zone. The presence of micro pop-in events was possibly linked to the microstructure and orientation of the enamel rods. From Figures 42.3a and 42.3e, it is observed that at the outer enamel regions, the enamel rods are oriented like fish scales (schematically shown in the inset of Figure 42.1f). In contrast, near to the DEJ zone, the enamel rods are seen to have longitudinal orientation (Figures 42.3d and 42.3h), as schematically shown in the inset of Figure 42.1f. When the indenter interacts with the second kind of situation, it may create the microcracking during the indentation that ultimately put a signature in the *P–h* plot (lower inset of Figure 42.1f). However, the possibility of microcracking during the nanoindentation measurement is reduced due to the more compact microstructure at the outer region of the enamel (Figures 42.3a and 42.3e and the lower inset of Figure 42.1f). Therefore, the chance of occurrence of micro pop-in events was almost absent in the case of the *P–h* plot at the outer region of the enamel (Figure 42.1f).

42.4 Conclusions

The nanohardness of tooth enamel increases as the distance from a region close to the dentin–enamel junction to the outer enamel zone is traversed. This increase is explained by a unified picture of the variations in nanohardness in terms of orientation-dependent variations in the extent of biomineralization as the distance from a region close to the dentin–enamel junction to the outer enamel zone is covered. In addition, a possible interaction between the nanoindenter and the enamel microstructure during the nanoindentation experiments is presented schematically. In Chapter 43, we shall further explore what happens if the loading rate is changed during the nanoindentation experiments. Eventually, it will be shown that such a situation has many interesting implications in terms of the basic physics of the deformation process in human dental enamel.

References

1. Meyers, M. A., P. Y. Chen, A. Y. M. Lin, and Y. Seki. 2008. Biological materials: Structure and mechanical properties. *Progress in Materials Science* 53:1–206.
2. Oliver, W. C., and G. M. Pharr. 1992. An improved technique for determining hardness and elastic modulus using load and displacement sensing indentation experiments. *Journal of Materials Research* 7:1564–83.

3. Jeng, Y. R., T. T. Lin, H. M. Hsu, H. J. Chang, and D. B. Shieh. 2011. Human enamel rod presents anisotropic nanotribological properties. *Journal of Mechanical Behaviour of Biomedical Materials* 4:515–22.

4. Wong, F. S. L., P. Anderson, H. Fan, and G. R. Davis. 2004. Microtomographic study of mineral concentration distribution in deciduous enamel. *Archives of Oral Biology* 49:937–44.

5. Frank, R. M., J. Capitant, and J. Goni. 1966. Electron probe studies of human enamel. *Journal of Dental Research* 45:672–82.

6. Teaford, M. F., M. M. Smith, and M. W. J. Ferguson. 2000. *Development, function and evolution of teeth*. Cambridge, UK: Cambridge University Press.

7. Habelitz, S., G. W. Marshall Jr., M. Balloch, and S. J. Marshall. 2002. Nanoindentation and storage of teeth. *Journal of Biomechanics* 35:995–98.

8. Biswas, N., A. Dey, S. Kundu, H. Chakraborty, and A. K. Mukhopadhyay. 2012. Orientational effect in nanohardness of functionally graded microstructure in enamel. *Journal of The Institute of Engineers India: Series D* 93:87–95.

9. Biswas, N., A. Dey, and A. K. Mukhopadhyay. 2012. Loading rate effect on nanohardness of human enamel. *Indian Journal of Physics* 86:569–74.

10. Biswas, N., A. Dey, and A. K. Mukhopadhyay. 2013. Micro-pop-in issues in nanoscale contact deformation resistance of tooth enamel. *ISRN Biomaterials* 2013 (545791): 1–6.

43

Slow or Fast Contact: Does it Matter for Enamel?

Nilormi Biswas, Arjun Dey, and Anoop Kumar Mukhopadhyay

43.1 Introduction

The tooth undergoes mastication, whereby it experiences changing loading rates. Hence, the effect of loading rate on the nanohardness of dental enamel nanocomposites is one of the most important aspects governing the reliable performance of a tooth in service as well as for in-depth scientific understanding of the mechanism of the contact events. Such information would be useful in the development of bio-inspired synthetic structures and their subsequent utilization in functional applications. So far, there has not been any significant systematic investigation on the influence of loading rate strictly on the nanohardness of tooth enamel. Therefore, our aim in this chapter is to study the effect of loading rate on the nanohardness and Young's modulus of human tooth enamel employing the novel nanoindentation technique.

43.2 Loading Rate Effect

The experimental samples and methods were similar to those described in Chapter 42 [1, 2]. The experiments were conducted at a constant load of 10^5 µN, and the loading time was varied from 0.4 to 100 s, with no holding time at the peak load, to obtain various loading rates in the range of 1×10^3 to 0.3×10^6 µN·s^{-1}. Further details have been given elsewhere [2]. The results are shown Figures 43.1–43.3. The nanohardness of the tooth enamel was enhanced with an increase in loading rate from 1,000 to 250,000 µN·s^{-1} (Figure 43.1a). The corresponding exploded view of a typical load versus depth (P–h) plot is shown in Figure 43.1d. To the best of our knowledge, this is the first ever experimental observation that there is an apparent increase in the nanohardness of human tooth enamel with an increase in loading rate.

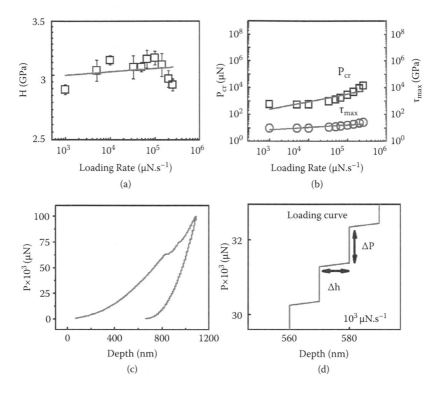

FIGURE 43.1
Variations of nanohardness and related parameters of human dental tooth enamel with load-ing rate: (a) nanohardness (*H*), (b) P_{cr}, τ_{max}, (c) *P–h* plot, and (d) exploded view of *P–h* plot. (Reprinted/modified with permission of Biswas, Dey, and Mukhopadhyay [7] from Indian Association for the Cultivation of Science (India) and Springer.)

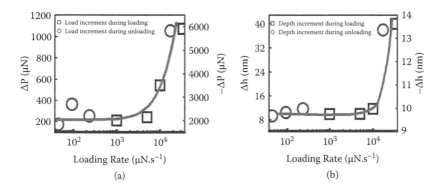

FIGURE 43.2
Variations of (a) load increment (Δ*P*) and load decrement (−Δ*P*) and (b) depth increment (Δ*h*) and depth decrement (−Δ*h*) as a function of loading rate in tooth enamel. (Reprinted/modi-fied with permission of Biswas, Dey, and Mukhopadhyay [7] from Indian Association for the Cultivation of Science (India) and Springer.)

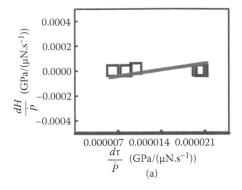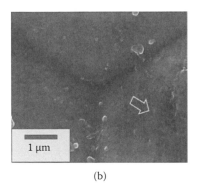

(a)   (b)

FIGURE 43.3
Influence of the rate of change of shear stress with loading rate $\left[\dfrac{d\tau}{\dot{P}}\right]$ on the rate of change of nanohardness $\left[\dfrac{dH}{\dot{P}}\right]$ with loading rate $\left(\dot{P}\right)$ and (b) FESEM photomicrograph of a typical nanoindent on tooth enamel at a loading rate of 10^3 µN·s^{-1}. (Reprinted/modified with permission of Biswas, Dey, and Mukhopadhyay [8] from Indian Association for the Cultivation of Science (India) and Springer.)

43.3 Evolution of Micro Pop-in Events

The typical load–depth data (Fig. 43.1c) showed evidence of the presence of multiple micro pop-in (Figure 43.2a) and multiple micro pop-out events (not shown here for the sake of brevity). This signifies the presence of plasticity initiation at the nanoscale of microstructure and/or microcrack generation during loading and/or unloading.

However, the magnitudes of load increments (ΔP) during the loading (Figure 43.2a) and those of load decrements ($-\Delta P$) during the corresponding unloading cycles (Figure 43.2a) were not exactly the same for the multiple micro pop-in and micro pop-out events. Both (ΔP) and ($-\Delta P$) increased rapidly in a nonlinear style with loading rates beyond 10^4 µN·s^{-1}. A similar trend was shown by the corresponding depth data (Figure 43.2b).

The critical load, P_{cr}, the point at which the first distinct positive displacement burst occurred, signifies the onset of plasticity at the nanoscale of the microstructure. This is what is meant by *incipient plasticity*. In addition, both P_{cr} and the corresponding associated maximum shear stress, τ_{max} [3, 4], active just underneath the nanoindenter, had a power law dependent enhancement with loading rate (Figure 43.1b). To calculate τ_{max} (10–30 GPa, depending on loading rate, Figure 43.1b), it was assumed [3, 4] that the contact between the triangular pyramidal Berkovich tip and the enamel can be approximately represented by a Hertzian contact between a spherical indenter and a flat plate.

These data (Figure 43.1b) suggested the interdependence between P_{cr}, τ_{max} and the apparent increase in nanohardness (Figure 43.1a) of human tooth enamel. It also highlighted the significant role of the pop-in events [5] in controlling the deformation process as a function of $\left(\dot{P} = \dfrac{dP}{dt} \right)$. It may be noted that the shear modulus (G) of human tooth enamel can be estimated as ≈31.5 GPa, assuming the Poisson's ratio as 0.27. Thus, the theoretical shear strength (τ_{theor}), which is estimated as ≈$G/5$–$G/10$, was found to be ≈3.15–6.3 GPa. Thus, τ_{theor} was much lower than τ_{max}.

Now, since τ_{max} is linearly related [3, 4] to P_{cr}, the loading-rate dependence of τ_{max} showed a trend similar to that of P_{cr} (Figure 43.1b). The linear relationship between $\left(\dfrac{d\tau_{max}}{\dot{P}} \right)$ and $\left(\dfrac{dH}{\dot{P}} \right)$ (Figure 43.3a) clearly suggested that the former controlled the later. Since the maximum shear stress active just underneath the nanoindenter was much higher than the theoretical shear strength of the enamel nanocomposite, the shear-induced microfracture (indicated by a hollow white arrow) happened just below the side surface of the nanoindent (Figure 43.3b).

43.4 Loading Rate versus Micro-/Nanostructure

With the increment in the loading rate value, the induced compressive strain became enhanced and, thus, led to a compressive stress zone in and around the nanoindentation cavity. To overcome this induced compressive stress at the next higher applied loading rate, greater critical load was necessary to initiate the onset of plasticity at the nanoscale of the microstructure.

It is well known that the enamel is a hybrid nanocomposite with a hierarchical architecture from the microscale to the nanoscale. As mentioned in Chapter 42, at the microscopic level it consists of aligned rods surrounded by a protein sheath that plays the role of an organic matrix. At the nanoscale level, each rod contains numerous HAp (hydroxyapatite) nanocrystals separated by a nanometer-thin protein layer. The rodlike HAp nanocrystals are about 50 nm in diameter, and they lean along the axis of the enamel prism.

In response to the load applied, the protein layer sheared to accommodate the deformation-induced strain and transferred the load among the neighboring mineral components. This repetitive mechanism of loading and load transfer followed by the unloading process most likely gave rise to the observed pop-in events during enamel nanoindentation.

The two other factors that might have added to the complete stretching of the protein matrix layer were (a) the extent of local sharing of the total strain between the organic matrix and the HAp nanocrystals within the purview of a tension shear chain (TSC) model [6] and (b) the process of stretching of the individual biomolecules in the protein matrix. Both of these factors could have contributed locally in an additive fashion.

At the lowest loading rate, the individual biomolecules in the protein matrix might have had just enough time to stretch in response to the applied loading rate. However, as the nanoindenter was withdrawn, the individual biomolecules in the protein matrix might not have had enough time to curl back, possibly due to the more open structure of the biomolecules and the inelastic component of the deformation. Such a process could explain why the overall elastic recovery for the enamel nanocomposite was so low. The lower nanohardness was evaluated at the lowest loading rate of 10^3 $\mu N \cdot s^{-1}$ at the given constant peak load of 10^5 μN. This was due to the fact that as the elastic recovery was less, the final depth of penetration was greater, and the projected contact area was thus on the higher side. This is a possible explanation for the lower magnitude of nanohardness.

However, at moderate and still higher loading rates, the time of contact was very small (≈ 0.4 s). The individual biomolecules in the protein matrix might thus have had just enough time to at least moderately stretch in response to the applied loading rates and curl back to a relatively much higher extent as the nanoindenter was withdrawn very quickly within the ≈ 0.4 s during the unloading cycle. This physical deformation process could have led to a situation such that the overall elastic recovery became much higher for the nanocomposite. As a consequence, the final depth of penetration became less, and hence the projected contact area was reduced, resulting in the measurement of higher nanohardness values as the loading rate was increased.

However, beyond a certain critical loading rate (10^4 $\mu N \cdot s^{-1}$, Figure 43.1a), the energy of the overall thermodynamic equilibrium for the physical deformation mechanism might have demanded the exhaustion of nearly all of the stored elastic energy due to the loading rate dependent nanoindentation process, and consequently the apparent increase in nanohardness would have had to be more or less saturated beyond that threshold loading rate [7].

43.5 Conclusions

This work presented, for the first time, a major effect of loading rate on the nanohardness of human tooth enamel. An apparent increment of about 8% was observed in nanohardness with an increase in loading rate from $\approx 1 \times 10^3$ to 0.3×10^6 $\mu N \cdot s^{-1}$. However, there was no effect of loading rate on the Young's modulus of human tooth enamel. The extent of this apparent increment in nanohardness with the loading rate was most likely controlled by the rate at which the maximum shear stress underneath the nanoindenter changed with variation of the loading rate. Such dependency of nanohardness on the rate of loading was most likely connected to the loading rate dependent unfolding and curling-back behavior of the organic macromolecules in the protein matrix. This physical deformation process controlled the local shear-stress accommodation and the attendant strain sharing at the nanoscale between the HAp nanocrystal rods and the protein matrix. Given this finding, it will be interesting to investigate the nanoindentation response of another natural nanocomposite material, such as a cortical bone. This is what we shall examine in Chapter 44.

References

1. Oliver, W. C., and G. M. Pharr. 1992. An improved technique for determining hardness and elastic modulus using load and displacement sensing indentation experiments. *Journal of Materials Research* 7:1564–83.
2. Biswas, N., A. Dey, S. Kundu, H. Chakraborty, and A. K. Mukhopadhyay. 2013. Mechanical properties of enamel nanocomposite. *ISRN Biomaterials* 2013 (253761): 1–15.
3. Packard, C. E., and C. A. Schuh. 2007. Initiation of shear bands near a stress concentration in metallic glass. *Acta Materialia* 55:5348–58.
4. Shang, H., T. Rouxel, M. Buckley, and C. Bernard. 2006. Viscoelastic behavior of a soda-lime-silica glass in the 293–833 K range by micro-indentation. *Journal of Materials Research* 21:632–38.
5. Mukhopadhyay, N. K., and P. Paufler. 2006. Micro- and nanoindentation techniques for mechanical characterisation of materials. *International Materials Reviews* 51:209–45.
6. Zhou, J., and L. L. Hsiung. 2006. Biomolecular origin of the rate-dependent deformation of prismatic enamel. *Applied Physics Letters* 89:051904.
7. Biswas, N., A. Dey, and A. K. Mukhopadhyay. 2012. Loading rate effect on nanohardness of human enamel. *Indian Journal of Physics* 86:569–74.

44

Anisotropy of Modulus in Cortical Bone

Arjun Dey, Himel Chakraborty, and Anoop Kumar Mukhopadhyay

44.1 Introduction

In the previous couple of chapters, we have discussed the nanomechanical properties of the hardest known biomaterials of the human body: tooth enamel. This structure is composed of a calcium phosphate compound or HAp-based composite materials. Further, it possesses a hierarchical structure ranging from the millimeter- to the angstrom-length scale. In this chapter, we will be discussing another biomaterial—human cortical bone—which also offers a hierarchical structure (Figure 44.1) like the tooth enamel. Natural bone is a composite material composed of organic compounds (mainly collagen) reinforced with inorganic compounds (HAp and other minerals). The most prominent structures seen at the nanoscale are the collagen fibers, surrounded and infiltrated by minerals. Bone builds its hierarchical architecture from these nanostructured building blocks. The detailed composition of bone differs, depending on species, age, dietary history, health status, and anatomical location. In general, however, the inorganic phase accounts for about 70% of the dry weight of bone and the organic matrix makes up the remainder. A critical survey of literature data [1–14] establishes clearly that we do not yet know with sufficient detail and accuracy the scientific reasons as to why and in which way the different compositional materials of bone are responsible for their varying nanomechanical properties. Moreover, to the best of our knowledge, the anisotropy in nanomechanical properties on cross section and plan section of the cortical bone has not yet been explored to a significant detail. The directional anisotropic properties of bone at the macrostructural length scale are well known. But we still need to know what happens at the microstructural length scale. This is what we want to explore in the present chapter. Eventually, the results will show us that the elastic properties of bone at local microstructural length scale do exhibit directional anisotropy that has not yet been explored to a significant detail.

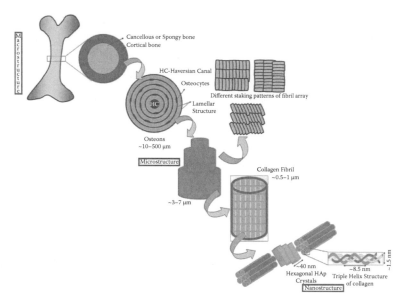

FIGURE 44.1
Schematic of the hierarchical structure of bone.

44.2 Microstructure

In the present work, dry bone samples from human broken fingers were collected from Railway Orthopaedic Hospital, Howrah, India. The samples were cut by a low-speed diamond saw and sectioned in both plan section and cross section. Bone samples were cold-mounted by embedding into a mixture of epoxy resin and hardener with desired ratio and subsequently cured for 24 hours (Figure 44.2). The exposed area was polished on special felt with diamond pastes ranging from 9 μm to 0.25 μm grit size prior to optical and scanning electron microscopy as well as the nanoindentation experiments. The optical micrographs of the human cortical bone microstructure are shown in both cross-sectional (i.e., at transverse section) (Figures 44.3a and 44.3b) and plan-sectional (Figure 44.3c) views [15]. The major components of the cortical bone, i.e., the lamella, Haversian canal, and osteon, have been marked in Figure 44.3a. A typical diameter of the osteon was found to be ≈75 μm, as shown in Figure 44.3b. The osteons, the base units of cortical bone, are generally circular and/or elliptical in shape. Lamellae have ring-like structures, which are stacked one after another. The Haversian canal, where the nutrition and essential fluids circulate, are also almost circular in nature. However, the diameters can differ, as was evident from the optical microscopy images shown in Figure 44.3. Scanning electron photomicrographs of the cortical bone are shown in Figure 44.4 [15].

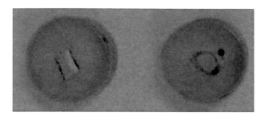

FIGURE 44.2
Cold-mounted polished cortical bone samples.

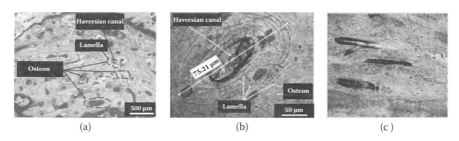

FIGURE 44.3
Optical photomicrographs of cortical bone: cross section at (a) 5× and (b) 50× magnifications, and (c) plan section at 20× magnification.

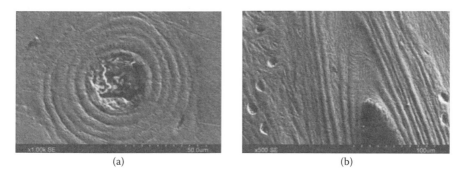

FIGURE 44.4
SEM photomicrographs of cortical bone: (a) cross section and (b) plan section.

44.3 Nanomechanical Behavior and Anisotropy

Nanoindentation experiments were carried out both on cross-sectional and plan-sectional sides, and optical photomicrographs were captured with the attached optical microscopy facility. Both Young's modulus (E) and hardness were recorded at a constant load of 100 mN. The typical load–depth plots (P–h) of the cortical bone are shown in Figure 44.5 [15]. The general nature of

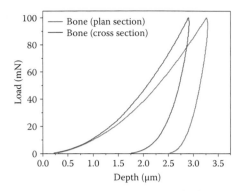

FIGURE 44.5 (See color insert.)
Typical *P–h* plots on both cross section and plan section of cortical bone sample.

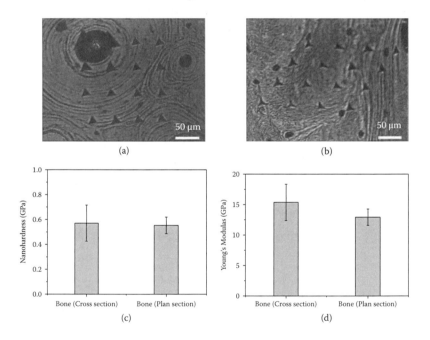

FIGURE 44.6
Nanoindentation array on human cortical bone: (a) cross section, (b) plan section, (c) nanohardness, and (d) Young's modulus.

these plots indicated the presence of an elastic-plastic deformation process, as expected for brittle materials. The final depth of penetration in the case of cross section was much larger than depth in the plan section. The typical Berkovich nanoindentation arrays on bone samples are shown for both the cross section (Figure 44.6a) and plan section (Figure 44.6b) [15].

Nanohardness and Young's modulus of the cortical bone for both the cross section and plan section are shown in Figures 44.6c and 44.6d, respectively [15]. The value of Young's modulus measured on the cross section was always higher than the modulus value measured on the plan section of the cortical bone. Therefore, a directional anisotropy was corroborated, possibly due to different structural orientations of HAp crystals and collagen fibers. However, the nanohardness data appeared to be insensitive to the direction of measurement.

44.4 Conclusions

Spatial and directional anisotropy of Young's modulus was found for the human cortical bone at the local microstructural length scale that was normally seen at the macrostructural length scale. Our next point of focus will be another natural nanomaterial—a fish scale—which has a spatially varying microstructure. We shall therefore try to understand in Chapter 45 how such local variations of chemical composition and microstructure of a fish scale can affect its nanomechanical properties.

References

1. Hoc, T., L. Henry, M. Verdier, D. Aubry, L. Sedel, and A. Meunier. 2006. Effect of microstructure on the mechanical properties of Haversian cortical bone. *Bone* 38:466–74.
2. Mullins, L. P., V. Sassi, P. E. McHugh, and M. S. Bruzzi. 2009. Differences in the crack resistance of interstitial, osteonal and trabecular bone tissue. *Annals of Biomedical Engineering* 37:2574–82.
3. Nomura, T., M. P. Powers, J. L. Katz, and C. Saito. 2007. Finite element analysis of a transmandibular implant. *Journal of Biomedical Research. Part B, Applied Biomaterials* 80:370–76.
4. Tang, B., A. H. Ngan, and W. W. Lu. 2007. An improved method for the measurement of mechanical properties of bone by nanoindentation. *Journal of Material Science: Materials in Medicine* 18:1875–81.
5. Rho, J. Y., and G. M. Pharr. 1999. Effects of drying on the mechanical properties of bovine femur measured by nanoindentation. *Journal of Material Science: Materials in Medicine* 10:485–88.
6. Rho, J. Y., T. Y. Tsui, and G. M. Pharr. 1997. Elastic properties of human cortical and trabecular lamellar bone measured by nanoindentation. *Biomaterials* 18:1325–30.
7. Rho, J. Y., M. E. Roy, T. Y. Tsui, and G. M. Pharr. 1999. Elastic properties of microstructural components of human bone tissue as measured by nanoindentation. *Journal of Biomedical Materials Research* 45:48–54.

8. Zysset, P. K., X. E. Guo, C. E. Hoffler, K. E. Moore, and S. A. Goldstein. 1999. Elastic modulus and hardness of cortical and trabecular bone lamellae measured by nanoindentation in the human femur. *Journal of Biomechanics* 32:1005–12.

9. Turner, C. H., J. Y. Rho, Y. Takano, T. Y. Tsui, and G. M. Pharr. 1999. The elastic properties of trabecular and cortical bone tissues are similar: Results from two microscopic measurement techniques. *Journal of Biomechanics* 32:437–41.

10. Tai, K., H. J. Qi, and C. Ortiz. 2005. Effect of mineral content on the nanoindentation properties and nanoscale deformation mechanisms of bovine tibial cortical bone. *Journal of Material Science: Materials in Medicine* 16:947–59.

11. Xu, J., J. Y. Rho, S. R. Misra, and Z. Fan. 2003. Atomic force microscopy and nanoindentation characterization of human lamellar bone prepared by microtome sectioning and mechanical polishing technique. *Journal of Biomedical Research. Part B, Applied Biomaterials* 67:719–26.

12. Pelled, G., K. Tai, D. Sheyn, Y. Zilberman, S. Kumbar, L. S. Nair, C. T. Laurencin, D. Gazit, and C. Ortiz. Structural and nanoindentation studies of stem cell-based tissue-engineered bone. *Journal of Biomechanics* 40:399–411.

13. Hengsberger, S., A. Kulik, and P. Zysset. 2001. A combined atomic force microscopy and nanoindentation technique to investigate the elastic properties of bone structural units. *European Cells and Materials* 1:12–17.

14. Novitskaya, E., P. Y. Chen, E. Hamed, J. Li, V. A. Uberda, I. Jasiuk, and J. McKittrick. 2011. Recent advances on the measurement and calculation of the elastic moduli of cortical and trabecular bone: A review. *Theoretical Applied Mechanics* 38:209–97.

15. Dey A., H. Chakraborty, and A. K. Mukhopadhyay. 2013. Unpublished work. Nanoindentation studies on human cortical bone.

45

Nanoindentation of Fish Scale

Arjun Dey, Himel Chakraborty, and Anoop Kumar Mukhopadhyay

45.1 Introduction

The scales of a fish act as its primary level of defense. Fish scale provides a protective barrier or shield from an enemy. The structure of a fish scale consists of a multilayered architecture in a functionally graded manner. Like the cases of the bone and tooth enamel discussed in previous chapters, fish scale also possesses a hierarchical structure. Jandt [1] reported on the concept of developing a biomimetic system to make body armor or impact-protection systems from one of Mother Nature's wonders: the scale of an ancient fish family called *Polypterus senegalus*. This particular type of fish scale consists of four different layers. The outer layer is made of ganoine, which is less compliant than the next inner layer called *dentin*. This layer can dissipate energy plastically, so it can bear load. Subsequently, the third layer is known as *isopedine*. The fourth and innermost layer consists of a bony lamellar structure. Further, the mineral content increases as we traverse from the inner layer to the outer layer. At the same time, the collagen content increases from the outer layer to the inner layer.

Christine Ortiz and coworkers at the Massachusetts Institute of Technology performed an in-depth nanoindentation study on scales of *Polypterus senegalus* [2]. They found that both hardness and Young's modulus were increasing from the bony side to the ganoine side. Ikoma et al. [3] reported the tensile strength and failure mechanism under strain for the fish scales of *Pagrus major*. However, in-depth nanomechanical properties of fish scales at the microstructural length are yet to be explored to a significant detail, particularly those from the Asian countries, e.g., India. Keeping in mind their unique properties like low density, high fracture toughness, and functionally graded structure, it is important to understand the mechanical behavior of such fish scales thoroughly. Therefore, in this chapter, we are concentrating on an Indian freshwater fish (Cyprinidae family) scale, which is a cycloid type. The microstructural and dielectric properties of such a cycloid type fish scale of *Catla catla* have already been investigated elsewhere [4]. In this chapter, we shall be emphasizing the nanoindentation study and how it correlates with the microstructural aspects.

45.2 Microstructure

In the present investigation, fresh fish scales of the cycloid type were collected from the Indian freshwater fish *Catla catla* (from the Ganges River). The scales were washed thoroughly with distilled water and dried properly. The fish scale samples were cut with a very-low-speed wafer slicing saw. The sample was cold-mounted by a proper mixture of epoxy resin and hardener, followed by 24 hours of curing. After that, the fish scale specimen was polished by subsequent application of diamond paste and alumina suspension. Mainly, the polishing process was done just prior to the nanoindentation study. The SEM photomicrograph of the top surface, i.e., the outermost surface of the fish scale, is shown in Figure 45.1a [5]. The apparent micropatterning has also been observed by others [4]. However, the rear or innermost side did not show the micropatterning (Figure 45.1b) [5].

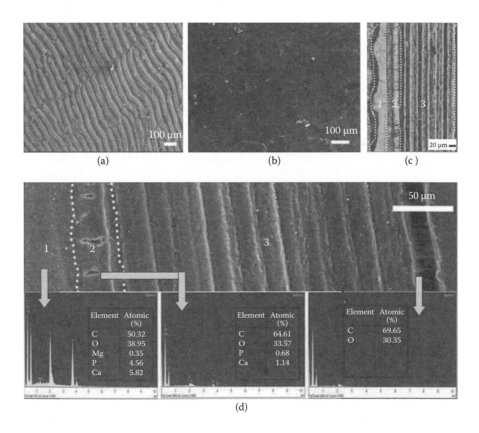

FIGURE 45.1

Microstructure of fish scale (a) outermost, (b) innermost, (c) cross-sectional surfaces, and (d) EDX spectra corresponding to stitched SEM images of cross-sectional surface.

Optical microscopy observations were conducted on the cross section of the fish scales using bright-field illumination where the image of any flat feature perpendicular to the incident light path appears to be bright or white. The different locations were identified and marked as 1, 2, and 3 (Figure 45.1c) [5]. Thus, the outermost surface was indexed as region 1; the middle surface was marked as region 2; and the innermost surface was marked as region 3. Figure 45.1d [5] illustrates the cross-sectional SEM photomicrographs of the same fish scale. It was also divided by region 1, region 2, and region 3 as in the optical microscopy data. The corresponding EDX data appended along with this figure showed that there was a continuous decrement of the mineral contents as we traverse from region 1 to region 3. The Ca/P ratio was only 1.27 in region 1 but rose to an almost HAp-like stoichiometric level of about 1.67 in region 2. However, in region 3, the cross-sectional region was devoid of both Ca and P.

45.3 Nanomechanical Behavior

The nanoindentation experiments were conducted on each region at a constant load of 10 mN. The typical load–depth (P–h) curves of different regions of the fish scale are shown Figure 45.2 [5]. The final depth of penetration was deliberately increased from region 1 to region 3. For the sake of comparison, the data obtained for the resin in which the sample had been mounted were also included. The lowest depth of penetration was obtained in region 1, which was biomineralization rich. The next

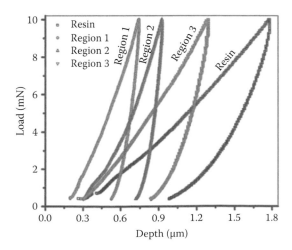

FIGURE 45.2
Load–depth curves of the different regions of fish scale.

highest depth of penetration was in region 2, which was biomineralization lean. The third highest depth of penetration was in region 3, which was absolutely devoid of any biomineralization. The highest depth of penetration, as expected, was in region 4, i.e., in the polymeric resin in which the sample was mounted.

The variation of nanohardness and Young's modulus at different regions of the fish scale are shown in Figures 45.3 and 45.4, respectively [5].

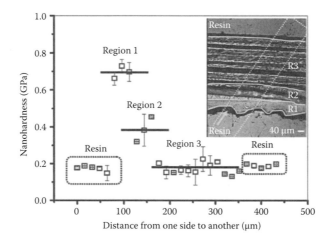

FIGURE 45.3
Variation of nanohardness at different regions of fish scale (inset: optical image of nanoindentation line array).

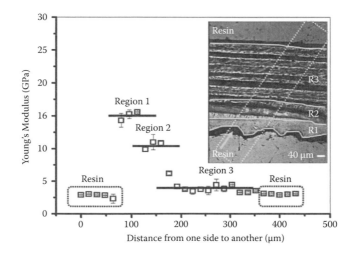

FIGURE 45.4
Variation of Young's modulus at different regions of fish scale (inset: optical image of nanoindentation line array).

The optical images of the nanoindentation line array are shown in the insets of Figures 45.3 and 45.4. Both nanohardness and Young's modulus decreased as we moved from region 1 to region 3. This was expected to happen due to the decrease in the extent of biomineralization, as seen in the EDX data presented in Figure 45.1d.

45.4 Conclusions

The microstructural and nanomechanical properties were correlated for the scale of a freshwater Indian fish *Catla catla*. A functionally graded microstructure was depicted where the extent of biomineralization continuously decreased as we traversed from the outer side to the inner side. The nanomechanical properties like nanohardness and Young's modulus also decreased in correspondence with the localized degradation of the Ca/P ratio.

To this point, we have studied the nanoindentation behavior of a wide variety of brittle solids like glass, alumina, ceramic matrix composites, C/C and C/C-SiC composites, thick coatings, thin films, and natural nanocomposites. Now, we shall shift our focus to some unique unresolved issues in the growing science and technology of nanoindentation technique. One such issue, and probably one of the most important, is the presence of indentation size effect and the occasional occurrence of a reverse indentation size effect in the experimentally measured data. Thus, in Chapter 46, our focus shall be on the indentation size effect (ISE) and reverse indentation size effect (RISE) in the nanoindentation responses of brittle solids.

References

1. Jandt, K. D. 2008. Biological materials: Fishing for compliance. *Nature Materials* 7:692–93.
2. Bruet, B. J. F., J. Song, M. C. Boyce, and C. Ortiz. 2008. Materials design principles of ancient fish armour. *Nature Materials* 7:748–56.
3. Ikoma, T., H. Kobayashi, J. Tanaka, D. Walsh, and S. Mann. 2003. Microstructure, mechanical, and biomimetic properties of fish scales from *Pagrus major*. *Journal of Structural Biology* 142:327–33.
4. Varma, K. B. R. 1990. Morphology and dielectric properties of fish scales. *Current Science* 59:420–22.
5. Dey, A., H. Chakraborty, and A. K. Mukhopadhyay. 2013. *Nanoindentation studies on scale of Katla katla fish*. Unpublished work.

Section 12

Some Unresolved Issues in Nanoindentation

46

Indentation Size Effect (ISE) and Reverse Indentation Size Effect (RISE) in Nanoindentation

Arjun Dey, Devashish Kaushik, Nilormi Biswas, Saikat Acharya, Riya Chakraborty, and Anoop Kumar Mukhopadhyay

46.1 Introduction

This chapter provides a comprehensive picture of indentation size effect (ISE) and reverse indentation size effect (RISE). With an increase in load and depth of penetration, if the hardness decreases, the phenomenon is called the *indentation size effect* (ISE). On the other hand, the opposite trend is a *reverse indentation size effect* (RISE).

For both micro- and nanoindentations, ISE can occur in metals, brittle solids, and nanocrystalline and quasi-crystalline materials [1–4]. Researchers are trying to understand this phenomenon. Indeed, a number of explanations have been proposed to understand ISE. These have been reviewed recently [3] and are, therefore, beyond the scope of a detailed discussion here. To mention briefly, though, some of the major proposals include the energy-balance concept, the strain-gradient plasticity theory, and the dislocation-nucleation theory [5–7]. Further, several factors may affect ISE. Some major issues identified here include the variation of contact surface, the friction between the material's surface and tip, the local microfracture processes, the presence or absence of residual stress, variations in the mechanisms of energy dissipation, etc. [8–13]. Researchers have also introduced several other concepts to explain ISE [14–16]. The most significant efforts in this regard are the minimum resistance on the surface, proportional specimen resistance (PSR), modified PSR concept, etc. [14–16]. We have already discussed the ISE in nanoindentation data of sintered coarse-grain alumina both before and after gas-gun shock loading (Chapters 19–22). Similarly, ISE was observed in the nanoindentation data of tape-cast alumina and monazite (Chapter 24). Besides the bulk brittle solids, ISE in heterogeneous ceramic coatings has not yet been explored in sufficient detail [17–19]. In this chapter, we shall describe how defects can influence the genesis of ISE in ceramic coatings. Next, we will offer a brief discussion of RISE. The conventional wisdom is that RISE may occur possibly due to residual stress in multiphase ceramics and thin films. Reports on RISE measured by the nanoindentation technique are not so common. Therefore, this should have caught our attention and, as will be shown here, did indeed gain our attention, at least as far as the experimental data are concerned.

Nevertheless, the most trusted and tested explanation of ISE in recent times is due to Nix and Gao [20], based on the dislocation theory of Taylor [21, 22]. They have linked ISE to the relative ease or difficulty of absorbing the localized strain gradient generated in a heavily plastically deformed small, confined volume that evolves during the nanoindentation process due to organization and reorganization of the dislocation network under the nanoindent. Their model [20] is represented as:

$$\left(\frac{H}{H_0} \right)^2 = 1 + \frac{h^*}{h} \tag{46.1}$$

where H is the measured hardness at a depth h, h^* is a characteristic length that depends on the properties of the test material and the indenter, and H_0 is the hardness at infinite depth.

46.2 ISE in HAp Coating

46.2.1 Nanoindentation at High Load

We have applied the celebrated Nix and Gao model in the nanoindentation data of a plasma-sprayed HAp coating. The experimental as well as the

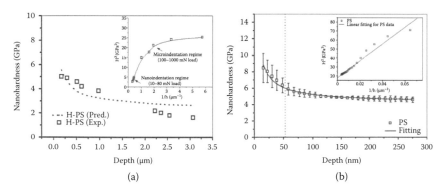

FIGURE 46.1
Measured and predicted hardness data as a function of depth up to (a) 1,000 mN load and (b) up to 10,000 μN load. Inset: square of indentation hardness, H^2, as a function of reciprocal depth, $1/h$, showing two slopes for plasma-sprayed HAp coating in (a) and a single slope in (b), PS = plan section. (Figure (a) is reprinted with permission of Dey et al. [19] from Springer.)

fitted data are shown in Figure 46.1a. Further, the data on H^2 as a function of reciprocal depth, $1/h$, is shown as an inset of Figure 46.1a. The fitting of the present data to Nix and Gao's model predicts H_0 as 2.37 GPa and h^* as 0.79 μm. The predicted value of H_0 appeared reasonable because the coating had characteristic porosities of ≈20%. However, if we take a critical look at the data, we can see that it indicates that there are two slopes signifying, for example, (a) a nanoindentation regime and (b) a microindentation regime, as pointed out in the inset of Figure 46.1a [19]. The slope for hardness data up to a depth of ≈0.75 μm is certainly differing from the data obtained at a depth of more than 1 μm. It is not a surprising phenomenon at all. Huang et al. [23] also reported a similar phenomenon in the case of MgO. This may be explained as a "scales of measurement" effect, where the region with the smaller slope is more strictly linked to the nanoindentation data regime obtained experimentally at, indeed, very small length scales in depth dimension. The region with the larger slope is more characteristically linked to the microindentation data at relatively larger length scale along the depth dimension.

We have opined elsewhere [19] that the phenomenon observed in the present data (Figure 46.1a) was mainly due to the contribution of the spatial defects such as micropores or microcracks besides the original ISE. The role of such microstructural defects (e.g., pores and cracks), termed as *characteristic microstructural flaws*, was to reduce the total solid load-bearing contact area. The lesser that such reduction occurred inside the indentation volume, the relatively higher would be the measured magnitude of nanohardness for those data obtained at lower loads. Therefore, we can suggest [18, 19] that from two important factors, there could be either an independent (yet collaborative) or interdependent contribution to the genesis of ISE in the MIPS-HAp coatings. These are (a) the localized enhancement in spatial density of flaws at higher depth and (b) associated reduction in solid load-bearing contact area inside the indentation volume. In the case of the ceramic coatings with characteristically heterogeneous microstructures in general at least, it is also a very difficult task to split the actual ISE due to the influence of the defect. Hence, we have introduced a new term, *apparent indentation size effect*, for the nanohardness data of the present plasma-sprayed HAp coatings.

We have a keen interest in further investigating the issue of characteristic defect length scale versus the probe length scale. Now it is plausible to argue that if we further decrease the nanoindentation load, then the defect influence zone for the newly created nanoindentation volume will be also reduced.

46.2.2 Nanoindentation at Ultralow Load

We have used a small load of 10,000 μN in a UBI Triboindenter, as mentioned in Chapter 9, on the same MIPS-HAp coating. The corresponding related data are plotted in Figure 46.1b. Here, the experimentally measured trend of nanohardness data plotted as a function of depth fitted well with the trend predicted from the Nix and Gao model. The fitting of the present data to the Nix and Gao model predicts H_0 as 4.2 GPa and h^* as 52.81 nm. The plot of H^2 versus $1/h$ (inset of Figure 46.1b) showed a linear relationship, as expected. Further, in the present case (Figure 46.1b), h^* was about 53 nm, which is much lower than the critical depth of $\approx$100 nm that was proposed earlier by Huang et al. [23]. This scenario suggests that when the data were deliberately restricted in the nanoindentation regime only, the match between theory and experiment was excellent. However, in the case of the data presented in Figure 46.1a (i.e., where the data were up to 1000 mN load), the results distinctly showed two regimes: the nano- and microindentation regimes.

46.3 ISE and RISE in AlN-SiC Composites

In the case of the pressureless sintered AlN-SiC composites, however, the data presented in Figures 46.2a–d show the presence of both ISE and RISE,

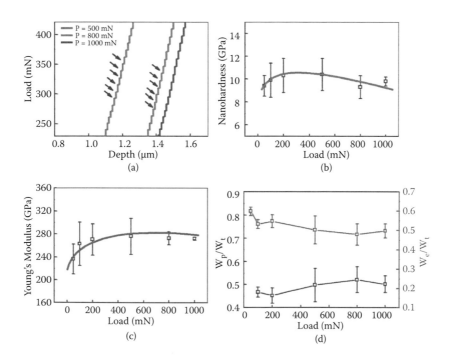

FIGURE 46.2(See color insert.)
The presence of both ISE and RISE depending upon load in AlN-SiC composites: variations of (a) load–depth (P–h) plot, (b) nanohardness, (c) Young's modulus, and (d) ratio of energies spent in elastic and plastic deformations—all as functions of the applied nanoindentation loads.

depending upon load. The typical *P–h* data plots showed the presence of multiple micro pop-in events here also, suggesting the presence of incipient plasticity at ultralow loads in this material (Figure 46.2a). Nanohardness in fact initially increased with load up to about 200 mN (Figure 46.2b). So there was a distinct presence of RISE here, akin to the case of Young's modulus up to the similar load range (Figure 46.2c). However, beyond 200 mN, the nanohardness decreased with indentation load in the AlN-SiC materials, thereby showing more of a conventional ISE pattern (Figure 46.2b). But, beyond 200 mN, the Young's modulus data appeared to have reached a plateau (Figure 46.2c). The corresponding data on variations of the energies spent in elastic and plastic deformations (Figure 46.2d) corroborated with each other as well as the nanohardness data (Figure 46.2b). Further work will definitely be necessary to understand such peculiar response of the AlN-SiC composites.

46.4 ISE in Dentin

Nevertheless, nanoindentation experiments conducted at a constant loading rate of 1 mN·s⁻¹ showed the presence of a remarkably strong ISE in the nanohardness and Young's modulus of a natural nanocomposite, namely the dentin region of human teeth (Figures 46.3a and 46.3b), which is a polymeric material that is supposed to exhibit a response more like a viscoelastic material than like a strong brittle material.

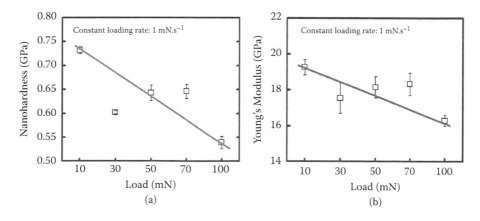

FIGURE 46.3
Presence of strong ISE in dentin region: variations of (a) nanohardness and (b) Young's modulus as functions of the applied nanoindentation loads.

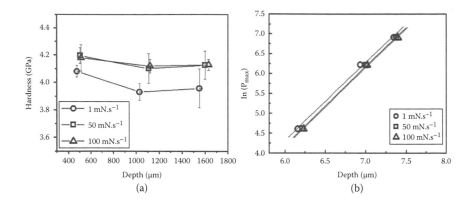

FIGURE 46.4

Presence of mild ISE at 1, 50, and 100 mN·s^{-1} loading rates in the SLS glass: variations of (a) nanohardness, H, with depth, h, and (b) $\ln(P_{max})$ with $\ln(h)$. (Reprinted with permission of Chakraborty, Dey, and Mukhopadhyay [24] from Springer.)

46.5 ISE in SLS Glass

On the other hand, the presence of mild ISE at three different loading rates of 1, 50, and 100 mN·s^{-1} was observed in soda-lime-silica (SLS) glass (Figure 46.4) [24]. For instance, the decrease of nanohardness (H) with depth (h) is shown in Figure 46.4a as a function of load. The plots of $\ln(P_{max})$ with $\ln(h)$ further confirmed the presence of ISE (Figure 46.4b). Fitting of the P–h data corresponding to loading rates of 1, 50, and 100 mN·s^{-1} to the well-known Mayer's law (Figure 46.4b) gave the values of the exponent (n), respectively, as 1.96, 1.98, and 1.99, which means that the Mayer's law exponent was still less than 2, and the presence of a mild ISE is definitely expected [16]. Despite numerous efforts to explain ISE in glass and ceramic, it is generally appreciated that it is yet to be unequivocally established as to which one of these is the best one [3]. Recent two-dimensional discrete dislocation modeling work [25] on ISE has shown that, compared to a case where the slip bands are oriented parallel to the surface, a less-pronounced decrease of the hardness with increasing indentation depth is observed for the nonparallel slip band orientation. The origin of the different effects was partly linked to the discrete nature of plasticity that operates in the domain of the nanometric scale regime of microstructure. In this connection, it may be mentioned that slip bands parallel to and at various angles to the surface were also noted in the present work (e.g., Chapters 10–12). However, a direct correlation between the orientation of slip bands and the mild ISE present in the data (Figures 46.4a and 46.4b), if any, is yet to be clearly understood. Therefore, the exact genesis of ISE in SLS glass still remains an interesting area of future study by other researchers. We also plan to address this issue more critically with further work in our future endeavors.

46.6 Conclusions

This chapter summarized the experimental observations related to the presence of an indentation size effect and a reverse indentation size effect in nanoindentation hardness data of MIPS-HAp coatings, alumina, AlN-SiC composites, dentin, and SLS glass. To the extent possible, attempts were made to look into the aspects related to the genesis of ISE in some of these materials. Another related issue is the signature of nanoscale plasticity initiation through micro pop-ins in the load–depth plots obtained from nanoindentation experiments of brittle solids. Therefore, in Chapter 47, we shall try to present a global viewpoint about the genesis of micro pop-ins in the nanoindentation behavior of brittle solids.

References

1. Ma, Q., and D. R. Clarke. 1995. Size dependent hardness of silver single crystals. *Journal of Materials Research* 10: 853–63.
2. Bull, S. J., T. F. Page, and E. H. Yoffe. 1989. An explanation of the indentation size effect in ceramics. *Philosophical Magazine Letters* 59:281–88.
3. Mukhopadhyay, N. K., and P. Paufler. 2006. Micro and nanoindentation techniques for mechanical characterisation of materials. *International Materials Reviews* 51:209–45.
4. Mukhopadhyay, N. K., J. Bhatt, A. K. Pramanik, B. S. Murty, and P. Paufler. 2004. Synthesis of nanocrystalline/quasicrystalline $Mg_{32}(Al,Zn)_{49}$ by melt spinning and mechanical milling. *Journal of Materials Science* 39:5155–59.
5. Bernhardt, E. O. 1941. On microhardness of solids at the limit of Kick's similarity law. [In German.] *Zeitschrift für Metallkunde* 33:135–44.
6. Nix, W. D., and H. Gao. 1998. Indentation size effects in crystalline materials: A law for strain gradient plasticity. *Journal of the Mechanics and Physics of Solids* 46:411–25.
7. Horstemeyer, M. F., M. I. Baskes, and S. J. Plimpton. 2001. Length scale and time scale effects on the plastic flow of FCC metals. *Acta Materialia* 49:4363–74.
8. Iost, A., and R. Bigot. 1996. Indentation size effect: Reality or artifact? *Journal of Materials Science* 31:3573–77.
9. Li, H., A. Gosh, Y. H. Han, and R. C. Bradt. 1993. The frictional component of the indentation size effect in low load microhardness testing. *Journal of Materials Research* 8:1028–32.

10. Swain, M. V., and M. Wittling. 1996. Indentation size effects for ceramics: Is there a fracture mechanics explanation? In *Fracture mechanics of ceramics*. Vol. 11, ed. R. C. Bradt, D. P. H. Hasselman, D. Munz, M. Sakai, and V. Y. Shevchenko, 379–87. New York: Plenum Press.

11. Gong, J., and Z. Guan. 2001. Load dependence of low-load Knoop hardness in ceramics: A modified PSR model. *Materials Letters* 47:140–44.

12. Gao, Y. X., and H. Fan. 2002. A micro-mechanism based analysis for size-dependent indentation hardness. *Journal of Materials Science* 37:4493–98.

13. Paternoster, C., A. Fabrizi, R. Cecchini, M. E. Mehtedi, and P. Choquet. 2008. Thermal stability of CrN_x nanometric coatings deposited on stainless steel. *Journal of Materials Science* 43: 3377–84.

14. Hays, C., and E. G. Kendall. 1973. An analysis of Knoop microhardness. *Metallography* 6:275–82.

15. Li, H., and R. C. Bradt. 1993. The microhardness indentation load/size effect in rutile and cassiterite single crystals. *Journal of Materials Science* 28:917–26.

16. Peng, Z., J. Gong, and H. Miao. 2004. On the description of indentation size effect in hardness testing for ceramics: Analysis of the nanoindentation data. *Journal of European Ceramic Society* 24:2193–2201.

17. Basu, D., C. Funke, and R. W. Steinbrech. 1999. Effect of heat treatment on elastic properties of separated thermal barrier coatings. *Journal of Materials Research* 14:4643–50.

18. Dey, A., A. K. Mukhopadhyay, S. Gangadharan, M. K. Sinha, D. Basu, and N. R. Bandyopadhyay. 2009. Nanoindentation study of microplasma sprayed hydroxyapatite coating. *Ceramics International* 35:2295–2304.

19. Dey, A., A. K. Mukhopadhyay, S. Gangadharan, M. K. Sinha, and D. Basu. 2009. Weibull modulus of nano-hardness and elastic modulus of hydroxyapatite coating. *Journal of Materials Science* 44:4911–18.

20. Nix, W. D., and H. Gao. 1998. Indentation size effects in crystalline materials: A law for strain gradient plasticity. *Journal of the Mechanics and Physics of Solids* 46:411–25.

21. Taylor, G. I. 1934. The mechanism of plastic deformation of crystals; Part I: Theoretical. *Proceedings of the Royal Society London A* 145:362–87.

22. Taylor, G. I. 1938. Plastic strain in metals. *Journal of the Institute of Metals* 62:307–24.

23. Huang, Y., F. Zhang, K. C. Hwang, W. D. Nix, G. M. Pharr, and G. Feng. 2006. A model of size effects in nano-indentation. *Journal of the Mechanics and Physics of Solids* 54:1668–86.

24. Chakraborty, R., A. Dey, and A. K. Mukhopadhyay. 2010. Loading rate effect on nanohardness of soda-lime-silica glass. *Metallurgical and Materials Transactions A* 41A:1301–12.

25. Kreuzer, H. G. M., and R. Pippan. 2007. Discrete dislocation simulation of nanoindentation: Indentation size effect and the influence of slip band orientation. *Acta Materialia* 55:3229–35.

47

Pop-in Issues in Nanoindentation

Riya Chakraborty, Arjun Dey, Manjima Bhattacharya, Nilormi Biswas, Jyoti Kumar Sharma, Devashish Kaushik, Payel Bandyopadhyay, Saikat Acharya, and Anoop Kumar Mukhopadhyay

47.1 Introduction

The load displacement curves obtained during a nanoindentation experiment may show discontinuities or steplike features at load levels typically well below 1 mN that characterize a sharp transition from purely elastic to plastic deformation. These are called *pop-in events*. The pop-in behavior is most often attributed to the nucleation of dislocations and the onset of incipient nanoscale plasticity. The reasons behind their genesis are still far from well understood. So what we propose to do is this: We first present the literature scenario of pop-ins in ceramics in particular and other materials in general. Then we look at our own experimental observations.

47.2 What is Known about Pop-ins?

Ceramics, as we already know, are hard and brittle solids. The pop-ins that occur in ceramics are proposed in the literature [1–9] to be linked to at least six different logics. One of these is the activation of rhombohedral twinning, whereby slip is initiated due to nucleation of homogeneous dislocations and the subsequent movement of these dislocations across favorably oriented slip planes [1–3]. Another possibility could be slip by punching out of dislocation bands parallel to specific crystallographic planes [4–6]. Yet another possible mechanism is multiplication of dislocation loops and then the movement of these higher dislocation density containing loops from plane to plane by cross-slips [7, 8], so much so that even pores and/or microcracks underneath the nanoindenter may also cause the pop-ins in ceramics [9]. Pop-ins have also been noted in semiconductors, e.g., silicon and many semiconducting thin films [10–13]. In the case of silicon, one major cause for pop-ins is the onset of pressure-induced phase change [10]. Two important parameters in this aspect are the critical load and the critical loading rate beyond which the pop-ins appear and below which they do not appear [11]. Their occurrence has also been linked to the genesis of maximum shear stress beneath the nanoindenter. When this stress is more than the theoretical shear strength of the material concerned, shear bands are formed inside the cavity, and the same corroborates with the pop-in formations [12]. As a matter of fact, even the presence of defects or cracks and their morphology at prospective locations can also give rise to the observed pop-in events [13]. Pop-ins are also observed in the cases of nanocrystalline materials, coatings, and multilayered structures [14–18]. Grain boundary defects may cause them to occur [14]. Similarly, cracking and coating delamination at the interface have been linked to pop-ins [15–17]. Of course, in these cases, the dislocation interaction still remains linked to the genesis of the pop-in behavior [18]. Pop-ins are very common to observe in bulk metallic glasses [19] and quasi-crystals [20]. In these cases, the deformations during pop-in physically occur mainly by discrete emission of shear bands at or below a critical load as well as a critical loading rate. It is also required that the activation of the maximum shear stress condition be fulfilled [19, 20]. In the case of the SLS glass, however, the genesis of pop-ins has been linked by us to the changes in the short range atomic arrangements at weak positions of the glass network structure, i.e., at the positions of network modifiers [21–23].

Other researchers [20] as well as our group [24] have opined that the situation is the most complicated in the case of the metallic materials. The main factors that could act as genesis points for pop-ins have been identified as the generation, amount, change of structure, movement, arrest, and repeated movement of dislocations [24]. Another important role is played by the slip rate dependence of interfacial friction. Here also a critical load and loading rate exist above which pop-in events do not happen, and they appear only when these critical conditions are met by the externally applied load and loading rates.

47.3 Pop-ins in Nanoindentation of Brittle Solids

47.3.1 Pop-ins in SLS Glass and Alumina

The experimental data presented in Figure 47.1 show the presence of micro pop-in events in the load–depth plots at constant load of 500 mN for the SLS glass slides at loading rates of 1, 10, and 500 mN·s^{-1} and for alumina at 1, 10, and 50 mN·s^{-1}. It may be noted from the data presented in Figure 47.1 that for both SLS glass and alumina, the recurrence of micro pop-in events is more evident at lower loading rates than at higher loading rates. Further, the variations in depth and load at the two consequent micro pop-in events were

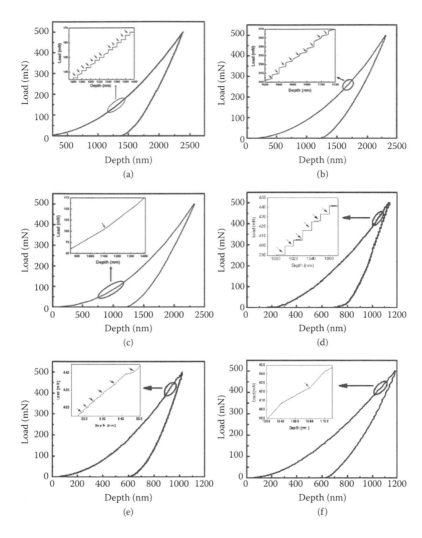

FIGURE 47.1
Examples of micro pop-ins in load–depth (*P–h*) plots at constant load of 500 mN for soda-lime-silica glass slides at loading rates of (a) 1 mN·s⁻¹, (b) 10 mN·s⁻¹, and (c) 500 mN·s⁻¹ and for alumina at loading rates of (d) 1 mN·s⁻¹, (e) 10 mN·s⁻¹, and (f) 50 mN·s⁻¹.

sensitive to the loading rate regimes as well as to material types. The presence of micro pop-in events in the load–depth plots of very thin microscope coverslips of SLS glass has been already reported by us [21–23]. Their presence has been discussed previously for both as-received (Chapters 10, 11, 16–18) and in as gas-gun shock-recovered (Chapters 19–22) conditions during static nanoindentation experiments on alumina. As discussed previously [21–23] (Chapters 10, 11, 16–18), the presence of shear bands is evident for the micro pop-in situations of both SLS glass (Figure 47.2a) and alumina (Figure 47.2b).

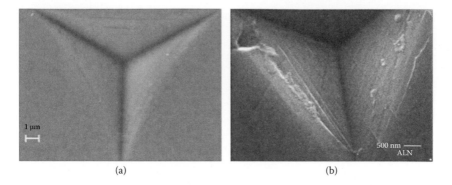

FIGURE 47.2
Shear bands in nanoindentation cavities at 500 mN loads at different loading rates in (a) SLS glass: FESEM, loading rate = 10 mN·s⁻¹ and (b) alumina: SEM, loading rate = 16.67 mN·s⁻¹.

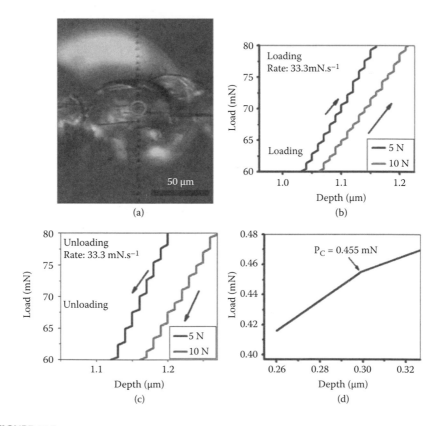

FIGURE 47.3
(a) Optical photomicrograph of the nanoindent made through the scratch grooves. Examples of micro pop-ins in *P–h* plots for nanoindentations made inside scratch grooves at 5 and 10 N loads in SLS glass during (b) loading and (c) unloading cycles. (d) Identification of critical load (P_c) value.

Figure 47.3a shows the optical photomicrographs of the nanoindentations through the scratch groove in an SLS glass slide. The nanoindents done at almost the middle of the scratch groove are shown by a circle. The exploded views of loading (Figure 47.3b) and unloading (Figure 47.3c) curves for the typical nanoindent marked with a circle in Figure 47.3a showed the micro pop-ins and pop-outs to be similar in nature as those observed in Figures 47.1a–c. The data presented in Figure 47.3d showed that the critical load (P_c) was calculated from the first nonlinearity observed in the loading curve of the nanoindentation experiment. This P_c value could be used to calculate the maximum shear stress (τ_{max}) active beneath the nanoindenter [25] by replacing P_{eff} with P_c in equation (1.18) in Chapter 1.

47.3.2 Why Pop-ins in SLS Glass?

We can try to provide some insight into how and why pop-ins do happen in an amorphous brittle solid like the present SLS glass [21–23]. It is well known that the range of atomic environments in a soda-lime-silica glass is such that some atoms can always reside in regions where the local topology is unstable. For instance, this could well be the positions of network modifiers (Na_2O and CaO). In such localized regions, the response to shear stress provided by the nanoindentation process could occur by atomic displacements and/or an elastic reshuffling of the atomic near-neighbors. Even if the actual fraction of atoms involved in such a process may be a small in number, the local strains caused by their displacements and anelastic reshuffling could be so large that their cumulative effect finally contributes to generate locally a significant amount of macroscopic strain that is relieved only by generation of shear deformation bands inside the nanoindentation cavity (Figure 47.2a). At lower loading rates, because of the long time available during the indentation process, the action of the first, the immediate next one, and all such consequent shear bands formed possibly came into play nearly sequentially one after another. Due to the sequential and discrete action of multiple shear bands, multiple serrations appear one after another in the *P–h* plot (Figures 47.1a–c). At higher loading rates, the loading as well as the unloading times were so small that only the overall average elastoplastic and plastic deformation processes came into play. In this case, the simultaneous operation of multiple shear bands forced a homogenization of the plastic deformation response of the glass, such that there were comparatively fewer shear bands located inside the nanoindentation cavity, as reflected in the number of serrations, which was comparatively much lower than those obtained at lower loading rates.

47.3.3 Why Pop-ins in Alumina Ceramic?

The formation of pop-ins in various alumina samples can be explained by the presence of an ultralow yet critical load, P_c, when the first micro pop-in event initiates the nanoindenter, which is supported by the surrounding microstructure that is under a huge compressive contact stress as well as shear stress at the vicinity of the tip. The initiation of P_c may cause a local relaxation of the surrounding microstructure by microcrack generation. When this happens, it not only aids in partially releasing the strain, but also reduces the local load-bearing contact area. Therefore, this step is followed by a small increase in nanoscale depth at P_c. It can happen because, over the small time scale involved, there is an instantaneous decrease in the available load-bearing contact area between the penetrating nanoindenter and the surrounding microstructure at the tip. As a consequence, the local compressive stress into the microstructure is momentarily enhanced, thereby leading to further displacement of the microstructure by the nanoindenter. As the depth of contact increases, so does the load-bearing contact area. This event causes a local drop of the compressive stress. Therefore, for initiation of the second micro pop-in event, the stress has to be higher than that at the immediate previous level. This can only happen if there is a minute increase in load for the same load-bearing contact area. As soon as the second micro pop-in event happens, the previously discussed cycle of events follows sequentially. This is how the series of micro pop-in events can happen.

47.3.4 Pop-ins in AlN-SiC Composites and Other Natural Biocomposites

It is further interesting to note that a similar situation can happen (Figures 47.4a and 47.4b) in the nanoindentation response of gas-pressure-sintered AlN-SiC composites [26]. Akin to the case of human dental enamels [24], a similar micro pop-in can be located in the nanoindentation load versus depth (*P–h*) plots of many natural nanocomposite biomaterials, such as human cortical bone (Chapter 44) and fish scales (Chapter 45). Some typical examples of micro pop-ins and micro pop-outs are presented in Figure 47.5 for outer and inner

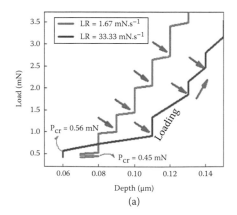

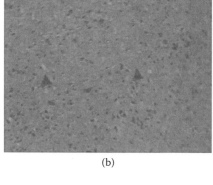

(b)

(a)

FIGURE 47.4

Example of micro pop-ins for AlN-SiC composites: (a) *P–h* plots at different loading rates, (b) nanoindentation array ($P = 1000$ mN, loading rate $= 333.3$ mN·s^{-1}, magnification 20×).

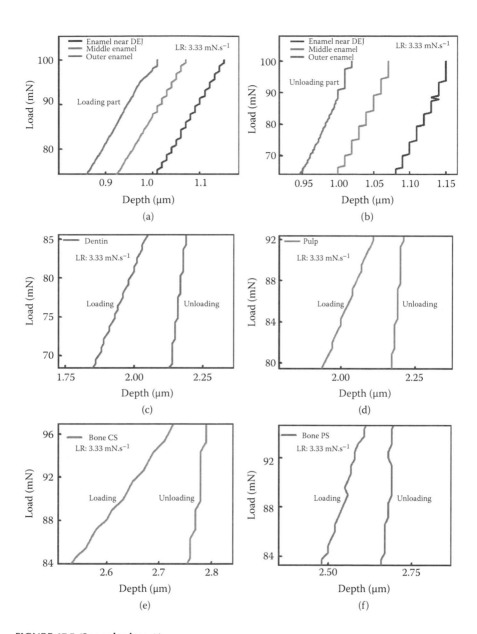

FIGURE 47.5 (See color insert.)
Examples of micro pop-ins and pop-outs occurring in load–depth (*P–h*) plots for outer and inner enamel and dentin–enamel junction regions during (a) loading and (b) unloading; for loading and unloading cycles in (c) dentin and (d) pulp regions of human teeth; and also for (e) cross and (f) plan sections of human cortical bones.

enamel and dentin–enamel junction regions during loading and unloading, for loading and unloading cycles in dentin pulp regions of human teeth, and also for cross and plan sections of human cortical bones.

47.3.5 Pop-ins in Tooth Enamel

At the nanoscale level, each dental enamel prism contains numerous HAp crystal rods of ≈50 nm diameter and oriented along the prism axis. A nanometer-thin organic layer separates these rods. Then, in response to the applied loading, the biopolymer matrix can shear to various extents to accommodate the strain due to the imposed deformation, and in the process the matrix can also transfer the load among the adjacent mineral components. Therefore, it follows that at the nanoscale level, a sequential process of (a) repetitive loading, (b) load transfer, (c) unloading, and (d) subsequent loading process will happen during the penetration and withdrawal of the nanoindenter into and out of the enamel microstructure. It has been proposed elsewhere [24] that these characteristic nanoscale deformation steps can give rise to micro pop-in events in enamel nanocomposites (Figures 47.5a and 47.5b).

Further, there can be two additional factors that may contribute to the observed micro pop-in events. These are: (a) the extent of local sharing of the total strain in between the protein matrix and the HAp crystal rods and (b) the process of stretching of individual biomolecules in the protein matrix, both of which can contribute in an additive fashion locally to the complete stretching of the protein matrix layer. However, a lot of further research work is needed to understand in detail the genesis of micro pop-in events in various brittle solids as well as the natural nanocomposites.

47.4 Conclusions

The major observation here is that the pop-ins can and do happen in brittle solids. The process is extremely sensitive to the ease or difficulty of nanoscale events, i.e., incipient plasticity initiation, homogeneous and/or heterogeneous dislocation initiation and the movement of these dislocations in specific and/or across preferential slip planes, and the applied load and the loading rate. A critical load and a critical loading rate may exist below which the propensity of pop-in events will be greater and above which the same would be less. Shear band formation can definitely occur in brittle solids when the local shear stress generated beneath the nanoindenter becomes greater than the shear strength of the material in question. The entire process of discrete shear band formation is extremely sensitive to both the load and loading rates applied as well as to the presence or absence of local, surface, or subsurface flaws. The mechanisms of pop-ins will definitely be different for amorphous solids, e.g., glass, ceramics, ceramic composites, and natural biomaterials. A lot of research will need to be conducted before the mechanisms of pop-in events in glass, ceramics, ceramic composites, and natural biomaterials are well understood. We have already seen that the issues of micro pop-ins and their occurrences are intimately related to the speed with which energy is transferred from the loading train into the brittle solids exposed under the nanoindenter. Therefore, in Chapter 48, we shall try to look at a more global pattern to understand the effects of loading rate on the nanoindentation responses of brittle solids.

References

1. Nowak, R., T. Sekino, and K. Niihara. 1996. Surface deformation of sapphire crystal. *Philosophical Magazine A* 74:171–94.
2. Mao, W. G., Y. G. Shen, and C. Luc. 2011. Nanoscale elastic-plastic deformation and stress distributions of the C plane of sapphire single crystal during nanoindentation. *Journal of the European Ceramic Society* 31:1865–71.
3. Mao, W. G., Y. G. Shen, and C Lu. 2011. Deformation behavior and mechanical properties of polycrystalline and single crystal alumina during nanoindentation. *Scripta Materialia* 65:127–30.
4. Samandari S. S., and K. A. Gross. 2009. Micromechanical properties of single crystal hydroxyapatite by nanoindentation. *Acta Biomaterialia* 5:2206–12.
5. Kucheyev S. O., J. E. Bradby, J. S. Williams, C. Jagadish, and M. V. Swain. 2002. Mechanical deformation of single-crystal ZnO. *Applied Physics Letters* 80:956–58.
6. Bradby J. E., S. O. Kucheyev, J. S. Williams, J. Wong-Leung, and M. V. Swain. 2002. Indentation induced damage in GaN epilayers. *Applied Physics Letters* 80:383–85.
7. Huang J., K. Xu, X. J. Gong, J. F. Wang, Y. M. Fan, J. Q. Liu, X. H. Zeng, G. Q. Ren, T. F. Zhou, and H. Yang. 2011. Dislocation cross-slip in GaN single crystals under nanoindentation. *Applied Physics Letters* 98 221906:1–3.
8. Guicciardi S., C. Melandri, and F. T. Monteverde. 2010. Characterization of pop-in phenomena and indentation modulus in a polycrystalline ZrB_2 ceramic. *Journal of the European Ceramic Society* 30:1027–34.
9. Rosa-Fox de la, N., V. M. Florez, J. A. T. Fernandez, M. Pinero, R. M. Serna, and L. Esquivias. 2007. Nanoindentation on hybrid organic/inorganic silica aerogels. *Journal of the European Ceramic Society* 27:3311–16.
10. Chang, L., and L. Zhang. 2009. Mechanical behaviour characterisation of silicon and effect of loading rate on pop-in: A nanoindentation study under ultra-low loads. *Materials Science and Engineering A* 506:125–29.
11. Mann, A. B., J. B. Pethica, W. D. Nix, and S. Tomiya. 1994. Nanoindentation of epitaxial films: A study of pop-in events. *Materials Research Society Symposium Proceedings* 356:271.
12. Kuan, S. Y., X. H. Du, H. S. Chou, and J. C. Huang. 2011. Mechanical response of amorphous ZrCuTi/PdCuSi nanolaminates under nanoindentation. *Surface and Coatings Technology* 206:1116–19.

13. Roa, J. J., E. Jimenez-Pique, T. Puig, X. Obradors, and M. Segarra. 2011. Nanoindentation of multilayered epitaxial YBa$_2$Cu$_3$O$_7$-δ thin films and coated conductors. *Thin Solid Films* 519:2470–76.
14. Richter, A., E. Czerwosz, P. Dłuzewski, M. Kozłowski, and M. Nowicki. 2009. Nanoindentation of heterogeneous carbonaceous films containing Ni nano-crystals. *Micron* 40:94–98.
15. He, J. Y., S. Nagao, H. Kristiansen, and Z. L. Zhang. 2012. Loading rate effects on the fracture of Ni/Au nano-coated acrylic particles. *Express Polymer Letter* 6:198–203.
16. Rabkin, E., J. K. Deuschle, and B. Baretzky. 2010. On the nature of displacement bursts during nanoindentation of ultrathin Ni films on sapphire. *Acta Materialia* 58:1589–98.
17. Jose, F., R. Ramaseshan, A. K. Balamurugan, S. Dash, A. K. Tyagi, and B. Raj. 2011. Continuous multicycle nanoindentation studies on compositionally graded Ti$_{1-x}$Al$_x$ multilayer thin films. *Materials Science and Engineering A* 528:6438–44.
18. Zhou, J., and L. L. Hsiung. 2006. Biomolecular origin of the rate-dependent deformation of prismatic enamel. *Applied Physics Letters* 89 (051904):1–3.
19. Packard, C. E., and C. A. Schuh. 2007. Initiation of shear bands near a stress concentration in metallic glass. *Acta Materialia* 55:5348–58.
20. Mukhopadhyay, N. K., and P. Paufler. 2006. Micro- and nanoindentation techniques for mechanical characterisation of materials. *International Materials Reviews* 51:209–45.
21. Chakraborty, R., A. Dey, and A. K. Mukhopadhyay. 2010. Loading rate effect on nanohardness of soda-lime-silica glass. *Metallurgical and Materials Transactions A* 41:1301–12.
22. Chakraborty, R., A. Dey, and A. K. Mukhopadhyay. 2010. Role of the energy of plastic deformation and the effect of loading rate on nanohardness of soda-lime-silica glass. *Physics and Chemistry of Glasses: European Journal of Glass Science and Technology Part B* 51:293–303.
23. Dey, A., R. Chakraborty, and A. K. Mukhopadhyay. 2011. Enhancement in nanohardness of soda–lime–silica glass. *Journal of Non-Crystalline Solids* 357:2934–40.
24. Biswas, N., A. Dey, and A. K. Mukhopadhyay. 2012. Loading rate effect on nanohardness of human enamel. *Indian Journal of Physics* 86:569–74.
25. Bandyopadhyay, P., and A. K. Mukhopadhyay. 2013. *Stress distribution in scratched glass*. Unpublished work.
26. Kaushik, D., N. Biswas, M. Bhattacharya, S. Acharya, and A. K. Mukhopadhyay. 2013. *Contact Deformation Behaviour of AlN-SiC Composites*. Unpublished work.

48

Effect of Loading Rate on Nanoindentation Response of Brittle Solids

Riya Chakraborty, Arjun Dey, Nilormi Biswas, Manjima Bhattacharya, Payel Bandyopadhyay, Jyoti Kumar Sharma, Devashish Kaushik, Saikat Acharya, and Anoop Kumar Mukhopadhyay

48.1 Introduction

The reports about loading rate effects on hardness of glass and ceramics are contradictory [1–5]. Microhardness of glass increased [2], decreased [3, 4], or remained independent of loading rate [5], while the nanohardness was slightly reduced [6], remained load independent [7–9], or was load dependent [10, 11]. The hardness of glass was slightly reduced [4, 12, 13], while the hardnesses of ceramics [14–16] were either enhanced with strain and loading rate variation or remained insensitive [17] to it. Recently, significant effects of loading rate on hardness of glass (see Chapters 10–12) [18–21], alumina (see Chapters 16–18) [22, 23], and natural biocomposites like human tooth (see Chapter 43) [24, 25] were reported by us. In each of these materials, there was an increase in the nanohardness of the material with the loading rates (see Chapters 10–12, 16–18, 43). Thus, some generic mechanism might have played a pivotal role in the enhancement of the nanohardness.

48.2 Loading Rate Effects in Brittle Solids: SLS Glass and Alumina

Some further examples of improvements in nanohardness and Young's modulus with loading rate during nanoindentation experiments are shown for SLS (soda-lime-silica) glass (Figures 48.1a and 48.1b) and 10 µm grain size alumina (Figures 48.1c and 48.1d). Figure 48.2 depicts the presence of a small number of shear bands inside nanoindentation cavities at $P = 500$ mN and 500 mN·s^{-1} loading rate for (a) an SLS glass slide and (b) 10 µm grain

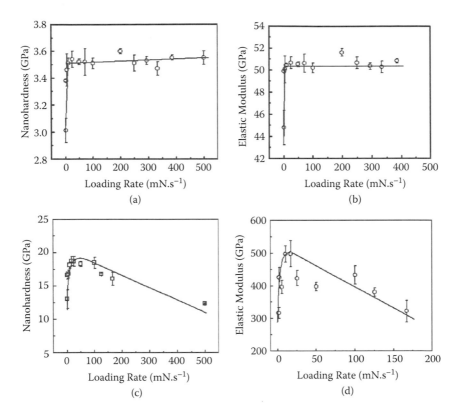

FIGURE 48.1
Variations with loading rate for nanohardness and Young's modulus of (a, b) SLS glass slide and (c, d) 10 µm grain size alumina.

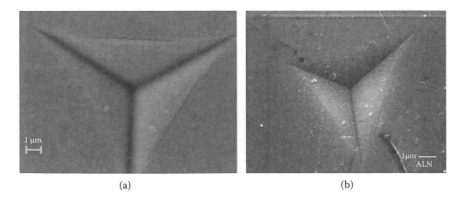

FIGURE 48.2
Shear bands inside nanoindentation cavities at $P = 500$ mN and 500 mN·s^{-1} loading rate: (a) FESEM, SLS glass slide and (b) SEM, 10 µm grain size alumina.

size alumina. In the case of the SLS glass, there is a steep rise in both nanohardness and Young's modulus at lower loading rate followed by a plateau (Figures 48.1a and 48.1b) at still higher loading rates. In the case of 10 μm grain size alumina, however, there was a threshold loading rate, up to which there is a steep rise in both nanohardness and Young's modulus at lower loading rate followed by a sharp fall at still higher loading rates (Figures 48.1c and 48.1d). Thus, the process of nanomechanical property enhancement was sensitive to material type and loading rate regime (see Chapters 10–12, 16–18).

48.2.1 Loading Rate Study on SLS Glass

During the nanoindentation of SLS glass at lower loading rates, the plastic deformation field and its effect are highly inhomogeneous inside the nanoindentation cavity. This is why there was a steep initial increment in nanohardness with loading rate, because the greater the number of slip bands formed, the larger was the amount of strain accommodated. At higher loading rates, the loading as well as the unloading times were so small that only the overall average elastoplastic and plastic deformation processes came into play. The simultaneous operation of multiple shear bands forced a homogenization of the plastic deformation response of the glass, such that there were comparatively fewer shear bands located inside the nanoindentation cavity, which was comparatively much lower than responses obtained at lower loading rates (see Figures 48.2a and 48.2b). As the number of shear bands formed was reduced, so (possibly) was the extent of shear flow and, thus, the resultant enhancement in nanohardness at loading rate higher than the threshold loading rate (TLR). This picture provides a rationale as to why the rate of increase of the nanohardness value was slowed or even degraded at loading rates higher than the TLR.

48.2.2 Loading Rate Study on Alumina

In the case of the coarse-grain alumina, the situation was completely different from that of glass. Here we have a brittle polycrystalline solid with a lot of grain boundaries. So how does it happen here that the nanohardness and Young's modulus increase with an increase in loading rate up to the TLR and then remains unchanged [22, 23] with further increase in loading rate or sharply degrades (Figures 48.1c and 48.1d) for loading rates higher than the TLR? We have argued elsewhere that the loading rate variation showed a consequent change in the intrinsic contact deformation resistance (P_c) of alumina, which ultimately led to an enhancement of the nanohardness (H) of the material [22, 23]. The effects were explained elsewhere by us (Chapters 16–18) [22, 23] also in terms of shear localization as well as possible formation of dislocation loops under the Berkovich nanoindenter. However, questions regarding why the nanohardness and Young's modulus of the

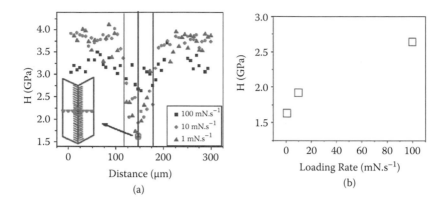

FIGURE 48.3
Nanohardness of SLS glass inside scratch groove as functions of (a) distance across scratch groove starting from the center and (b) loading rate.

10 μm grain size alumina sharply degraded beyond the TLR are, as of now, far from well understood. Hence, situations like this should demand more dedicated attention in future research.

48.2.3 Loading Rate Study inside Scratch Groove in SLS Glass

It was interesting to note that both nanohardness and Young's modulus of the SLS glass degrade from their respective values measured at the surface by about 30% to 60% inside the scratch groove (Figure 48.3a), particularly at the deepest point or base of the groove, as indicated by the arrow and the inset in Figure 48.3a. What is most interesting to note is that even these degraded values of nanohardness were loading rate dependent (Figure 48.3b). At a loading rate of 1 mN·s⁻¹, the nanohardness was about 1.5 GPa, but it quickly increased to about 2.75 GPa when measured at a loading rate of 100 mN·s⁻¹. Since the glass is the same SLS glass as mentioned previously (Chapters 10–12; Figure 48.1), similar reasoning may be applicable to explain the enhancement in nanohardness.

48.2.4 Loading Rate Study on AlN-SiC Composites

It was interesting to note that both the nanohardness (Figure 48.4a) and Young's modulus (Figure 48.4b) of AlN-SiC composites also can be enhanced with loading rate. Such enhancement of nanomechanical properties can be explained in terms of the variations of the critical load (P_c) (Figure 48.3c) and shear stress (τ_{max}) (Figure 48.3d) with loading rates.

48.2.5 Loading Rate Study on Tooth Enamel

Similar logic was shown to be valid when the nanohardness of the tooth enamel nanocomposite was enhanced up to 8% with an increase in

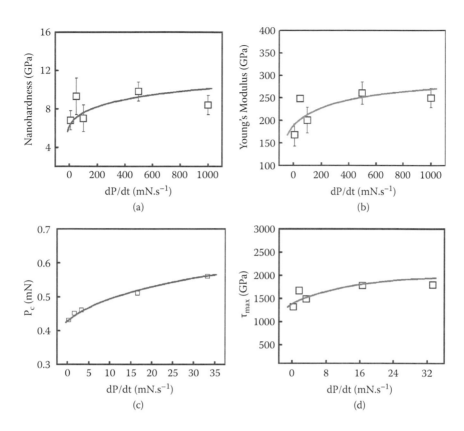

FIGURE 48.4
Effect of loading rate (dP/dt) variations on nanomechanical properties of AlN-SiC composites:
(a) nanohardness, (b) Young's modulus, (c) critical load (P_c), and (d) maximum shear stress
(τ_{max}).

loading rate [24]. The typical load-depth data (see Chapters 43 and 47)
showed evidence of serrations, which signify the presence of plasticity
initiation at the nanoscale of microstructure. The critical load, P_c, at which
the first distinct positive displacement burst occurred signifies the onset
of plasticity at the nanoscale of the microstructure. In addition, both P_c
and the corresponding associated maximum shear stress, τ_{max}, active just
underneath the nanoindenter had a power law dependent enhancement
with loading rate. This suggested the interdependence between P_{cr}, τ_{max},
and the apparent increase in nanohardness of human tooth enamel with
loading rates [24, 25]. Similar to the case of the enamel nanocomposite part
of the human teeth, the nanohardness (Figure 48.5a) and Young's modulus (Figure 48.5b) of the dentin part also increased with an increase in
loading rate to a plateau value, before degrading at still higher loading
rates [26]. Further studies will be necessary to understand the genesis of
such behaviors.

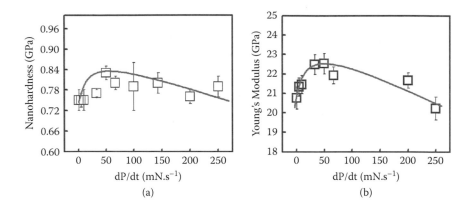

FIGURE 48.5
Effect of loading rate (dP/dt) variations on (a) nanohardness and (b) Young's modulus in dentin region of human tooth.

48.3 Conclusions

The generic nature of the loading rate's effect on nanomechanical properties of different materials was summarized. Some plausible explanations were offered for the effect of loading rate on nanohardness enhancement in SLS glass, 10 and 20 µm grain size alumina, AlN-SiC composites, and the enamel as well as dentin regions of human tooth. So far, we have not discussed anything about the possibilities of utilizing the nanoindentation technique to evaluate the residual stress in ceramic thick coatings. In Chapter 49 we shall illustrate the successful utilization of the nanoindentation technique to evaluate the residual stress in bioactive, porous, heterogeneous ceramic coatings.

References

1. Mao, W. G., Y. G. Shen, and C. Lu. 2011. Nanoscale elastic–plastic deformation and stress distributions of the C plane of sapphire single crystal during nanoindentation. *Journal of the European Ceramic Society* 31:1865–71.
2. Gunnasekera, S. P., and D. G. Holloway. 1973. Effect of loading time and environment on the indentation hardness of glass. *Physics and Chemistry of Glasses* 14:45–52.
3. Fairbanks, C. J., R. S. Polvani, S. M. Wiederhorn, B. J. Hockey, and B. R. Lawn. 1982. Rate effects in hardness. *Journal of Materials Science Letters* 1:391–93.
4. Yoshioka, M., and N. Yoshioka. 1995. Dynamic process of Vickers indentation made on glass surfaces. *Journal of Applied Physics* 78:3431–37.

5. Ainsworth, L. 1954. The diamond pyramid hardness of glass in relation to the strength and structure of glass. *Journals: Society of Glass Technology* 38:479–500.
6. Hohne, L., and C. Ullner. 1995. How does indentation velocity influence the recording hardness value? *VDI Berichte* 1194:119–28.
7. Eriksson, C. L., P. L. Larsson, and D. J. Rowcliffe. 2003. Strain-hardening and residual stress effects in plastic zones around indentations. *Materials Science and Engineering A* 340:193–203.
8. Kese, K. O., Z. C. Li, and B. Bergman. 2005. Method to account for true contact area in soda-lime glass during nanoindentation with the Berkovich tip. *Materials Science and Engineering A* 404:1–8.
9. Peng, Z., J. Gong, and H. Miao. 2004. On the description of indentation size effect in hardness testing for ceramics: Analysis of the nanoindentation data. *Journal of the European Ceramic Society* 24:2193–2201.
10. Gong, J., H. Miao, and Z. Peng. 2004. On the contact area for nanoindentation tests with Berkovich indenter: Case study on soda-lime glass. *Materials Letters* 58:1349–53.
11. Kese, K. O., and Z. C. Li. 2006. Semi-ellipse method for accounting for the pileup contact area during nanoindentation with the Berkovich indenter. *Scripta Materialia* 55:699–702.
12. Grau, P., G. Berg, H. Meinhard, and S. Mosch. 1998. Strain rate dependence of the hardness of glass and Meyer's law. *Journal of the American Ceramic Society* 81:1557–64.
13. Marsh, D. M. 1964. Plastic flow in glass. *Proceedings of the Royal Society of London Series A, Mathematical and Physical Sciences* 279:420–35.
14. Evans, G., and T. R. Wilshaw. 1977. Dynamic solid particle damage in brittle materials. *Journal of Materials Science* 12:97–116.
15. Chaudhri, M. M., J. K. Wells, and A. Stephens. 1981. Deformation and fracture of simple ionic crystals at very high rates of strain. *Philosophical Magazine A* 43:643–64.
16. Marshall, D., A. Evans, and Z. Nisenholz. 1983. Measurement of dynamic hardness by controlled sharp-projectile impact. *Journal of the American Ceramic Society* 66:580–85.
17. Anton, R. J., and G. Subhash. 2000. Dynamic Vickers indentation of brittle materials. *Wear* 239:27–35.
18. Quinn, G. D., P. J. Patel, and I. Lloyd. 2002. Effect of loading rate upon conventional ceramic microindentation hardness. *Journal of Research of the National Institute of Standards and Technology* 107:299–306.
19. Chakraborty, R., A. Dey, and A. K. Mukhopadhyay. 2010. Loading rate effect on nanohardness of soda-lime-silica glass. *Metallurgical and Materials Transactions A* 41A:1301–12.
20. Chakraborty, R., A. Dey, and A. K. Mukhopadhyay. 2010. Energy issues in loading rate effect on nanohardness of soda-lime-silica glass. *Physics and Chemistry of Glasses: European Journal of Glass Science and Technology Part B* 51:293–303.
21. Dey, A., R. Chakraborty, and A. K. Mukhopadhyay. 2011. Enhancement in nanohardness of soda-lime-silica glass. *Journal of Non-Crystalline Solids* 357:2934–40.
22. Bhattacharya, M., R. Chakraborty, A. Dey, A. K. Mandal, and A. K. Mukhopadhyay. 2012. Improvement in nanoscale contact resistance of alumina. *Applied Physics A* 107:783–88.

23. Bhattacharya, M., R. Chakraborty, A. Dey, A. K. Mandal, and A. K. Mukhopadhyay. 2013. New observations in micro-pop-in issues in nanoindentation of coarse grain alumina. *Ceramics International* 39:999–1009.
24. Biswas, N., A. Dey, and A. K. Mukhopadhyay. 2012. Loading rate effect on nanohardness of human enamel. *Indian Journal of Physics* 86:569–74.
25. Biswas, N., A. Dey, and A. K. Mukhopadhyay. 2013. Micro-pop-in issues in nanoscale contact deformation resistance of tooth enamel. *ISRN Biomaterials* 2013 (545791): 1–6.
26. Biswas, N., A. Dey, and A. K. Mukhopadhyay. 2013. *Effect of loading rate on nanohardness of enamel.* Unpublished work.

49

Measurement of Residual Stress by Nanoindentation Technique

Arjun Dey and Anoop Kumar Mukhopadhyay

49.1 Introduction

This chapter discusses the efficacy of nanoindentation technique toward the measurement of residual stress. Processing of ceramic materials involved sintering, thermal-spray process, etc., all of which require high temperature. We shall be concentrating here only on the effect of the plasma-spray technique for the fabrication of HAp coating, as discussed in Chapter 9. We know that quenching and thermal mismatch-induced stresses are two prime contributors to the residual stress in plasma-sprayed ceramic coatings [1]. Basically, in plasma spraying, the flowable powder granules/particles are fed from the powder feeder into the plasma jet and then propelled at very high speed ($\approx$200 m·s^{-1}) through the plasma jet toward the metallic substrate. The molten ceramic droplets then spread as splats in the shape of a pancake over metals. As cooling is done in an open-air atmosphere, an extremely rapid heat-transfer process starts between the stack of the ceramic splats and the metal substrate. Therefore, it is understood that the heat rejection is directly dependent on thermal diffusivity of the respective metal substrates [2]. The residual stress is generated in the ceramic coating due to the sudden quenching from high temperature to room temperature and the mismatch of the coefficient of thermal expansion (CTE) between the metal and the ceramic coating (e.g., MIPS-HAp).

We recently presented a collection of literature data on residual stress of HAp and related composite coatings showing that the numerical magnitude of residual stress may vary from as low as $\approx$5 MPa to as high as $\approx$450 MPa [2]. Moreover, the nature of the stress also varied in a random manner. Researchers have employed several methods like XRD, materials-removal techniques, piezospectroscopy, neutron diffraction, analytical treatment, etc., to evaluate residual stress of HAp coatings. However, each technique has its own pros and cons. In the case of XRD-based techniques, the sample is assumed to be a defect-free and ideally isotropic bulk material, which is not strictly true.

On the other hand, the spatial resolution of XRD-based methods is much superior to the material-removal methods. However, residual stresses in thin and thick coatings as well as bulk materials have recently been measured by nanoindentation technique [3–7]. Compared to the other methods, e.g., piezospectroscopy, X-ray diffraction, neutron diffraction, etc., the nanoindentation-based technique is less tedious and time consuming. So, we decided to apply the same to evaluate the residual stress of the MIPS-HAp coatings.

49.2 Measurement of Residual Stress by Nanoindentation: Concept

The method was first introduced by Suresh and Giannakopoulos [3]. Figures 49.1a–c show the nanoindentation processes on (a) a stress-free coating without the substrate, (b) a bonded coating with residual compressive stress, and (c) a bonded coating with residual tensile stress along with substrate. The presence of a residual compressive stress will offer a resistance to the penetration of the indenter into the coating. This results in lesser residual depth of penetration in comparison to what would actually occur in a stress-free coating. In the same way, the presence of residual tensile stress will assist the penetration of the indenter into the coating. This therefore shows greater residual depth of penetration in comparison to what would actually occur in

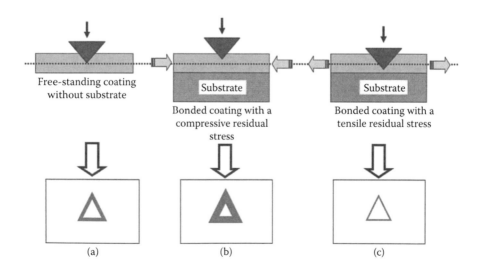

(a)   (b)   (c)

FIGURE 49.1
Schematic of nanoindentation processes on (a) a free-standing coating without substrate, (b) a bonded coating with compressive stress, and (c) a bonded coating with tensile stress.

a stress-free coating. Thus, for given fixed indentation load, the residual depth of penetration—and hence the projected contact area—shall be lower in a coating on a given substrate with residual compressive stress (Figure 49.1b) in comparison to those of a stress-free coating (Figure 49.1a) without the substrate. Similarly, the residual depth of penetration (for the same given fixed load)—and hence the projected contact area—shall be higher in a coating (on the same given substrate) with residual tensile stress (Figure 49.1c) in comparison to those of a stress-free coating (Figure 49.1a) without the substrate.

Suresh and Giannakopoulos [3] showed that the ratio of projected contact area is related to residual tensile and compressive stress by equations (49.1) and (49.2), respectively:

$$\frac{A_t}{A_0} = \frac{h_t}{h_0} = \left(1 - \frac{\sigma_t \cdot \sin\beta}{H_t}\right)^{-1} \tag{49.1}$$

and

$$\frac{A_c}{A_0} = \frac{h_c}{h_0} = \left(1 + \frac{\sigma_c \cdot \sin\beta}{H_c}\right)^{-1} \tag{49.2}$$

where σ is the residual stress of the coating material and A is the projected area of bonded coating with tensile (t) or compressive (c) residual stress. Further, A_0 signifies the projected area of free-standing coating (i.e., coating not bonded to the substrate). Here, h means the final depth of penetration, and h_0 is the corresponding final depth of penetration of free standing coating. Hardness of the coating is represented as H, and β is the complementary angle to the semi-apex angle of the indenter. In the present case, 10 mN load was employed on the substrate-bonded HAp coatings as well as the free-standing (i.e., unbonded) HAp coatings.

49.3 Evaluation of Residual Stress by Nanoindentation of HAp Coating

Further, we measured the residual stress on a HAp coating for the first time ever [2] by employing the nanoindentation technique described here. The residual stresses of magnitude ≈22 MPa (compressive) and ≈11 MPa (tensile) were estimated for both the microplasma-sprayed HAp coating on SS316L (HACS) and Ti-6Al-4V (HACT). Later, we evaluated the efficacy of the present method by comparing it to the XRD-based method, which is explained in detail elsewhere [2]. The XRD-based method gave a residual compressive stress of ≈11 MPa for HACS; however, a residual tensile stress of ≈12 MPa was obtained for HACT. Thus, both the order of magnitude and the nature of the residual stress data evaluated by the nanoindentation technique were comparable to the data derived from the XRD-based technique, thereby confirming the efficacy of the novel nanoindentation technique for the present MIPS-HAp coatings.

The A_0 and the corresponding H values of the free-standing HAp coating were ≈8.4 × 10^5 nm^2 and ≈4.31 GPa, respectively, while HACS showed respective A_c and H values of about ≈8.2 × 10^5 nm^2 and 4.96 GPa [2]. On the other hand, HACT showed respective A_c and H values of about 8.5 × 10^5 nm^2 and 3.77 GPa [2]. Thus, a decrease in the projected area is a signature of the presence of residual compressive stress, whereas the increase in projected area represents the presence of a residual tensile stress. The CTE (α) of sintered and dense bulk HAp is $\alpha_{HAp} = 11 \times 10^{-6}$ K^{-1}. It is higher than the α value of SS316L substrate, which is $\alpha_{SS} = 18 \times 10^{-6}$ K^{-1}. Similarly, the α value of Ti-6Al-4V, which is $\alpha_{TA} = 9 \times 10^{-6}$ K^{-1}, is less than that of the HAp phase. Therefore, the state of residual stress was expected to be compressive in nature for the HACS and tensile in nature for the HACT.

49.4 Conclusions

In this chapter, the efficacy of the nanoindentation technique in evaluating the residual stress of a MIPS-HAp coating on both SS316L and Ti-6Al-4V substrates was established beyond doubt. The data obtained from the nanoindentation technique matched well with those obtained from the XRD-based technique. Thus, the estimation of residual stress by the novel nanoindentation technique can be adopted for other materials in the future. As it is obvious that the nanoindentation techniques are extremely sensitive to local microstructural variations, it is of utmost importance to address the reliability issues of the nanomechanical properties data evaluated by the nanoindentation technique. This is exactly what we have presented in Chapter 50.

References

1. Matejicek, J., and S. Sampath. 2001. Intrinsic residual stresses in single splats produced by thermal spray processes. *Acta Materialia* 49:1993–99.
2. Dey, A., and A. K. Mukhopadhyay. 2014. Evaluation of residual stress in microplasma sprayed hydroxyapatite coating by nanoindentation. *Ceramics International*. Vol. 1, Issue 1, Part A, pp. 1263–1272. http://www.sciencedirect.com/science/article/pii/S0272884213007487.

3. Suresh, S., and A. E. Giannakopoulos. 1998. A new method for estimating residual stresses by instrumented sharp indentation. *Acta Materialia* 46:5755–67.
4. Bouzakis, K. D., G. Skordaris, J. Mirisidis, S. Hadjiyiannis, J. Anastopoulos, N. Michailidis, G. Erkens, and R. Cremer. 2003. Determination of coating residual stress alterations demonstrated in the case of annealed films and based on a FEM supported continuous simulation of the nanoindentation. *Surface and Coatings Technology* 174–175:487–92.
5. Bouzakis, K. D., and N. Michailidis. 2004. Coating elastic–plastic properties determined by means of nanoindentations and FEM-supported evaluation algorithms. *Thin Solid Films* 469–470:227–32.
6. Taylor, C. A., M. F. Wayne, and W. K. S. Chiu. 2003. Residual stress measurement in thin carbon films by Raman spectroscopy and nanoindentation. *Thin Solid Films* 429:190–200.
7. Chudoba, T., N. Schwarzer, V. Linss, and F. Richter. 2004. Determination of mechanical properties of graded coatings using nanoindentation. *Thin Solid Films* 469–470:239–47.

50

Reliability Issues in Nanoindentation Measurements

Arjun Dey and Anoop Kumar Mukhopadhyay

50.1 Introduction

In this chapter we discuss the reliability issue of the nanomechanical properties data for brittle solids as obtained by the nanoindentation technique. Whether the relevant data are on hardness, modulus, toughness, energy, etc., more often than not there is a significant degree of scatter. The point is: Why is it so? Is it real? Is it an error inherent to the experimental technique? Can we explain it?

We believe that this scatter can, at times, carry very useful information on a case-to-case basis. From a very simple yet fundamental standpoint, a few things are very relevant and very important to understand, at the very beginning. It does not really matter what fabrication technique we choose to make a brittle solid in amorphous or bulk or thin film or coating form or, for that matter, to make a corresponding composite, e.g., a ceramic matrix composite. It will always remain nearly impossible to make the same absolutely flawless and truly isotropic composite, as is often proposed in the standard theoretical framework of elasticity theory. When we are performing nanoindentation, the heterogeneity at the local microstructural length scale interacts with the indenter tip, which has a small nanometer-radius as the tip penetrates a given sample. Therefore, any microstructural discontinuity and/or heterogeneity in local microstructural length scale will leave the corresponding fingerprint in the measured data. Thus, we need a way to measure the reliability of the nanoindentation data.

Our objective here is to determine whether the well-established Weibull statistical model can be used to assess the reliability of the nanoindentation data. In 1951, Weibull introduced the applicability of this statistical method to understand primarily the reliability issues in the strength data of ceramics [1]. Afterwards, the method was applied to several indentation-derived mechanical properties like hardness, modulus, fracture toughness, etc.

This statistical method has been widely employed to determine the characteristic values for the micro-/nanohardness and Young's modulus of heterogeneous brittle coatings like thermal barrier coatings [2–6], microplasma-sprayed HAp coatings [7, 8], micro arc oxidation ceramic coatings [9], bulk composites like C/C and C/SiC composites [10, 11], and multilayered ceramic matrix composites for solid oxide fuel cell applications [12].

50.2 The Weibull Statistical Distribution

The two-parameter Weibull statistical distribution function is one of the most popular [2–12]. We shall confine our focus here on the expression of the probability of survival (p) for a given parameter (x) as:

$$p = 1 - e^{\left(-\frac{x}{x_0}\right)^m} \tag{50.1}$$

where x_0 is the scale parameter, and m is known as the Weibull modulus, which is dimensionless. Actually, m reflects the degree of reliability in the experimentally found data within the distribution. When m is high, the scatter is low, so the reliability is high for the data. On the other hand, when m is low, the scatter is high, so the reliability is low yet countable or dependable for the data. Further, to calculate Weibull modulus, the data should be arranged in ascending order. Then, we can write the survival probability of the ith observation as:

$$p = \frac{(i - 0.5)}{N} \tag{50.2}$$

where N denotes the total number of observations.

Now, if we take for two consecutive times the natural logarithm of both sides of equation (50.2), then the equation becomes:

$$\ln\left[\ln\left\{\frac{1}{(1-p)}\right\}\right] = m\left[\ln(x) - \ln(x_0)\right] \tag{50.3}$$

Thus, by fitting the dataset by least squares regression, m and x_0 can be easily found. The value of m can be obtained from the slope of the fitted straight line. Similarly, x_0 can be attained from the y-intercept of the fitted straight line. Ultimately, equating the quantity $\ln\left[\ln\left\{\frac{1}{(1-p)}\right\}\right]$ to zero and placing the values of m and x_0 in equation (50.3), the characteristic value (x_{ch}) of the given parameter, x, can be calculated. The quantity x may be hardness, Young's modulus, fracture toughness, etc. The values of m and x_{ch} have an enormous engineering and technological significance, as they provide the designer with a unique and dependable value of the required parameter. The probability of occurrence for the characteristic value is 63.2%. Thus, it can also provide an idea about the uniformity of the material and the reliability of the data.

Recently, Yang [13] has opined that the Weibull model is a powerful tool for predicting long-term stability as well as designing optimal biomaterials. He asserts that the prime advantage of the Weibull model is that it provides reasonably accurate failure analysis and failure forecasts while using only a small volume of samples to present relevant data in useful graphical plots. We believe that these data plots are highly useful in identifying the engineering design values for relevant properties. In this chapter, we discuss how the Weibull model was successfully employed in three typical cases. The first is a plasma-sprayed HAp coating with a very heterogeneous microstructure. The second is a bulk C/C composite, and the third is a C/SiC composite. The two composites comprised several phases and different fiber orientations. The processing details are discussed in Chapter 9.

50.3 Weibull Analysis for HAp Coating

The Weibull moduli (m) values of the nanohardness and Young's modulus data increased with load (Figures 50.1a and 50.1b) and were generally larger than 3. An m value larger than 3 is always good, especially for biomedical applications, as reported by Yang [13]. Therefore, the present plasma-spray coating is reliable enough for the purpose of further biomedical applications. This observation has been linked [8] to the load dependent length scale of interaction phenomena between the penetrating nanoindenter and the characteristic microstructural scale flaws and defects of the MIPS-HAp coating (Figure 50.1c). At relatively lower loads, the nanoindent size almost scaled with the size of microstructural defects (Figure 50.1c, upper part). The nanoindent itself was small enough and therefore was most likely to have also negotiated many intersplat boundaries. These are the typical weak, defect-populated zones of the coating. Therefore, such interactions had possibly led to a relatively higher degree of scatter in both nanohardness and Young's modulus data. Thus the m values were a little low (Figures 50.1a and 50.1b).

However, at relatively higher loads on the plan section, the nanoindent size was much bigger. Therefore, it covered a much larger number of splats

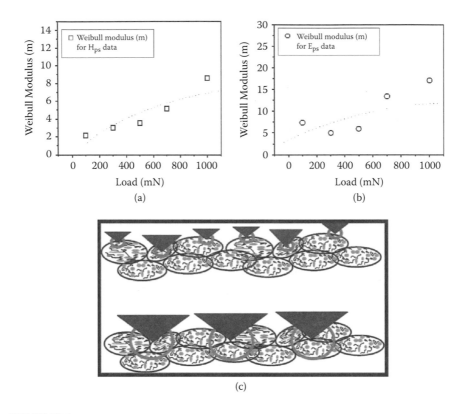

FIGURE 50.1
Load dependencies of Weibull modulus for (a) the nanohardness (*H*), (b) the Young's modulus (*E*) data on plasma-sprayed HAp coating, and (c) schematic explanation of load dependencies. PS = plan section, (Reprinted (modified) with permission from Dey et al. [8].)

and now had a chance to negotiate with only a small number of intersplat boundaries (Figure 50.1c, lower part). Thus the extent of interaction with intersplat boundaries was spatially diminished. This was why a relatively lower scatter occurred in the data, which explains why the *m* value was high (Figure 50.1b).

It has been proposed further [8] that, at higher loads, the randomly oriented interactions might have also cancelled out each other's effect. If this happens, it will also reduce scatter and hence enhance reliability. A reduction in data scatter should be reflected in relatively higher *m* values. The credence to such a picture was also borne out from the fact that the present experimental data showed an increasing trend of Weibull moduli with load (Figures 50.1a and 50.1b) for both nanohardness and Young's moduli data, evaluated by the nanoindentation experiments with a Berkovich nanoindenter. Similar observations were also reported [14] for atmospheric plasma-sprayed thermal barrier coatings.

50.4 Weibull Analysis for C/C and C/SiC Composites

In the next two typical illustrative examples, Weibull distribution plots are shown for two-dimensional (2-D) C/C and C/SiC composites with perpendicular and parallel fiber orientations for both nanohardness (H) and Young's modulus (E) in Figures 50.2 and 50.3, respectively [11]. Here, a load-controlled depth sensitive nanoindentation technique was employed with a Berkovich tip at an ultralow load of 10 mN. The m values (1.2–1.5 for H and 1.5–2.4 for E) were very low, as expected. Similar low m values were also reported by Kanari et al. [10]. Thus, a high scatter of data has been reported. We also found that our m values were dependent on fiber orientation [11]. The values of m were characteristically low simply because both of these composites had highly heterogeneous microstructures. The microstructures of the respective matrices had both dense and undensified regions, reinforcement phases, numerous interfaces between fiber and matrix phases, and surface defects such as cracks and pores of different sizes and shapes. These microstructural inhomogeneities had a major role on the reliability of the nanomechanical properties such as hardness, Young's modulus, etc., evaluated at the microstructural length scale. Further, it has been shown elsewhere [11] that the characteristic values of the mean contact pressure, relative stiffness, relative spring-back, and indentation energies can be easily derived using a similar technique to provide reliable engineering design data.

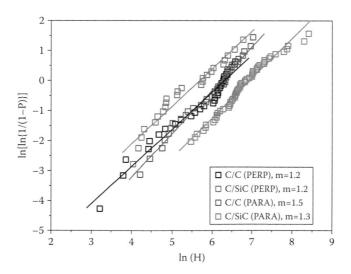

FIGURE 50.2 (See color insert.)
Typical examples of Weibull statistical fittings for the nanohardness (H) data on C/C and C/SiC composite with fiber orientation of perpendicular and parallel fashions. (Reprinted (modified) with permission from Sarkar et al. [11].)

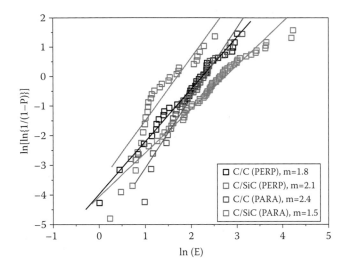

FIGURE 50.3 (See color insert.)
Typical examples of Weibull statistical fittings for the Young's modulus (*E*) data on C/C and
C/SiC composite with fiber orientation of perpendicular and parallel fashions. (Reprinted
(modified) with permission from Sarkar et al. [11].)

50.5 Conclusions

This chapter discusses the application of the Weibull statistical method
to interpret nanoindentation data such as hardness and modulus in
ceramic coatings and bulk composite materials. The load dependencies of the Weibull moduli were also considered and explained. It is clear
that this method is useful in understanding how the heterogeneity/uni-
formity of the microstructure affects the measured nanoindentation data.
We now understand the importance of these reliability issues in the field
of design engineering for microstructures. The next step is to consider the
effects of how a substrate's particular mechanical properties can contribute
to the measured nanomechanical properties of a ceramic thin film. This issue
is discussed in Chapter 51.

References

1. Weibull, W. 1951. A statistical distribution function of wide applicability. *Journal
   of Applied Mechanics* 18:293–97.
2. Lugscheider, E., K. Bobzin, S. Barwulf, and A. Etzkorn. 2001. Mechanical properties of EB-PVD-thermal barrier coatings by nanoindentation. *Surface and
   Coatings Technology* 138:9–13.

3. Guo, S., and Y. Kagawa. 2004. Effect of loading rate and holding time on hardness and Young's modulus of EB-PVD thermal barrier coating. *Surface and Coatings Technology* 182:92–100.
4. Guo, S., and Y. Kagawa. 2006. Effect of thermal exposure on hardness and Young's modulus of EB-PVD yttria-partially-stabilized zirconia thermal barrier coatings. *Ceramics International* 32:263–70.
5. Zhou, H., F. Li, B. He, J. Wang, and B. Sun. 2007. Air plasma sprayed thermal barrier coatings on titanium alloy substrates. *Surface and Coatings Technology* 201:7360–67.
6. Basu, D., C. Funke, and R. W. Steinbrech. 1999. Effect of heat treatment on elastic properties of separated thermal barrier coatings. *Journal of Materials Research* 14:4643–50.
7. Dey, A., A. K. Mukhopadhyay, S. Gangadharan, M. K. Sinha, D. Basu, and N. R. Bandyopadhyay. 2009. Nanoindentation study of microplasma sprayed hydroxyapatite coating. *Ceramics International* 35:2295–2304.
8. Dey, A., A. K. Mukhopadhyay, S. Gangadharan, M. K. Sinha, and D. Basu. 2009. Weibull modulus of nano-hardness and elastic modulus of hydroxyapatite coating. *Journal of Materials Science* 44:4911–18.
9. Dey, A., R. U. Rani, H. K. Thota, A. K. Sharma, P. Bandyopadhyay, and A. K. Mukhopadhyay. 2013. Microstructural, corrosion and nanomechanical behaviour of ceramic coatings developed on magnesium AZ31 alloy by micro arc oxidation. *Ceramics International* 39:313–20.
10. Kanari, M., K. Tanaka, S. Baba, and M. Eto. 1997. Nanoindentation behavior of a two-dimensional carbon-carbon composite for nuclear applications. *Carbon* 35:1429–37.
11. Sarkar, S., A. Dey, P. K. Das, A. Kumar, and A. K. Mukhopadhyay. 2011. Evaluation of micromechanical properties of carbon/carbon and carbon/carbon-silicon carbide composites at ultra-low load. *International Journal of Applied Ceramic Technology* 8:282–97.
12. Dey, T., A. Dey, P. C. Ghosh, M. Bose, A. K. Mukhopadhyay, and R. N. Basu. 2014. Influence of microstructure on nano-mechanical properties of SOFC single cell in pre- and post-reduced conditions. *Materials and Design* 53:182–91.
13. Yang, C.-W. Forthcoming. Development of hydrothermally synthesized hydroxyapatite coatings on metallic substrates and Weibull's reliability analysis of its shear strength. *International Journal of Applied Ceramic Technology*. http://onlinelibrary.wiley.com/doi/10.1111/ijac.12145/abstract.
14. Malzbender, J., and R. W. Steinbrech. 2003. Determination of the stress-dependent stiffness of plasma-sprayed thermal barrier coatings using depth-sensitive indentation. *Journal of Materials Research* 18:1975–84.

51

Substrate Effect in Thin Film Measurements

Arjun Dey, I. Neelakanta Reddy, N. Sridhara, Anju M. Pillai, Anand
Kumar Sharma, Rajib Paul, A. K. Pal, and Anoop Kumar Mukhopadhyay

51.1 Introduction

In this chapter, we mainly concentrate on the effects of a substrate's mechanical properties on the measured nanomechanical properties of a ceramic thin film. Usually, contact depths of less than 10% of the film/coating thickness are recommended in nanoindentation tests on thin films and coatings [1]. Otherwise, the mechanical properties of the substrate may affect the measured nanoindentation data of the thin film or coating. In terms of both modeling and experimental work, a significant amount has been devoted to the evaluation of nanomechanical properties of the soft thin films on hard substrate having different architecture combinations, e.g., metallic thin film on glass or metallic thin film on ceramic substrates [1–8]. Evaluation of the plastic-zone concept has been introduced for an Al/Si system [2]. Tsui and Pharr [3] reported that indentation pileup in the Al films on glass is significantly affected by the mechanical properties of the glass substrate. Further, the unloading curve of the $P–h$ plot of soft Al film was significantly influenced by the mechanical properties of the hard glass substrate, which ultimately had a major effect on the nanoindentation data analysis. Even empirical models applicable only to Al/Si and W/Si systems [4] have been developed. Attempts have also been made [5–7] to explain the nanomechanical responses of both the compliant thin films on stiff substrates (e.g., Al/sapphire) and the stiff films on compliant substrates (e.g., W/Si and W/glass). Further, King [8] has modified the Doerner and Nix [4] solution, which is applicable for all film/substrate systems. The modified equation is given by [8]:

$$\frac{1}{E_r} = \frac{1 - \nu_{\text{indenter}}^2}{E_{\text{indenter}}} + \frac{1 - \nu_{\text{film}}^2}{E_{\text{film}}} \left(1 - e^{-\frac{\alpha T}{A}} \right) + \frac{1 - \nu_{\text{substrate}}^2}{E_{\text{substrate}}} \left(e^{-\frac{\alpha T}{A}} \right) \tag{51.1}$$

where the first part indicates the contribution of the indenter, and the second and third parts represent the contributions of the thin film and the substrate, respectively. In equation (51.1), A is the square root of the projected contact area, and T is the thickness of the film where α is a scale parameter that is different for different indenter geometries. In another work by Saha and Nix [6], King's model [8] has been applied successfully to Al/glass and W/sapphire systems to determine and quantify the substrate effect. While the literature contains rich resources on the effect of the substrate in nanoindentation data of both ductile and hard thin films on brittle substrates [1–8], there are not many reports on the nanomechanical responses of ceramic/ceramic and ceramic/metal systems. That's why in the present case we have judiciously chosen (a) DLC/quartz glass (e.g., a hard/hard system) and (b) alumina/SS (e.g., a hard/soft system) combinations to look at their respective nanomechanical responses.

51.2 Substrate Effect in Nanocomposite DLC Thin Films

Nanocrystalline gold incorporated DLC (AuDLC) composite films as well as pure DLC (diamond-like carbon) thin films were developed on quartz glass. The details of processing are described in Chapters 9 and 41. The films had thickness of about 70–120 nm. The AuDLC nanocomposite films were deposited with 50%, 60%, 70%, and 80% of argon in the argon–methane mixture. The corresponding thin films shall be denoted here as AuDLC50, AuDLC60, AuDLC70, and AuDLC80. The DLC thin films were also similarly deposited with 50%, 60%, 70%, and 80% of argon in the argon–methane mixture. The corresponding thin films shall be denoted here as DLC50, DLC60, DLC70, and DLC80. In the present investigation, partial unloading mode was applied at a peak load of 1000 µN by using a Berkovich indenter.

For the AuDLC nanocomposite thin films, the nanohardness and Young's modulus data are shown as a function of depth in Figures 51.1a and 51.1b, respectively. Similar data for the DLC films are shown in Figures 51.1c and 51.1d. At typical ultralow depths of about 20–25 nm, both the nanohardness and Young's modulus values of AuDLC nanocomposite films register apparently higher values before settling off to a much lower plateau value at relatively higher depths of penetration. Similar behavioral patterns could also be observed for nanohardness and Young's modulus of DLC films. The decrease of nanohardness with increase in the load as well as depth was also observed by Jian, Fang, and Chuu [9] for DLC film at very low depths of 10–50 nm. A similar observation was made by Bharathy et al. [10] for titanium-doped DLC films up to a depth of about 35 nm. However, beyond a certain critical depth ($\approx$35 nm) in the cases of both DLC and nanocomposite AuDLC films, both nanohardness and Young's modulus data were again sluggishly increased. This behavior clearly indicates the influence of the mechanical properties of the quartz glass substrate on the measured nanohardness and Young's modulus data. Therefore, to avoid the substrate effect on the nanoindentation measurements, the nanohardness and Young's moduli data should

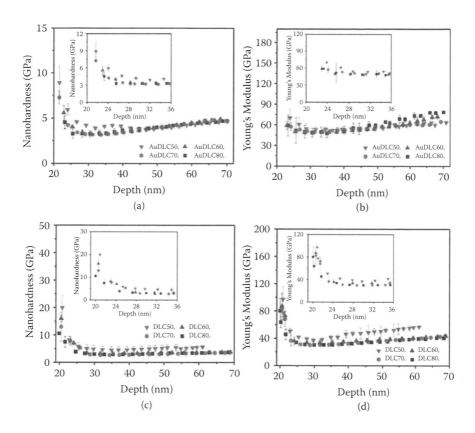

FIGURE 51.1
(a) Nanohardness and (b) Young's modulus as a function of depth for AuDLC. (c) Nanohardness and (d) Young's modulus as a function of depth for DLC. Inset: blown-up view of nanohardness and Young's modulus as a function of depth up to 35 nm.

better be examined as a function of depth up to 35 nm for both the DLC and AuDLC nanocomposite films. In fact, this is what has been shown as a blown-up picture as the insets of the respective figures. It should be noted that all the data plotted as insets in Figures 51.1a–d, respectively, showed decreasing trend with increase in depth, as expected.

51.3 Substrate Effect in Alumina Film

The other experiments were done on the alumina/SS304 system at a low load of 1.5 mN and then at a hundred times higher load of 150 mN. The alumina film of thickness ≈1.5 μm was grown by a direct rf magnetron sputtering method. Typical $P–h$ plots of the alumina film of thickness ≈1.5 μm

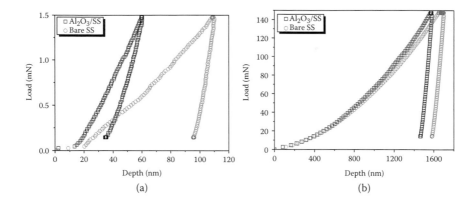

FIGURE 51.2
Typical *P–h* plots of alumina film of thickness ~1.5 μm on SS304 and bare SS304 at (a) lower load of 1.5 mN and (b) higher load of 150 mN.

on SS304 and bare SS304 are shown in Figures 51.2a and 51.2b for both a very low load of 1.5 mN and a relatively high load of 150 mN, respectively. When we kept the maximum depth of penetration to about 60 nm, we got a reasonably good result like hardness of ≈17 GPa and modulus of ≈270 GPa. At such depth, the maximum depth of penetration is well below 10% of the film thickness (≈150 nm). So, here the influence of the mechanical properties of the substrate did not come into the picture at all. The corresponding *P–h* plots of ductile SS and brittle alumina are also depicted here in Figure 51.2a. Here, the *P–h* plot of alumina is showing an elastoplastic deformation, as expected for a brittle ceramic, whereas that of the SS304 indicates a more plastic deformation pattern, as expected for a ductile and soft material. Now, if we increase the load to 150 mN when the corresponding depth goes up to ≈1500 nm, the nature of the *P–h* plot of the same 1.5 μm alumina thin film is entirely transformed from an elastoplastic to a plastic pattern. The reason is that we have gone now deliberately to a depth that is at least 10 times the depth within which the measurement should have been done. So, in this particular case, the mechanical properties of the substrate have contributed a significant influence on the measured nanoindentation data of the alumina thin film. Now, if we look at the data, we get hardness of 2.2 GPa and modulus of ≈175 GPa measured at the nanoindentation load of 150 mN for the alumina thin film, whereas at the same load, the bare SS304 substrate had hardness of ≈2.6 GPa and modulus of ≈200 GPa. Therefore, the data obtained at 150 mN load are not reliable for the present alumina thin film. Hence, it is obvious that to get reliable nanoindentation data of thin films/coatings, the measurement should always be done at less than 10% of the thickness of the film/coating. Otherwise, an appropriate model should be devised to eliminate the influence of the mechanical properties of the substrate from the measured nanoindentation data of the thin film/coating.

51.4 Conclusions

We have discussed how the mechanical properties of the substrate can significantly influence the nanoindentation-derived data of thin films. Experimental data presented for both DLC as well gold-doped DLC and alumina thin films showed a strong effect of the mechanical properties of the substrate on the nanoindentation data measured at relatively higher indentation depth. To avoid the same, therefore, it is advisable that the measurement be done at less than 10% of the thickness of the relevant film/coating. In Chapter 52 we shall focus on the scope for future research.

References

1. Chen, X., and J. J. Vlassak. 2001. Numerical study on the measurement of thin film mechanical properties by means of nanoindentation. *Journal of Materials Research* 16:2974–82.
2. Kramer, D. E., A. A. Volinsky, N. R. Moody, and W. W. Gerberich. 2001. Substrate effects on indentation plastic zone development in thin soft films. *Journal of Materials Research* 16:3150–57.
3. Tsui, T. Y., and G. M. Pharr. 1999. Substrate effects on nanoindentation mechanical property measurement of soft films on hard substrates. *Journal of Materials Research* 14:292–301.
4. Doerner, M. F., D. S. Gardner, and W. D. Nix. 1986. Plastic properties of thin films on substrates as measured by submicron indentation hardness and substrate curvature techniques. *Journal of Materials Research* 1:845–51.
5. Han, S. M., R. Saha, and W. D. Nix. 2006. Determining hardness of thin films in elastically mismatched film-on-substrate systems using nanoindentation. *Acta Materialia* 54:1571–81.
6. Saha, R., and W. D. Nix. 2002. Effects of the substrate on the determination of thin film mechanical properties by nanoindentation. *Acta Materialia* 50:23–38.
7. Cao, Y., S. Allameh, D. Nankivil, S. Sethiaraj, T. Otiti, and W. Soboyejo. 2006. Nanoindentation measurements of the mechanical properties of polycrystalline Au and Ag thin films on silicon substrates: Effects of grain size and film thickness. *Materials Science and Engineering A* 427:232–40.
8. King, R. B. 1987. Elastic analysis of some punch problems for a layered medium. *International Journal of Solids and Structures* 23:1657–64.
9. Jian, S. R., T. H. Fang, and D. S. Chuu. 2004. Nanoindentation investigation of amorphous hydrogenated carbon thin films deposited by ECR-MPCVD. *Journal of Non-Crystalline Solids* 333: 291–95.
10. Bharathy, P. V., D. Nataraj, P. K. Chu, H. Wang, Q. Yang, M. S. R. N. Kiran, and D. Mangalaraj. 2010. *Applied Surface Science* 257:143–50.

52

Future Scope of Novel Nanoindentation Technique

Arjun Dey and Anoop Kumar Mukhopadhyay

52.1 Introduction

In recent years, the nanoindentation technique has been applied not only to bulk materials such as metals, ceramics, polymers, composites, coatings, and thin films, but also to hard/soft tissues, biological materials, nanomaterials, in situ deformation and recrystallization, high-temperature nanomechanical behavior, etc. Now look into a simple trend: If we search for the keyword *nanoindentation* in Google Scholar, we can find there are about 52,000 entries. Now, if we restrict this search to the last ten years, the number of hits is about 22,700, which is almost 50% of the global result of 50,200. This indicates that interest regarding nanoindentation is undoubtedly increasing in recent years at a very rapid rate.

52.2 Nanoindentation on Biological Materials and Nanostructures

A series of recent reviews [1–7] have shown in very impressive manner how the nanoindentation technique has become very useful in evaluating the nanomechanical properties of biomolecules [1, 3], biological materials [2, 7], living cells [3], and mineralized hard tissues [4–6]. A recently published book by Oyen [8] is dedicated to the application of nanoindentation for biological materials. Further, the implementation of nanoindentation for nanomaterials and nanostructures like nanodots, nanopillars, nanotowers, nanobelts, nanotubes, quantum dots, and atomic layer depositions (ALDs) are well documented in the recent review by Palacio and Bhushan [9]. Moreover, the use of the nanoindentation technique for the study of in situ deformation behavior is also very well summarized here by Palacio and Bhushan [9].

52.3 In Situ Nanoindentation and Picoindentation

A more specific review work on in situ nanoindentation has been recently published by Nili et al. [10], wherein they showed a complete picture of nanomechanical deformation inside SEM and TEM, high-temperature nanoindentation, and picoindentation (where load level is of the order of few nanonewtons). The issue of nano/picoindentation revealed by SEM is documented in review articles by Ghisleni et al. [11] and Gianola et al. [12]. Further, the critical issues of importance during a study of nanoindentation inside the TEM have been reviewed by Kiener et al. [13].

52.4 High Temperature Nanoindentation

In the case of high-temperature nanoindentation, the challenges are more like oxidation of the sample, temperature drift, cooling, etc. These issues have been summarized by Duan and Hodge [14]. The diamond tip is probably a more critical issue, as diamond possesses high thermal conductivity. Consequently, the correctness of measurement at a specific temperature is still questionable. The high-temperature nanoindentation work had most likely started from 1994 when a work by Suzuki and Ohmura showed utilization of microindentation at comparatively lower temperature on silicon [15]. As of today, it is possible to conduct nanoindentation experiments up to temperatures of only 600°C [16–20]. Such a temperature would be good enough to look at the nanoindentation behavior of amorphous brittle solids like the SLS (soda-lime-silica) glass and the polymers. However, this temperature will need to be raised further if high-temperature nanoindentation work of real relevance is to be conducted on structural and functional ceramics, including the bioceramics.

52.5 Properties other than Hardness and Modulus: A Direct Measurement

It is well known that depth-sensitive indentation or nanoindentation techniques generally have been applied to evaluate hardness and modulus. However, some of the other related mechanical properties like fracture toughness, residual stress, adhesion strength of films, and fatigue can also be quantified using this technique. Although fracture toughness and residual stress topics have already been discussed in Chapters 35 and 49, we want to take a final look at them for the future expansion of the nanoindentation technique, especially for brittle solids.

52.5.1 Fracture Toughness

The measurement of fracture toughness by the nanoindentation method is critically reviewed in publications by Zhang and Zhang [21, 22]. Fracture toughness can be evaluated by direct radial crack-length measurement of nanoindents from the built-in optical/SPM facility or by postindentation SEM of nanoindents. The direct radial crack measurement method in nanoindentation was employed first by Pharr, Harding, and Oliver [23] to measure the fracture toughness of several brittle materials like glass, quartz, silicon, sapphire, and silicon nitride. Further, many researchers, including us (in Chapter 35), have adopted this method to evaluate the fracture toughness of brittle solids, more specifically coatings. The same has been extended to thin films. The best advantage here is that the measured fracture toughness is of relevance to the microstructural length scale itself. The other advantage is that the specimen size requirement is much smaller compared to what would be required to prepare a sample for single-edge/chevron-notched beam to be broken by three-point/four-point fixtures in a universal testing machine. However, problems may arise in the case of shallow indentation depths, where the radial crack is very difficult to locate even under an electron microscope [24–26]. In this case, Li, Bhushan, and coworkers [24–26] applied energy-release-based fracture mechanics to evaluate the fracture toughness of brittle solids. Here, the discontinuity (large pop-in) observed in load versus depth (P–h) plots due to through-thickness cracking is assumed to represent the energy released due to the cracking process. Generally, the cube corner tip has been preferred [24–26] for evaluation of fracture toughness by this method, whereas the Vickers or Berkovich tip is preferred [21–23] for the direct radial crack measurement method. The question that we would leave with the reader is whether it is possible to extend these two methods to evaluate the microstructural lengthscale-relevant fracture toughness of hard natural biomaterials like bone and teeth. The success, if it happens, will also pave the way toward extending or modifying these methods to be able to measure the fracture toughness of the whole genre of soft natural biomaterials like the nerve, the eyeball, the fingertips, and the skin by the nanoindentation technique.

52.5.2 Residual Stress

Residual stress can also be quantified by applying the novel nanoindentation technique. Jang [27] recently reviewed the work about the estimation of residual stress by instrumented indentation. A very simple but effective nanoindentation method for evaluating residual stress was shown both theoretically and experimentally for the very first time by Suresh and

Giannakopoulos [28]. Recently, the present authors followed this method to evaluate the residual stress of ceramic coatings, as described in Chapter 49. Further, many researchers [29–31] have tried to develop nanoindentation-based residual stress models. However, most of the aforesaid models are modifications or adaptations of the Suresh and Giannakopoulos proposal [28]. The question we want to ask here is whether the nanoindentation technique can be used to measure the residual stress in a particulate or fiber-reinforced ceramic matrix composite (CMC).

52.5.3 Adhesion Strength

Adhesion strength of the thin film is generally measured by the nano-scratch test. The pioneer researchers in this field are Bhushan and his coworkers, who have put forth enormous effort to understand the scientific principles that are operative in such situations, and these works are already well documented in several books [32–35]. Here, the critical load that can delaminate the top coat can be identified by the kink that appears in the plot of lateral or tangential force versus scratch length. This critical load is taken as a measure of the adhesion strength of the coating or thin film. However, any deviation of slopes or the appearances of pop-ins (due to coating peel-off) in the loading curve of *P–h* plots during the conventional nanoindentation on thin films can also identify the critical load point, beyond which the adhesion will be lost. Both Stone et al. [36] and Gupta and Bhushan [37] have utilized this method to calculate adhesion strength/critical load point where both Berkovich and conical indenters were used. Recently, Borrero-Lopez et al. [38] also showed how the conical indenter with a nearly spherical tip can be utilized to measure adhesion strength of amorphous carbon thin film. In their work [38], the film failure loads were determined from the positions of the pop-ins in the loading part of the nanoindentation curves. In another related work, Manso-Silvan et al. [39] also successfully implemented this method on a much thicker coating (≈5 μm) of calcium phosphate prepared by the aerosol-gel technique. Further, the extrapolation of this proposition can be utilized to evaluate interfacial fracture toughness, as illustrated by Mehrotra and Quinto [40] and critically reviewed by Volinsky, Moody, and Gerberich [41]. The open question here is whether this method can be extended to measure the adhesion strength and/or fracture toughness of biological interfaces in both natural and synthetic biomaterials.

52.5.4 Nanofatigue

Fatigue testing at the scale of microstructure is also gaining popularity nowadays. A harmonic force is required to perform nanofatigue testing. The continuous stiffness measurement (CSM) technique can be utilized using sinusoidal-shaped load cycles at higher frequencies. The change in contact stiffness indicates the fatigue behavior, as the contact stiffness depends on the deformation behavior of the test coupons. Several researchers reported fatigue behavior by the depth sensitive indentation or micro-/nanoindentation technique. For instance, Li and Bhushan [42] have utilized this technique for MEMS/NEMS devices. Wu, Shull, and Berriche [43] have taken advantage of this technique to study the fatigue behavior of submicron-thick carbon films. Similarly, Li and Bhushan [44] have used this method to study the nanofatigue for overcoats used in magnetic storage devices. However, there is not yet any theoretical work that has been initiated to understand and simulate the nanoindentation fatigue behavior of brittle solids. More experimental work and theoretical modeling would be most welcome in this emerging area of application for the nanoindentation technique. Another interesting area that has not yet received much attention is the nanoindentation fatigue study of the natural biomaterials.

References

1. Kurland, N. E., Z. Drira, and V. K. Yadavalli. 2012. Measurement of nanomechanical properties of biomolecules using atomic force microscopy. *Micron* 43:116–28.
2. Bowen, W. R., R. W. Lovitt, and C. J. Wright. 2000. Application of atomic force microscopy to the study of micromechanical properties of biological materials. *Biotechnology Letters* 22:893–903.
3. Kasas, S., and G. Dietler. 2008. Probing nanomechanical properties from biomolecules to living cells. *Pflugers Archiv-European Journal of Physiology* 456:13–27.
4. Lewis, G., and J. S. Nyman. 2008. The use of nanoindentation for characterizing the properties of mineralized hard tissues: State-of-the-art review. *Journal of Biomedical Materials Research Part B, Applied Biomaterials* 87B:286–301.
5. Thurner, P. J. 2009. Atomic force microscopy and indentation force measurement of bone. *Wiley Interdisciplinary Reviews: Nanomedicine and Nanobiotechnology* 1:624–49.
6. He, L. H., and M. V. Swain. 2008. Understanding the mechanical behaviour of human enamel from its structural and compositional characteristics. *Journal of the Mechanical Behavior of Biomedical Materials* 1:18–29.
7. Ebenstein D. M., and L. A. Pruitt. 2006. Nanoindentation of biological materials. *Nanotoday* 1:26–33.
8. Oyen, M. L., ed. 2010. *Handbook of nanoindentation: With biological applications.* Singapore: Pan Stanford Publishing.
9. Palacio, M. L. B., and B. Bhushan. 2013. Depth-sensing indentation of nanomaterials and nanostructures. *Materials Characterization* 78:1–20.
10. Nili, H., K. Kalantar-zadeh, M. Bhaskaran, and S. Sriram. 2013. In situ nanoindentation: Probing nanoscale multifunctionality. *Progress in Materials Science* 58:1–29.

11. Ghisleni, R., K. Rzepiejewska-Malyska, L. Philippe, P. Schwaller, and J. Michler. 2009. In situ SEM indentation experiments: Instruments, methodology, and applications. *Microscopy Research and Technique* 72:242–49.

12. Gianola, D. S., A. Sedlmayr, R. Mönig, C. A. Volkert, R. C. Major, E. Cyrankowski, S. A. S. Asif, O. L. Warren, and O. Kraft. 2011. In situ nanomechanical testing in focused ion beam and scanning electron microscopes. *Review of Scientific Instruments* 82:063901.

13. Kiener, D., K. Durst, M. Rester, and A. M. Minor. 2009. Revealing deformation mechanisms with nanoindentation. *Journal of the Minerals, Metals and Materials Society* 61:14–23.

14. Duan, Z., and A. Hodge. 2009. High-temperature nanoindentation: New developments and ongoing challenges. *Journal of the Minerals, Metals and Materials Society* 61:32–36.

15. Suzuki, T., and T. Ohmura. 1996. Ultra-microindentation of silicon at elevated temperatures. *Philosophical Magazine* A74:1073–84.

16. Lund, A. C., A. M. Hodge, and C. A. Schuh. 2004. Incipient plasticity during nanoindentation at elevated temperatures. *Applied Physics Letters* 85:1362.

17. Volinsky, A. A., N. R. Moody, and W. W. Gerberich. 2004. Nanoindentation of Au and Pt/Cu thin films at elevated temperatures. *Journal of Materials Research* 19:2650–57.

18. Komvopoulos, K., and X.-G. Ma. 2005. Pseudoelasticity of martensitic titanium-nickel shape-memory films studied by in situ heating nanoindentation and transmission electron microscopy. *Applied Physics Letters* 87:263108.

19. Schuh, C. A., J. K. Mason, and A. C. Lund. 2005. Quantitative insight into dislocation nucleation from high-temperature nanoindentation experiments. *Nature Materials* 4:617–21.

20. Schuh, C. A., C. E. Packard, and A. C. Lund. 2006. Nanoindentation and contact-mode imaging at high temperatures. *Journal of Materials Research* 21:725–36.

21. Zhang, S., and X. Zhang. 2012. Toughness evaluation of hard coatings and thin films. *Thin Solid Films* 520:2375–89.

22. Zhang, X., and S. Zhang. 2012. Rethinking the role that the "step" in the load–displacement curves can play in measurement of fracture toughness for hard coatings. *Thin Solid Films* 520:3423–28.

23. Pharr, G. M., D. S. Harding, and W. C. Oliver. 1993. Measurement of fracture toughness in thin films and small volumes using nanoindentation methods. In *Mechanical properties and deformation behavior of materials having ultra-fine microsctructures*, ed. M. A. Nastasi, D. M. Parkin, and H. Gleiter, 449. Heidelberg: Springer.

24. Li, X., D. Diao, and B. Bhushan. 1997. Fracture mechanisms of thin amorphous carbon films in nanoindentation. *Acta Materialia* 45:4453–61.

25. Li, X., and B. Bhushan. 1998. Measurement of fracture toughness of ultra-thin amorphous carbon films. *Thin Solid Films* 315:214–21.

26. Li, X., and B. Bhushan. 1999. Evaluation of fracture toughness of ultra-thin amorphous carbon coatings deposited by different deposition techniques. *Thin Solid Films* 355/356:330–36.

27. Jang, J. 2009. Estimation of residual stress by instrumented indentation: A review. *Journal of Ceramic Processing Research* 10:391–400.

28. Suresh, S., and A. E. Giannakopoulos. 1998. A new method for estimating residual stresses by instrumented sharp indentation. *Acta Materialia* 46:5755–67.

29. Carlsson, S., and P. L. Larsson. 2001. On the determination of residual stress and strain fields by sharp indentation testing; Part I: Theoretical and numerical analysis. *Acta Materialia* 49:2179–91.

30. Carlsson, S., and P. L. Larsson. 2001. On the determination of residual stress and strain fields by sharp indentation testing; Part II: Experimental investigation. *Acta Materialia* 49:2193–2203.

31. Lee, Y. H., and D. Kwon. 2003. Measurement of residual-stress effect by nanoindentation on elastically strained (100) W. *Scripta Materialia* 49:459–65.

32. Bhushan, B., and B. K. Gupta. 1991. *Handbook of tribology: Materials, coatings and surface treatments.* New York: McGraw-Hill.

33. Bhushan, B. 1999. *Handbook of micro/nanotribology.* Boca Raton, FL: CRC Press.

34. Bhushan, B. 1999. *Modern tribology handbook.* Vols. 1 and 2. Boca Raton, FL: CRC Press.

35. Bhushan, B. 2001. *Modern tribology handbook.* Vols. 1 and 2. Boca Raton, FL: CRC Press.

36. Stone, D., W. R. LaFontaine, P. Alexopoulos, T.-W. Wu, and C.-Y. Li. 1988. An investigation of hardness and adhesion of sputter-deposited aluminum on silicon by utilizing a continuous indentation test. *Journal of Materials Research* 3:141–47.

37. Gupta, B. K., and B. Bhushan. 1995. Micromechanical properties of amorphous carbon coatings deposited by different deposition techniques. *Thin Solid Films* 270:391–98.

38. Borrero-Lopez, O., M. Hoffman, A. Bendavid, and P. J. Martin. 2009. Reverse size effect in the fracture strength of brittle thin films. *Scripta Materialia* 60:937–40.

39. Manso-Silvan, M., M. Langlet, C. Jimenez, M. Fernandez, and J. M. Martinez-Duart. 2003. Calcium phosphate coatings prepared by aerosol-gel. *Journal of the European Ceramic Society* 23:243–46.

40. Mehrotra, P. K., and D. T. Quinto. 1985. Techniques for evaluating mechanical properties of hard coatings. *Journal of Vacuum Science and Technology A* 3:2401–5.

41. Volinsky, A. A., N. R. Moody, and W. W. Gerberich. 2002. Interfacial toughness measurements for thin films on substrates. *Acta Materialia* 50:441–66.

42. Li, X., and B. Bhushan. 2003. Fatigue studies of nanoscale structures for MEMS/NEMS applications using nanoindentation techniques. *Surface and Coatings Technology* 163/164:521–26.

43. Wu, T. W., A. L. Shull, and R. Berriche. 1991. Microindentation fatigue tests on submicron carbon films. *Surface and Coatings Technology* 47:696–709.

44. Li, X., and B. Bhushan. 2002. Nanofatigue studies of ultrathin hard carbon overcoats used in magnetic storage devices. *Journal of Applied Physics* 91:8334.

Conclusions

Here we shall present a brief overview of what we have learnt so far as we breeze through the different chapters on the nanoindentation of brittle solids. In Chapter 1 we introduced the important aspects of relevance in contact deformation of brittle solids and the related mathematical formalisms, whereas the elastic and the elastoplastic contact mechanics are detailed in Chapter 2. The genesis of indentation testing from 16th century up to modern times was sketched in Chapter 3 while the basic concepts of hardness and modulus of solids and the typical different approaches used to evaluate them remained the focus of Chapter 4. In Chapter 5, we demonstrated that in both the ex-situ and in-situ measurement scenarios of many advanced structural, functional, biological, cellular, and composite materials nanoindentation is the only useable tool for evaluation of the system's mechanical integrity at microstructural or sub-microstructural length scale.

Chapter 6 provided very basic ideas about the nanoindentation technique and those relevant models, which are most commonly practiced. The nominal details about how to calculate the contact area from depth measurements and how to account for piling up or sinking in issues in such measurements are given in Chapter 7. In Chapter 8, we tried to understand how a nanoindenter actually works in practice, discuss the essential components of a nanoindenter, and explain how these different components work in coordination with each other. We have also touched upon the issues of the varieties of types of load and displacement sensors and their respective working principles, as well as their relative limitations of range and resolutions. Chapter 9 contained a very brief overview of the preparation of the huge variety of brittle materials that we utilized and the relevant nanoindentation machines, and presented the machines utilized for microstructural characterizations.

It has been shown that depending on the peak load (P_{max}) and loading rate (LR) combination, e.g., 100 to 1000 mN, LR – (1–1000 mN.s^{-1}, Chapter 10); or 10,000 μN, LR – (10–20000 μN.s^{-1}, Chapter 11) the nanohardness of very thin (~330 μm) soda lime silica (SLS) glass can be significantly enhanced, e.g., by about 5 to 70%. For a given P_{max}, at loading rates lower than the threshold loading rate (TLR) a profuse presence of shear bands in and around the nanoindentation cavity was noted; but their presence decreased at loading rates higher than the TLR. It is suggested that the position of the shear flow–induced shear band formation would be governed by the local, short-range atomic arrangements in the glass, e.g., the positions of local weakness provided by the network modifiers. The applied loading provided enough shear stress just beneath the tip of the nanoindent to cause shear-induced flow and/or deformation in the SLS glass at positions of local weakness provided by the network modifiers. At loading rates higher than the TLR though, the forced homogenization of the plastic deformation response of the glass caused a reduction in the extent of shear flow that limited the resultant enhancement in nanohardness because most of the stored elastic energy got already exhausted. It was shown further in Chapter 12 that the relative amount of energy spent in plastic deformation (W_p/W_t) percent reached its maximum at the same TLR values. Further, the inelastic deformation (IED) parameter had the maximum values at these same TLR values, and decreased thereafter slowly with further increase in loading rate. These results suggested that the maximum magnitude of IED parameter occurred when the relative amount of energy spent in plastic deformation during the nanoindentation process was also the maximum, thereby contributing to the maximum rate of change in enhancement of nanohardness of the present glass with loading rate at the TLR values.

The severity of dynamic contact-induced damages and their interactions at both surface and subsurface regions of the present SLS glass increased with the applied normal load (Chapter 13), but decreased with speed of scratching (Chapter 14). Three distinct roughly semicircular subsurface damage zones with different damage characteristics were present in the current SLS glass. Further, for a given peak load, the width and depth of the scratch grooves, wear volume, and wear rate decreased with scratching speeds following inverse power law dependencies. These could easily be explained by a mathematical model proposed by the present authors; but the shear stress component played a major role in damage initiation. In addition, it has been shown in Chapter 15 that depending on the load and scratch speed combination, the nanomechanical properties can be degraded by about 30 to 70% due to the presence of microcracks inside the scratch grooves of the SLS glass.

As a typical illustrative example for a polycrystalline, coarse grain ceramic, the nanohardness (H~20.4–21.3 GPa) and Young's modulus (E~400 GPa) of a dense (~99.8%), coarse grain (~10 μm) alumina were measured as a function of load (10–1000 mN) by means of the nanoindentation technique (Chapter 16). The presence of a slight indentation size effect (ISE) in the present data was explained successfully through the application of the well-established Nix and Gao model that predicted H_0 = 19.93 GPa and h* = 9 nm. Subsequent extension of this approach provided the very first experimental observation (Chapters 17, 18) that for a high density (95% of theoretical) coarse grain (~20 μm) alumina ceramic the intrinsic contact deformation resistance, i.e., the critical load (P_c) against the initiation of nanoscale plasticity events, can increase with the loading rates following a power law dependence with a positive exponent. Here also, the maximum shear stress generated just underneath the nanoindenter had magnitude much higher than the theoretical shear strength of alumina.

Chapters 19 to 22 provided us with an idea of how the entire scenario would be affected if the given coarse grain ceramic was already highly strained, e.g., due to a high strain rate impact at a pressure much higher than its own Hugoniot elastic limit (HEL). It was shown for the very first time that there is a strong ISE present in the 10 μm grain sized alumina shocked up to 6.5 GPa pressure by a gas gun (Chapters 19, 20), which was otherwise mild in the same alumina. The strong ISE was linked to the presence of highly localized micro-tensile stresses in the suitably oriented grains of the damaged microstructure of shock recovered alumina. In the presence of such tensile stress at the same comparable load, the depth of penetration was more, thereby giving rise to the observed ISE. At the same time, presence of slip lines, deformation bands, and microcracks inside the nanoindentation cavity led to more frequent occurrences of multiple "micro pop-in" events in the load depth plots obtained during the loading cycles in the nanoindentation experiments conducted on the damaged microstructure of the shock-recovered alumina.

When the shock pressure was enhanced to about 12 GPa, the drop in nanohardness was very significant (Chapter 21). There were more frequent multiple micro pop-in events in the P-h plots of the alumina samples shocked up to 12 GPa pressure by the same gas gun. The experimental observation of ISE was the strongest in 12 GPa shocked alumina samples, stronger in the 6.5 GPa shocked alumina samples, and strong in the As-sintered alumina samples (Chapter 22). The Nix and Gao model could successfully explain the observed ISE in all the alumina samples. This fact suggests that a generic mechanism was responsible. Further, the generic mechanism was linked to the relative frequency of the multiple micro pop-in events at the microstructural and/or nanostructural length scale and the extent of stochastic interaction with the spatial distribution between the pre-existing plus shock-induced defects and the nanoindentation zone of influence depending upon the loads as well as the loading rates applied during the nanoindentation experiments. It is understandable that better insights into these aspects can really help us to design more damage-tolerant polycrystalline ceramics.

The nanoindentation responses of ceramic matrix composites (CMCs) were presented in Chapters 23 to 25. As far as the CMCs are concerned, the average nanohardness and Young's modulus and the extent of anisotropy of the aforesaid properties were the highest for the LSI- derived 2D C/C-SiC composites along the direction parallel to that of the C-fabric stacking as shown in Chapter 23. The critical examination of the relative stiffness and relative spring-back data explained why the extents of anisotropy in nano-mechanical properties and elastic recovery were the highest in the C/C-SiC composites compared to those of the C/C composites. The characteristic high scatter in the experimentally measured nanoindentation data was possibly linked to the characteristic heterogeneous microstructure of the composites. The nanoindentation behavior of tape-cast multilayered sintered alumina and lanthanum phosphate composites were summed up in Chapter 24. However, the nanohardness and Young's modulus values of the HAp-mullite composites were inferior to those of the HAp matrix phase as discussed in Chapter 25, possibly due to the presence of relatively softer grain boundary phases in the former. In contrast, an opposite trend was observed in the case of other HAp based composite, e.g. HAp-calcium titanate which is mainly developed to increase the electrical properties of pure HAp monolith.

Similarly, the nanoindentation responses and damage evolution mechanisms of a wide variety of functional materials and ceramics are dealt with in Chapters 26 to 31 of Section 8. For instance, Chapter 26 focused on the pop-out in P-h plots of silicon a phase that transforms under nanoindentation-induced pressure, while Chapter 27 elaborated on the nanomechanical behavior of ZTA where the tetragonal zirconia can phase transform under appropriate stress to the monoclinic phase. It was further shown in Chapter 27 that the incorporation of zirconia in alumina matrix degraded both nanohardness and Young's modulus of ZTA. In Chapter 28, the nanoindentation work on PZN-BT-PT and PZN-BT materials has shown that the PZN-BT-PT actuator had more elastic energy and higher nanohardness and Young's modulus compared to that of the PZN-BT actuator because its Ti^{4+} content was more than that of the PZN-BT actuator and it is in B-site of the distorted perovskite structure (ABO_3). As $Ti^{4+}(d^0)$ is in body center position with octahedral coordination, the Ti-O bond became rigid enough to restrict the plastic deformation. Hence, the presence of more Ti in PZN-BT-PT caused the relative hike in the elastic energy dissipation with a corresponding decrease in plastic deformation energy. These results also corroborated with the results obtained from the polarization studies of these two systems. Due to inherent rigidity of the Ti-O bond, charge separation and hence polarization of the PZN-BT-PT actuator was lower than that of the PZN-BT actuator ceramics. Further, possibly for the very first time ever, the nanoindenatation response of a nanostructured green magnetoelectric multiferroic material, e.g., nano bismuth ferrite (NBFO), has been presented in Chapter 29.

The nanohardness value of the NBFO pellet decreased at a rate of only 0.26 GPa.μN^{-1} when the load was increased by an order of magnitude, e.g., from 100 μN to 1000 μN. This may be an indication of a highly strain-tolerant microstructure. The decrease in nanohardness and Young's modulus were further explained with elastic recovery and plastic deformation energy concepts. Further, in both pre- and post-reduction conditions, the mechanical properties of each individual component layer of a planar single solid oxide fuel cell (SOFC), e.g., the NiO/Ni-YSZ anode, the 8YSZ electrolyte, and the LSM cathode, are evaluated at microstructural length scales by nanoindentation techniques in Chapter 30.

The electrolyte was the strongest of the three components in both pre- and post-reduction conditions. It was elaborated further that very high degree of "characteristic" scatter present in the nanomechanical properties data measured for both the cathode and the anode layers was due to their characteristically heterogeneous and nanoporous microstructures. Since a lot of glass is used in sealing different components of an effective stack of SOFCs used for power generation, the nanomechanical properties of magnesium lanthanum alumino borosilicate and barium magnesium borosilicate–based sealant glass-ceramics were also investigated in order to explore their suitability for SOFC sealing applications (Chapter 31). In general, the nanomechanical properties of the glass ceramics from the magnesium lanthanum alumino borosilicate system were much better than those from the magnesium barium borosilicate system.

Section 9, encompassing Chapter 32 to Chapter 37, actually has provided an answer to the question of whether it is possible to utilize the nanoindentation technique to investigate the nanoscale contact deformation behavior of ceramic coatings which are thick, porous, highly microcracked, and possess truly heterogeneous microstructures. It has been shown that the same is really possible to be achieved. Two coatings were investigated. One is a bioactive ceramic coating, e.g., microplasma-sprayed hydroxyapatite (MIPS-HAp) coatings (Chapters 32–36). The other is a protective ceramic oxide coating formed on magnesium alloy by micro arc oxidation, e.g., MAO coatings (Chapter 37) on AZ31B Mg alloys (~3% Al, 1% Zn, and 96% Mg).

In the case of the bioactive thick ceramic coatings it has been proposed that the variations in the extent of interaction of the indenter with average size defects across the depth of MIPS-HAp coatings may cause an ISE (Chapter 32). It is evident from this picture that there is an intrinsic presence of stochastic processes, which will have a statistical distribution, guided by the local microstructural heterogeneity. That is why in Chapter 33 we have made an attempt to study the Weibull modulus of nanohardness and elastic modulus of the hydroxyapatite coatings. The results show that the typical Weibull modulus ("m") values were about 2–8 for the nanohardness data, but increased to about 5–15 for the Young's modulus measured in the plan section of MIPS-HAp coatings. It was noted further in Chapter 34 that at comparable loads, the nanohardness (H_{cs}) of a cross section was usually higher than those (H_{ps}) measured on the plan section of the coating (i.e., $H_{cs} > H_{ps}$), thereby showing a characteristic anisotropy. The anisotropy was linked to the larger volume percent porosity as well as higher spatial density of planar defects, pores, and cracks in the plan section over those in the cross section. In addition, it has been depicted here for the very first time that in spite of having ~11 vol% porosity the MIPS-HAp coatings deposited on SS316L substrates had a high fracture toughness of about 0.6 MPa·m$^{0.5}$ (Chapter 35). Further, the extensive SEM study of the nanoindents also showed that crack blunting at micro- and macro-pore, localized secondary/multiple cracking, partial local delamination, crack branching and crack deflection/bifurcation provided various means of energy dissipation for achieving higher toughness in the present MIPS-HAp coatings. For the same coating, after immersion for 14 days in the simulated body fluid (SBF) solution, the nanohardness and Young's modulus values and coefficient of friction values were only slightly lower than those of the MIPS-HAp coating in the "as-sprayed" condition, suggesting the possibility of the major dominance of the nascent apatite deposition over the dissolution process (Chapter 36). Most important point to note was that there was no large-scale delamination or coatings peel-off in scratch tests which proved the stability of the coating after immersion in SBF environment.

On the other hand, the characteristic values of nanohardness and Young's modulus of the MAO coatings were evaluated through the Weibull approach to be ~3 GPa and ~90 GPa, respectively (Chapter 37). Further, sealing of MAO samples showed only a marginal improvement of Young's modulus while the nanohardness remained as good as that of the unsealed coating. However, when exposed to the corrosive environment, the sealed coatings performed much better than the unsealed one, as expected.

The cases of the soft, e.g., $Mg(OH)_2$, and hard, e.g., TiN, Al_2O_3, and metal doped and un-doped DLC thin ceramic films in terms of nanoindentation behavior as well as nanotribological behavior, have been briefly exposed in Chapter 38 to Chapter 41 which comprise Section 10. By virtue of a strain-tolerant layered microstructure the soft magnesium hydroxide thin films deposited by an inexpensive green chemical deposition technique on SLS glass substrates films showed nanohardness ~0.3 GPa and Young's modulus ~3 GPa, which provided the support to the viewpoint that the nanostructured 1.5 μm thick, phase pure magnesium hydroxide film may be useful for protective as well as structural applications (Chapter 38). On the other hand, the hard TiN thin films showed strong ISE up to ~40 nm; afterwards, the data showed depth independency. Further work is needed to understand the basic reasons behind such peculiar behavior. In all of these experiments, however, the depth of experiments was strictly kept well below 10% of the film thickness to avoid any influence of the mechanical properties of the substrate on the measured nanomechanical properties of the TiN or other thin films (Chapters 38, 39). Like TiN, the alumina thin films also have many important applications and associated intricacies of nanoindentation-induced damage evolution process. Therefore, the nanoindentation behavior of alumina thin films for space applications was explored in Chapter 40. The continuous stiffness measurement mode (CSM) nanoindentation study gave indentation size effect or ISE in both nanohardness and modulus of the rf magnetron sputtered alumina thin films.

Similarly, the metal nanoparticle encapsulated diamond-like carbon (M-DLC) thin films have also evolved in recent times as a new class of materials for advanced tribological applications. The AuDLC nanocomposite films showed lower nanohardness and elastic moduli compared to those of the DLC films as expected, because metal incorporation increased the ratio of sp^2/sp^3 hybridization, affected through the increase in the percentage of argon from 50 to 80% (Chapter 41). The most interesting observation was that, in terms of tribological property, the AuDLC films behaved almost similar to the DLCs.

So far we have not studied the aspect of structure–nanomechanical property relationship of the natural biomaterials, e.g., human teeth, bone, fish scale, etc. developed by Mother Nature over uncountable times and ages in such a way that they possess nanoscale to macroscale functionally graded microstructure. Thus, the results presented in Section 11 which spanned Chapters 42 to 45 have focused on the nanoindentation behavior of ceramic-based natural hybrid nanocomposites.

Increase of the nanohardness as the distance from a region close to the dentin enamel junction (DEJ) to the outer enamel zone is traversed (Chapter 42) was explained by a unified picture of the variations in nanohardness in terms of orientation-dependent variations in the extent of biomineralization as the distance from a region close to the DEJ to the outer enamel zone is covered. In addition, it has been shown for the very first time that about 8% apparent increments had occurred in nanohardness (Chapter 43) of the dental enamel nanocomposites with increase in loading rate from ~1×10^3 to 0.3×10^6 $\mu N.s^{-1}$, possibly due to the loading rate–dependent unfolding and curling back behavior of the organic macromolecules in the protein matrix. This physical deformation process had finally controlled the local shear stress accommodation and attendant strain sharing at the nanoscale between the HAp nanocrystal rods and the protein matrix. It is further interesting to note that both spatial and directional anisotropy of Young's modulus was found for the human cortical bone at local microstructural length scale which was not normally seen at the macrostructural length scale (Chapter 44).

In Chapter 45, the correlation of the nanomechanical properties with the functionally graded microstructure was established for a sweet water Indian fish scale, e.g., that of *Catla Catla*. A functionally graded structure was depicted where mineralization was continuously decreased when we traverse from the outer side to the inner side. The nanomechanical properties like nanohardness and Young's modulus were also decreasing in correspondence with the Ca/P ratio.

After studying the nanoindentation behavior of a wide variety of brittle solids including glass, alumina, ceramic matrix composites, C/C and C/C-SiC composites, thick coatings, thin films, and natural nanocomposites, we have indeed shifted our focus to some unique unresolved issues in the field of nanoindentation technique. Thus, the experimental observations were presented that related to presence of indentation size effect and reverse indentation size effect in the nanohardness data of MIPS-HAp coatings, alumina, AlN-SiC composites, dentin, and SLS glass. To the extent possible, attempts were made to look into the aspects related to the genesis of ISE in some of these materials (Chapter 46).

Another related issue is the signature of nanoscale plasticity initiation through "micro pop-in" in the load-depth plots obtained from nanoindentation experiments of brittle solids. The major observation here (Chapter 47) is that the pop-ins in brittle solids were extremely sensitive to the ease or difficulty of nanoscale, i.e., incipient plasticity initiation, homogeneous and/or heterogeneous dislocation initiation, their movements in specific and/or across preferential slip planes, the applied load, and the loading rate. A critical load and a critical loading rate were found to exist below which its propensity will be over and above which the same would be lesser. Shear band formation had always occurred in brittle solids when the local shear stress generated beneath the nanoindenter becomes greater than the shear strength of the material in question. The entire process of discrete shear band formation was found to be extremely sensitive to both load and loading rates applied, as well as to the presence or absence of local, surface or subsurface flaws. Much research needs to be conducted before the mechanisms of pop-in events in glass, ceramics, ceramic composites, and natural biomaterials are well understood.

The generic nature of loading rate's effect on nanomechanical properties of different materials was summarized in Chapter 48. Plausible explanations were searched out for the effect of loading rate on nanohardness enhancement in SLS glass, 10 and 20 μm grain size alumina, AlN-SiC composites, and enamel as well as dentin regions of human tooth.

In Chapter 49, for the first time we have successfully utilized the nanoindentation technique to evaluate the residual stress for plasma-sprayed HAp coatings on both SS316L and Ti-6Al-4V substrates. As it is almost obvious that nanoindentation techniques are extremely sensitive to local microstructural variations, it is of utmost importance to address the reliability issues of the nanomechanical properties data evaluated by the nanoindentation technique. Therefore, in Chapter 50, the Weibull statistical method has been successfully employed for evaluation of the characteristic values of the nanohardness and Young's modulus in ceramic coatings and bulk composite materials. The load dependencies of the Weibull's moduli were also explained.

In Chapter 52, we discussed the influence of the mechanical properties of the substrate on the film's/coating's nanomechanical properties data. From the experimental data it could be shown that a strong substrate effect can exist at higher depths of DLC, AuDLC, and alumina thin films. Finally, the future scope of work needed to expand the application zones of the nanoindentation technique for both brittle solids and biomaterials was very briefly touched upon in Chapter 52.

Common Abbreviations

8YSZ	8 mol% yttria-stabilized zirconia
AFM	atomic force microscopy
A/LP/A	alumina/lanthanum phosphate/alumina
ARA	as-received alumina
AuDLC	gold nanoparticle encapsulated diamond-like carbon
C/C	carbon fiber reinforced in carbon matrix composites
CFRP	carbon fiber reinforced polymer
CMC	ceramic matrix composite
C/SiC	carbon fiber reinforced in silicon carbide matrix
CSM	continuous stiffness measurement
CTE	coefficient of thermal expansion
CVD	chemical vapor deposition
DEJ	dentin enamel junction
DLC	diamond-like carbon
E	Young's modulus
EDX	energy dispersive X-ray spectroscopy
FESEM	field emission scanning electron microscopy
FTIR	Fourier transform infrared spectroscopy
G	shear modulus
GFRP	glass fiber reinforced polymer
H	hardness/nanohardness
HAp	hydroxyapatite
HEL	Hugoniot elastic limit
ICP-AES	inductively coupled plasma atomic emission spectroscopy
IED	inelastic deformation
IFT	indentation fracture toughness
ISE	indentation size effect
LP	lanthanum phosphate
LR	loading rate
LSI	liquid silicon infiltration
LSM	lanthanum strontium manganite
MAO	micro arc oxidation
MAPS	macroplasma spraying
M-DLC	metal nanoparticle-encapsulated diamond-like carbon
MEMS	microelectromechanical systems
MIPS	microplasma spraying
MLCC	multilayer ceramics matrix composite
MM	magnetoelectric multiferroic
NBFO	nano bismuth ferrite
NEMS	nanoelectromechanical systems

P–h	load versus depth
PSR	proportional specimen resistance
PZT	lead zirconate titanate
rf	radio frequency
RISE	reverse indentation size effect
SA	shocked alumina
SBF	simulated body fluid
SDAF	shock-deformed alumina fragment
SEM	scanning electron microscopy
SLS	soda-lime-silica
SOFC	solid oxide fuel cell
SPM	scanning probe microscopy
TEM	transmission electron microscopy
TLR	threshold loading rate
XRD	X-ray diffraction
YSZ	yttria-stabilized zirconia
ZTA	zirconia-toughened alumina

Index